Lie Groups and
Lie Algebras
for Physicists

Lie Groups and Lie Algebras for Physicists

Ashok Das
Susumu Okubo

University of Rochester, USA

HINDUSTAN
BOOK AGENCY

World Scientific

NEW JERSEY · LONDON · SINGAPORE · BEIJING · SHANGHAI · HONG KONG · TAIPEI · CHENNAI

Published by

World Scientific Publishing Co. Pte. Ltd.

5 Toh Tuck Link, Singapore 596224

USA office: 27 Warren Street, Suite 401-402, Hackensack, NJ 07601

UK office: 57 Shelton Street, Covent Garden, London WC2H 9HE

British Library Cataloguing-in-Publication Data
A catalogue record for this book is available from the British Library.

LIE GROUPS AND LIE ALGEBRAS FOR PHYSICISTS

ISBN 978-981-4603-27-0

Printed in India, bookbinding made in Singapore.

To

Mary Okubo

Preface

Group theory was discovered by Évariste Galois in the 19th century for the case of finite dimensional symmetric group. It has been successfully generalized subsequently for the infinite dimensional case (i.e. Lie group and Lie algebra) by Sophus Lie. Before the advent of quantum mechanics in the early 20th century, group theory was thought by many physicists to be unimportant in the study of physics. It is indeed of some interest to note the following anecdote narrated by Freeman Dyson (as quoted in Mathematical Apocrypha, p 21, by S. A. Kranz), "In 1910, the mathematician Oswald Veblen (1880-1960) - a founding member of the Institute for Advanced Study - and the physicist James Jean (1877-1946) were discussing the reform of the mathematics curriculum at Princeton University. Jeans argued that they 'may as well cut out group theory,' for it 'would never be of any use to physics'." The real fundamental change in thinking truly occurred with the development of quantum mechanics. It was soon realized that a deep knowledge of group theory and Lie algebra in the study of of angular momentum algebra is crucial for real understanding of quantum mechanical atomic and nuclear spectral problems. At present, group theory permeates problems in practically every branch of modern physics. Especially in the study of Yang-Mills gauge theory and string theory the use of group theory is essential. For example, we note that the largest exceptional Lie algebra E_8 appears in heterotic string theory and also in some one-dimensional Ising model in statistical mechanics, some of whose predictions have been experimentally verified recently.

Our goal in this monograph is to acquaint (mostly) graduate students of physics with various aspects of modern Lie group and Lie algebra. With this in view, we have kept the presentation of the material in this book at a pedagogical level avoiding unnecessary mathematical rigor. Furthermore, the groups which we will discuss in this book will be mostly Lie groups which are infinite dimensional. We will discuss finite groups only to the extent that they will be necessary for the development of our discussions. The interested readers on this topic are advised to consult many excellent text books on the subject. We have not tried to be exhaustive in the references. Rather, we have given only a few references to books (at the end of each chapter) that will be easier to read for a student with a physics background. We assume that the readers are familiar with the material of at least a two semester graduate course on quantum mechanics as well as with the basics of linear algebra theory.

We would like to thank Dr. Brenno Ferraz de Oliveira (Brazingá) for drawing all the Young tableau diagrams in chapter 5.

A. Das and S. Okubo
Rochester

Contents

CHAPTER 1

Introduction to groups

In this chapter, we introduce the concept of a group and present in some detail various examples of commonly used groups in physics. This is helpful in establishing the terminology as well as the notations commonly used in the study of groups which will also be useful in further development of various ideas associated with groups.

1.1 Definition of a group

Let us start with the formal definition of a group G as follows:

$(G1)$: For any two elements a and b in a group, a product is defined in G satisfying

$$ab = c \in G, \quad \forall a, b \in G. \tag{1.1}$$

$(G2)$: The group product is associative so that

$$(ab)c = a(bc)(\equiv abc), \quad \forall a, b, c \in G. \tag{1.2}$$

$(G3)$: The group has a unique identity (unit) element $e \in G$ such that

$$ea = ae = a, \quad \forall a \in G. \tag{1.3}$$

This implies that

$$ee = e \in G. \tag{1.4}$$

$(G4)$: Any element $a \in G$ has a unique inverse element $a^{-1} \in G$ so that

$$aa^{-1} = a^{-1}a = e. \tag{1.5}$$

Any set of elements G satisfying all the axioms $(G1)$-$(G4)$ is defined to be a group. On the other hand, a set of elements which satisfies only the first three axioms $(G1)$-$(G3)$, but not $(G4)$, is known as a semi-group. (More rigorously, a semi-group is defined as the set of elements which satisfy only $(G1)$-$(G2)$. However, one can always add the identity element to the group since its presence, when an inverse is not defined, is inconsequential (see (1.3)-(1.4)) and we will adopt this definition commonly used in physics.)

Some comments are in order here. We note that the definition of a group does not require that the product rule satisfy the commutativity law $ab = ba$. However, if for any two arbitrary elements of the group, $a, b \in G$, the product satisfies $ab = ba$, then the group G is called a commutative group or an Abelian group (named after the Norwegian mathematician Niels Henrik Abel). On the other hand, if the product rule for a group G does not satisfy commutativity law in general, namely, if $ab \neq ba$ for some of the elements $a, b \in G$, then the group G is known as a non-commutative group or a non-Abelian group. Furthermore, it is easy to see from the definition $(G4)$ of the inverse of an element that the inverse of a product of two elements satisfies

$$(ab)^{-1} = b^{-1}a^{-1} \neq a^{-1}b^{-1}, \tag{1.6}$$

in general, unless, of course, the group G is Abelian. Equation (1.6) is easily checked in the following way

$$
\begin{aligned}
(ab)(b^{-1}a^{-1}) &= a(b(b^{-1}a^{-1})), &&\text{by } (1.2) \text{ with } c = b^{-1}a^{-1}, \\
&= a((bb^{-1})a^{-1}), &&\text{by } (1.2), \\
&= a(ea^{-1}), &&\text{by } (1.5), \\
&= aa^{-1}, &&\text{by } (1.3), \\
&= e, &&\text{by } (1.5). \tag{1.7}
\end{aligned}
$$

Similarly, it is straightforward to verify that $(b^{-1}a^{-1})(ab) = e$.

In order to illustrate the proper definitions in a simple manner, let us consider the following practical example from our day to day life. Let "a" and "b" denote respectively the operations of putting on a coat and a shirt. In this case, the (combined) operation "ab" would correspond to putting on a shirt first (b) and then putting on a coat (a) whereas the (combined) operation "ba" would denote putting on a coat first and then a shirt. Clearly, the operations are not commutative, namely, $ab \neq ba$. If we now introduce a third operation "c" as corresponding to putting on an overcoat, then the law of associativity of the operations (1.2) follows, namely, $(ca)b = c(ab) = cab$ and corresponds to putting on a shirt, a coat and an overcoat in that order. It now follows that the operation "bb" denotes putting on two shirts while "$b(bb) = (bb)b = bbb$" stands for putting on three shirts etc. The set of these operations would define a semi-group if we introduce the identity (unit) element e (see (1.3)) to correspond to the operation of doing nothing. However, this does not make the set of operations a group for the following reason. We note that we can naturally define the inverses "a^{-1}" and "b^{-1}" to correspond respectively to the operations of taking off a coat and a shirt. It follows then that $a^{-1}a = e = b^{-1}b$ as required. However, we also note that the operation "aa^{-1}" is not well defined, in general, unless there is already a coat on the body and, therefore, $aa^{-1} \neq e$, in general (but $a^{-1}a = e$ always). As a result, the set of operations so defined do not form a group. Nonetheless, (1.6) continues to be valid, namely, $(ab)^{-1} = b^{-1}a^{-1} \neq a^{-1}b^{-1}$, whenever these operations are well defined. In fact, since ab corresponds to putting on a shirt and then a coat, if we are already dressed in this manner, the inverse $(ab)^{-1}$ would correspond to removing these clothes (come back to the state prior to the operation ab) which can only be done if we first take off the coat and then the shirt. This leads to $(ab)^{-1} = b^{-1}a^{-1}$. Furthermore, let us note that if "d" denotes the operation of putting on a trouser, then the operation of putting on a shirt and a trouser are commutative, namely, $db = bd$ and so on.

1.2 Examples of commonly used groups in physics

In this section, we would discuss some of the most commonly used groups in physics. This would also help to set up the conventional notations associated with such groups.

1.2.1 Symmetric group S_N. Although most of the commonly used groups in physics are infinite dimensional (Lie) groups, let us begin with a finite dimensional group for simplicity. Let us consider a set of N distinct objects labelled by $\{x_i\}, i = 1, 2, \cdots, N$ and consider all possible distinct arrangements or permutations of these elements. As we know, there will be $N!$ possible distinct arrangements (permutations) associated with such a system and all such permutations (operations) form a group known as the symmetric group or the permutation group of N objects and is denoted by S_N. Since the number of elements in the group is finite, such a group is known as a finite group and the number of elements of the group ($n!$) is known as the order of the group S_N. Permutation groups are relevant in some branches of physics as well as in such diverse topics as the study of the Rubik's cube. Note that every twist of the Rubik's cube is a rearrangement of the faces (squares) on the cube and this is how the permutation group enters into the study of this system. There are various possible notations to denote the permutations of the objects in a set, but let us choose the most commonly used notation known as the cycle notation. Here we denote by (ij) the operation of permuting the objects x_i and x_j in the set of the form

$$(ij): \quad x_i \leftrightarrow x_j, \quad i, j = 1, 2, \cdots, N, \tag{1.8}$$

with all other objects left unchanged. In fact, the proper way to read the relation (1.8) is as

$$(ij): \quad x_i \to x_j \to x_i. \tag{1.9}$$

This brings out the cyclic structure of the cycle notation (parenthesis) and is quite useful when the permutation of more objects are involved. It is clear from this cyclic structure that under a cyclic permutation

$$(ij) = (ji), \tag{1.10}$$

and when permutations of three objects are involved we have

$$(ijk): \quad x_i \to x_j \to x_k \to x_i, \tag{1.11}$$

and

$$(ijk) = (kij) = (jki) \, (\neq (jik) = (kji) = (ikj)). \tag{1.12}$$

Let us clarify the group structure of S_N in the case of $N = 3$, namely, let us consider the set of three objects (elements) $\{x_i\} = (x_1, x_2, x_3)$ where $i = 1, 2, 3$. With the cycle notation we note that the 3! distinct permutations of these elements can be denoted by the set of operations $\{P\}$ with the elements in the set given by

$$P_1 = e = (1), P_2 = (12), P_3 = (23), P_4 = (13),$$

$$P_5 = (123), P_6 = (132), \tag{1.13}$$

where we have introduced the identity element $P_1 = e = (1)$ to correspond to doing no permutation (every object is left unchanged under this operation). The product rule for the permutations can now be understood in the sense of operations. For example,

$$P_2 P_3 = (12)(23), \tag{1.14}$$

implies carrying out the permutation (23) first followed by the permutation (12). We note that under this combined operation

$$x_1 \xrightarrow{(23)} x_1 \xrightarrow{(12)} x_2,$$

$$x_2 \xrightarrow{(23)} x_3 \xrightarrow{(12)} x_3,$$

$$x_3 \xrightarrow{(23)} x_2 \xrightarrow{(12)} x_1, \tag{1.15}$$

so that the combined operation is equivalent to

$$P_2 P_3 = (12)(23) = (123) = P_5 \in \{P\}, \tag{1.16}$$

and the group is closed under multiplication (see (1.1)). Physically this can be understood from the fact that since $\{P\}$ represents all possible distinct arrangements of the objects in the set, any two arrangements would lead to an element of the set. Following (1.15), we can also show that

$$P_3 P_2 = (23)(12) = (132) = P_6 \in \{P\}, \tag{1.17}$$

which shows that the product rule for the permutations is not commutative

$$P_2 P_3 \neq P_3 P_2. \tag{1.18}$$

The associativity of the product of permutations (1.2) can also be easily verified. For example, we note that

$$(P_2 P_3) P_4 = P_5 P_4 = (123)(13) = (23) = P_3,$$

$$P_2 (P_3 P_4) = (12)((23)(13)) = (12)(123) = (23) = P_3, \tag{1.19}$$

so that associativity holds and we have

$$(P_2 P_3) P_4 = P_2 (P_3 P_4) = P_2 P_3 P_4. \tag{1.20}$$

The identity element $P_1 = e$ of the group corresponding to no permutation is clearly unique and satisfies (see (1.3))

$$P_\alpha P_1 = P_1 P_\alpha = P_\alpha, \quad \alpha = 1, 2, \cdots, 6. \tag{1.21}$$

Finally, the inverse of the permutations is determined easily as follows. For any pair permutation (ij), we see that doing the permutation twice brings it back to the original arrangement of the objects. Therefore, we have

$$P_2 P_2 = P_3 P_3 = P_4 P_4 = e = P_1, \tag{1.22}$$

and we can identify (identity is its own inverse, $(P_1)^{-1} = P_1$)

$$(P_2)^{-1} = P_2, \quad (P_3)^{-1} = P_3, \quad (P_4)^{-1} = P_4. \tag{1.23}$$

On the other hand, for permutations involving all three objects we see that (we are using (1.16) and (1.17) as well as (1.22))

$$P_5 P_6 = P_2 P_3 P_3 P_2 = P_2 P_2 = P_1, \tag{1.24}$$

which determines that P_6 is the inverse of P_5 and *vice versa*. This shows that S_3 indeed satisfies all the postulates of a group and a similar proof can be extended to the case of S_N as well.

After the discussion of this simple finite dimensional group, let us next move onto infinite dimensional (Lie) groups (named after the Norwegian mathematician Sophus Lie) which are more commonly used in the study of modern physics.

1.2.2 One dimensional translation group T_1. Let x denote a real variable (namely, $x \in \mathbb{R}$ and it can be either a spatial coordinate or the time coordinate, for example). For any real constant parameter a, we define an operation $T(a)$ which acting on the coordinate x simply translates it by an amount a such that

$$T(a)x = x + a, \quad \forall x \in \mathbb{R}. \tag{1.25}$$

This is to be interpreted in one of the following two ways. First, we can think of $T(a)$ as the active transformation

$$T(a) : \quad x \xrightarrow{T(a)} x' = x + a. \tag{1.26}$$

Alternatively, we can interpret $T(a)$ as a linear operator satisfying (1.25) with the action of $T(a)$ leaving any constant parameter b unchanged (a linear operator does not act on constants, only on vectors), namely,

$$T(a)b = b. \tag{1.27}$$

In either case, the combined effect of two translations can now be determined. Viewed as an active transformation, for example, we can write

$$T(a)T(b): \quad x \xrightarrow{T(b)} x' = x + b, \quad \text{followed by}$$

$$x' \xrightarrow{T(a)} x'' = x' + a = x + b + a, \tag{1.28}$$

which, in the notation of (1.25), corresponds to

$$(T(a)T(b))\,x \equiv T(a)\,(T(b)x) = (T(b)x) + a = x + b + a$$
$$= x + (b + a) = T(b + a)x = T(a + b)x. \tag{1.29}$$

One can represent the transformations in (1.28) diagrammatically as shown in Fig. 1.1. On the other hand, considered in terms of linear operators, it is easy to check with the use of (1.27) that

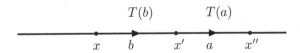

Figure 1.1: The combined operation of two translations $T(a)T(b)$ where a translation $T(b)$ is followed by another translation $T(a)$.

$$(T(a)T(b))\,x \equiv T(a)\,(T(b)x) = T(a)\,(x + b) = T(a)x + T(a)b$$
$$= x + a + b = T(a + b)x = T(b + a)x, \tag{1.30}$$

which coincides with (1.29).

We note that we may identify $T(a)$ with the differential operator

$$T(a) = \mathbb{1} + a\,\frac{\mathrm{d}}{\mathrm{d}x}, \tag{1.31}$$

acting on x so that

$$T(a)x = \left(\mathbb{1} + a\,\frac{\mathrm{d}}{\mathrm{d}x}\right)x = x + a,$$

$$T(a)b = \left(\mathbb{1} + a\,\frac{\mathrm{d}}{\mathrm{d}x}\right)b = b. \tag{1.32}$$

Here $\mathbb{1}$ denotes the identity operator. It follows now that

$$
\begin{aligned}
T(a)T(b)x &= \left(\mathbb{1} + a\frac{\mathrm{d}}{\mathrm{d}x}\right)\left(\mathbb{1} + b\frac{\mathrm{d}}{\mathrm{d}x}\right)x \\
&= \left(\mathbb{1} + (a+b)\frac{\mathrm{d}}{\mathrm{d}x} + ab\frac{\mathrm{d}^2}{\mathrm{d}x^2}\right)x \\
&= \left(\mathbb{1} + (a+b)\frac{\mathrm{d}}{\mathrm{d}x}\right)x = T(a+b)x, \qquad (1.33)
\end{aligned}
$$

consistent with (1.30) (or (1.29)).

Equations (1.29) and (1.30) show that the product $T(a)T(b)$, the combined operation of two translations, is well defined and corresponds also to a translation, namely,

$$
T(a)T(b) = T(b+a) = T(a+b) = T(b)T(a), \quad \forall a, b \in \mathbb{R}. \quad (1.34)
$$

This defines the product rule for the operation (see (1.1)) which, in this case, is commutative. The associativity of the product (1.2) is easily verified from

$$
\begin{aligned}
(T(a)T(b))T(c) &= T(a+b)T(c) = T((a+b)+c) \\
&= T(a+b+c), \\
T(a)(T(b)T(c)) &= T(a)T(b+c) = T(a+(b+c)) \\
&= T(a+b+c), \qquad (1.35)
\end{aligned}
$$

where (1.34) has been used and (1.35) leads to

$$
(T(a)T(b))\,T(c) = T(a)\,(T(b)T(c)). \qquad (1.36)
$$

The unit element e (see (1.3)) in this case can be identified with

$$
e = T(0), \qquad (1.37)
$$

so that it satisfies

$$
eT(a) = T(0)T(a) = T(a) = T(a)T(0) = T(a)e. \qquad (1.38)
$$

Finally, the inverse of a translation $T(a)$ by an amount a can be defined as a translation by an amount $(-a)$

$$
(T(a))^{-1} = T(-a), \qquad (1.39)
$$

so that (see (1.5))

$$T(a)\,(T(a))^{-1} = T(a)T(-a) = T(0) = e$$

$$= T(-a)T(a) = (T(a))^{-1}\,T(a). \tag{1.40}$$

Thus, we see that the set of translations $T_1 = \{T(a)\}$ with the parameter of translation $a \in \mathbb{R}$ (namely, $-\infty < a < \infty$) satisfy all the properties $(G1)$-$(G4)$ defining a group. This is a continuous group since the parameter of translation can take any continuous value in $\mathbb{R}$ and is known as the one dimensional translation group. Furthermore, since the product rule for this group is commutative (see (1.34)), this is an example of an Abelian group. To conclude this subsection, we note that if the parameters of translation were restricted to a finite interval $L_1 \le a \le L_2$, then the set of translations will not define a group since two successive translations may not remain in the interval.

1.2.3 One dimensional unitary group $U(1)$. Let z be a complex variable and let us consider the phase transformation

$$z \to z' = e^{i\theta}\,z, \tag{1.41}$$

where θ is a real constant angular parameter $0 \le \theta < 2\pi$. We can write the phase transformation (1.41) in terms of operators also as (see also (1.25))

$$U(\theta)z = e^{i\theta}\,z. \tag{1.42}$$

Following the discussion in section 1.2.2 we can show from (1.41) (or (1.42)) that the product

$$U(\theta_1)U(\theta_2) = U(\theta_1 + \theta_2) = U(\theta_2)U(\theta_1), \tag{1.43}$$

is well defined if we identify $U(\theta + 2\pi) = U(\theta)$ (since $e^{2\pi i} = 1$) whenever $\theta_1 + \theta_2 = \theta + 2\pi \ge 2\pi$. We note that in this case we can identify († denotes the Hermitian conjugate, in this one dimensional case, only the complex conjugate)

$$U^{\dagger}(\theta) = U^{-1}(\theta) = U(-\theta), \tag{1.44}$$

namely,

$$U^{\dagger}(\theta)z = U^{-1}(\theta)z = U(-\theta)z = e^{-i\theta}\,z, \tag{1.45}$$

so that the operator $U(\theta)$ is unitary

$$U(\theta)U^{\dagger}(\theta) = e = U^{\dagger}(\theta)U(\theta). \tag{1.46}$$

The set of phase transformations $U(1) = \{U(\theta),\ 0 \le \theta < 2\pi\}$ forms a group called the one dimensional unitary group $U(1)$ (also written as U_1). We see from (1.43) that it is a commtative group. Let us also note that we can restrict the range of the parameter of transformation to $0 \le \theta < 2\pi$ since $e^{2\pi i} = 1$ and this has to be contrasted with the parameters of translations T_1 discussed earlier. Correspondingly, $U(1)$ is known as a compact group (parameters have a compact range) as opposed to T_1 which is called a non-compact group. We will discuss these concepts in more detail in section 1.3.

1.2.4 $U(N)$ and $SL(N)$ groups. We can generalize the group of one dimensional phase transformations $U(1)$ discussed in section 1.2.3 to N dimensions as follows. Let A denote a $N \times N$ unitary matrix, namely,

$$
A = \begin{pmatrix}
\alpha_{11} & \alpha_{12} & \cdots & \alpha_{1N} \\
\alpha_{21} & \alpha_{22} & \cdots & \alpha_{2N} \\
\vdots & \vdots & \vdots & \vdots \\
\alpha_{N1} & \alpha_{N2} & \cdots & \alpha_{NN}
\end{pmatrix},
\tag{1.47}
$$

where the matrix elements $\alpha_{ij},\ i,j = 1, 2, \cdots, N$ are complex parameters such that

$$
AA^\dagger = E_N = A^\dagger A,
\tag{1.48}
$$

Here $E_N\ (= \mathbb{1}_N)$ denotes the $N \times N$ unit (identity) matrix so that the unitarity condition (1.48) leads to the condition

$$
\sum_{k=1}^{N} \alpha_{ik}\alpha_{jk}^* = \delta_{ij} = \sum_{k=1}^{N} \alpha_{ki}^* \alpha_{kj},
\tag{1.49}
$$

where δ_{ij} denotes the familiar Kronecker delta. For $i = j$, equation (1.49) yields

$$
\sum_{k=1}^{N} |\alpha_{ik}|^2 = 1 = \sum_{k=1}^{N} |\alpha_{ki}|^2,
\tag{1.50}
$$

which, in turn, implies that

$$
|\alpha_{ik}| \le 1, \quad i, k = 1, 2, \cdots, N.
\tag{1.51}
$$

Let us further note that

$$
\det A^\dagger = \left(\det A^T\right)^* = (\det A)^*, \quad \det E_N = 1,
\tag{1.52}
$$

where $(\dagger, T, *)$ denote respectively Hermitian conjugate, transpose and complex conjugate of the matrix. With these, we can take the determinant of both sides in (1.48) to obtain

$$\det A \, (\det A)^* = |\det A|^2 = 1,$$

$$\text{or,} \quad |\det A| = 1, \tag{1.53}$$

so that A^{-1} always exists and can be identified, as in (1.44), with (see (1.48))

$$A^{-1} = A^{\dagger}, \tag{1.54}$$

and it follows that A^{-1} is also unitary. For example,

$$A^{-1} \left(A^{-1}\right)^{\dagger} = A^{\dagger} \left(A^{\dagger}\right)^{\dagger} = A^{\dagger} A = E_N, \tag{1.55}$$

which follows from (1.48). Similarly, if A and B are both unitary matrices, then their matrix product AB also is. Finally, the $N \times N$ unit (identity) matrix E_N is unitary. Thus, we conclude that the set of all $N \times N$ unitary matrices $U(N) = \{A\}$ form a group known as the unitary group $U(N)$ (or U_N).

In a similar manner, if we consider all $N \times N$ matrices A satisfying only the condition

$$\det A = 1, \tag{1.56}$$

we can show that they also form a group. From the matrix product, it follows that, for any two matrices satisfying (1.56), we have $\det(AB) = \det A \det B = 1$ since each of the matrices has unit determinant, namely, $\det A = 1 = \det B$. Furthermore, we note that $\det A^{-1} = (\det A)^{-1} = 1$ and $\det E_N = 1$. (Note that since the matrices have unit determinant, the inverses exist.) This group is known as the special linear group in N dimensions and is denoted by $SL(N)$ (or SL_N). If the set of $N \times N$ matrices $\{A\}$, in addition to satisfying (1.56), also satisfy the unitary condition (1.48), the resulting group is known as the special unitary group in N dimensions and is denoted by $SU(N)$ (or SU_N) which is a subgroup of both SL_N and U_N. (The notion of a subgroup will be discussed later.) For $N \geq 2$, these groups are not Abelian (non-Abelian).

The $U(N)$ group is of interest in the operator algebra of harmonic oscillators in quantum mechanics. Let a_i and $a_j^{\dagger}$ with $i, j = 1, 2, \cdots, N$ denote respectively the annihilation and the creation operators for a N dimensional oscillator system satisfying

$$[a_i, a_j] = 0 = [a_i^{\dagger}, a_j^{\dagger}], \quad [a_i, a_j^{\dagger}] = \delta_{ij}. \tag{1.57}$$

Here $[a, b] = ab - ba$ denotes the commutator of two linear operators a and b. Let us next consider the linear canonical transformation of the annihilation and the creation operators in (1.57) of the form

$$a_i \rightarrow b_i = \sum_{j=1}^{N} \alpha_{ij} a_j,$$

$$a_i^\dagger \rightarrow b_i^\dagger = \sum_{j=1}^{N} a_j^\dagger \alpha_{ij}^*, \tag{1.58}$$

where α_{ij} denote constant parameters. It follows now that if the parameters α_{ij} satisfy the unitarity condition (1.49), the transformed operators satisfy

$$[b_i, b_j] = 0 = [b_i^\dagger, b_j^\dagger], \quad [b_i, b_j^\dagger] = \delta_{ij}, \tag{1.59}$$

for $i, j = 1, 2, \cdots, N$. The last relation in (1.59) (which is the only one that needs checking), for example, can be seen as

$$[b_i, b_j^\dagger] = \left[\sum_{k=1}^{N} \alpha_{ik} a_k, \sum_{\ell=1}^{N} a_\ell^\dagger \alpha_{j\ell}^* \right] = \sum_{k,\ell=1}^{N} \alpha_{ik} \alpha_{j\ell}^* [a_k, a_\ell^\dagger]$$

$$= \sum_{k,\ell=1}^{N} \alpha_{ik} \alpha_{j\ell}^* \delta_{k\ell} = \sum_{k=1}^{N} \alpha_{ik} \alpha_{jk}^* = \delta_{ij}, \tag{1.60}$$

where we have used (1.57) in the intermediate step as well as (1.49) in the last step.

Since the $N \times N$ matrix A defined by (1.48) (or equivalently by (1.49)) is unitary, we conclude that the canonical commutation relations (1.57) for the creation and annihilation operators of the harmonic oscillator are invariant under the unitary group of transformations $U(N)$. More importantly, the Hamiltonian operator (normal ordered although the presence of a zero point energy does not modify the result) for the harmonic oscillator system defined by

$$H = \sum_{i=1}^{N} a_i^\dagger a_i, \tag{1.61}$$

is also invariant under $U(N)$. This is easily checked from the fact that (see (1.58))

$$\sum_{i=1}^{N} b_i^\dagger b_i = \sum_{i=1}^{N} \left(\sum_{j=1}^{N} a_j^\dagger \alpha_{ij}^* \right) \left(\sum_{k=1}^{N} \alpha_{ik} a_k \right)$$

$$= \sum_{j,k=1}^{N} a_j^\dagger a_k \sum_{i=1}^{N} \alpha_{ik} \alpha_{ij}^*$$

$$= \sum_{j,k=1}^{N} a_j^\dagger a_k \delta_{jk} = \sum_{j=1}^{N} a_j^\dagger a_j = H, \qquad (1.62)$$

where we have used (1.49) in the intermediate step.

The results in (1.59) and (1.62) can be seen more easily as follows. Let us arrange the two sets of N annihilation operators into column matrices as

$$\hat{a} = \begin{pmatrix} a_1 \\ a_2 \\ \vdots \\ a_N \end{pmatrix}, \quad \hat{b} = \begin{pmatrix} b_1 \\ b_2 \\ \vdots \\ b_N \end{pmatrix}. \qquad (1.63)$$

Correspondingly the two sets of creation operators $\hat{a}^\dagger$ and $\hat{b}^\dagger$ would be denoted by row matrices. With this notation, the transformation (1.58) can be written in the matrix form (see also (1.47))

$$\hat{b} = A\hat{a}, \quad \hat{b}^\dagger = \hat{a}^\dagger A^\dagger, \qquad (1.64)$$

and the relations (1.59) and (1.62) follow if the transformation matrix A is unitary, namely, if it satisfies (1.48). For example, we note that (the summation over all modes is automatically contained in the matrix multiplication)

$$\hat{b}^\dagger \hat{b} = \hat{a}^\dagger A^\dagger A \hat{a} = \hat{a}^\dagger \mathbb{1} \hat{a} = \hat{a}^\dagger \hat{a} = H, \qquad (1.65)$$

The invariance of the Hamiltonian of a dynamical system under a transformation defines a symmetry of the system as we will discuss later.

1.2.5 $O(N)$ **and** $SO(N)$ **groups.** In the earlier example of the unitary group studied in section 1.2.4 we assumed the $N \times N$ matrix A to be a complex unitary matrix. On the other hand, if we restrict further the matrix A to be a real matrix, namely,

$$A^* = A, \tag{1.66}$$

then (1.48) leads to

$$AA^T = E_N = A^T A, \tag{1.67}$$

which follows from the fact that $A^\dagger = (A^*)^T = A^T$ for real matrices. Here A^T denotes the transpose of the matrix A, namely,

$$(A^T)_{ij} = A_{ji}, \tag{1.68}$$

so that explicitly we can write (see (1.47))

$$A^T = \begin{pmatrix} \alpha_{11} & \alpha_{21} & \cdots & \alpha_{N1} \\ \alpha_{12} & \alpha_{22} & \cdots & \alpha_{N2} \\ \vdots & \vdots & \vdots & \vdots \\ \alpha_{1N} & \alpha_{2N} & \cdots & \alpha_{NN} \end{pmatrix}. \tag{1.69}$$

We note that the restriction on the matrix elements following from (1.67) has the form

$$\sum_{k=1}^{N} a_{ik}a_{jk} = \delta_{ij} = \sum_{k=1}^{N} a_{ki}a_{kj}, \tag{1.70}$$

which can be compared with (1.49) (for real elements, $\alpha_{ij}^* = \alpha_{ij}$). In particular, for $i = j$, relation (1.70) leads to

$$\sum_{k=1}^{N} (a_{ik})^2 = 1 = \sum_{k=1}^{N} (a_{ki})^2, \quad i = 1, 2, \cdots, N, \tag{1.71}$$

so that the elements of the matrix A are restricted to have values $-1 \leq a_{ij} \leq 1$ for any $i, j = 1, 2, \cdots, N$.

We can readily verify as in the last section that the set of all $N \times N$ real matrices satisfying (1.67) forms a group which is called the real orthogonal group in N dimensions and is denoted by $O(N)$ (or O_N) (the matrices A satisfying (1.67) are known as orthogonal matrices). Furthermore, taking the determinant of both sides of (1.67), we note that since $\det A^T = \det A$, we have

$$(\det A)^2 = 1,$$

or, $\quad \det A = \pm 1. \tag{1.72}$

If we require the orthogonal matrices A to satisfy the additional condition

$$\det A = 1, \tag{1.73}$$

then it can be shown that all such matrices define the real special orthogonal group in N dimensions which is denoted by $SO(N)$ (or SO_N). Sometimes it is also known as the group of rotations in N dimensions identified as R_N. This nomenclature can be understood as follows. Let us consider the N dimensional Euclidean space labelled by the coordinates

$$\mathbf{x} = \begin{pmatrix} x_1 \\ x_2 \\ \vdots \\ x_N \end{pmatrix}. \tag{1.74}$$

Rotations define the maximal symmetry (which leaves the origin invariant) of such a space and the length of a vector in this space defined as

$$\mathbf{x}^T \mathbf{x} = x_1^2 + x_2^2 + \cdots + x_N^2, \tag{1.75}$$

is invariant under rotations. Let us note that if we define rotations through $N \times N$ matrices R as

$$\mathbf{x} \rightarrow \mathbf{x}' = R\mathbf{x}, \tag{1.76}$$

it follows from (1.75) that the length remains invariant, namely,

$$\mathbf{x}'^T \mathbf{x}' = \mathbf{x}^T R^T R \mathbf{x} = \mathbf{x}^T \mathbf{x},$$

$$\text{or,} \quad (x_1')^2 + (x_2')^2 + \cdots + (x_N')^2 = x_1^2 + x_2^2 + \cdots + x_N^2, \tag{1.77}$$

only if

$$R^T R = E_N. \tag{1.78}$$

Comparing (1.78) with (1.67) we recognize that the rotation matrices R in N dimensional Euclidean space correspond to orthogonal matrices and correspondingly the $O(N)$ group is also identified with the rotation group R_N. The difference between the $O(N)$ and the $SO(N)$ groups lies in the fact that the latter group does not include "mirror reflections" (which render the matrices A to have $\det A = -1$). (Note that the length of a vector defined in (1.75) is clearly invariant under "mirror reflections" $x_i \rightarrow -x_i$ for some of the coordinates.) Correspondingly $SO(N)$ is also known as the group of proper rotations. We note here that the rotation group SO_3 (without "mirror reflections") plays an important role in the study of the Schrödinger equation with central potentials in general and the group SO_4 clarifies the origin of the "accidental degeneracy" in the case of the Hydrogen atom (which we discuss in chapter 3).

A group that is very important in the study of relativistic systems is the Lorentz group. It is the maximal symmetry group (which leaves the origin fixed) of the four dimensional space-time Minkowski manifold (space) just as the three dimensional rotations (which leave the origin invariant) define the maximal symmetry group of the three dimensional Euclidean manifold (space) (and are important in non-relativistic phenomena). In the case of the four dimensional space-time manifold, if we define a coordinate four vector as (t denotes the time coordinate, $\mathbf{x}$ the three dimensional spatial coordinate vector and c represents the speed of light)

$$
\begin{aligned}
x^\mu &= (x^0, \mathbf{x}) \\
&= (x^0, x^1, x^2, x^3) \\
&= (ct, x^1, x^2, x^3), \quad \mu = 0, 1, 2, 3,
\end{aligned} \tag{1.79}
$$

then the length invariant under a Lorentz transformation (which includes three dimensional rotations as well as Lorentz boosts), is given by

$$
c^2 t^2 - \mathbf{x}^2 = c^2 t^2 - (x^1)^2 - (x^2)^2 - (x^3)^2, \tag{1.80}
$$

and is known as the invariant length of the Minkowski space. Denoting the coordinate four vector by a four component column matrix (see (1.74))

$$
x = \begin{pmatrix} x^0 \\ \mathbf{x} \end{pmatrix} = \begin{pmatrix} ct \\ x^1 \\ x^2 \\ x^3 \end{pmatrix}, \tag{1.81}
$$

and introducing a 4×4 constant diagonal matrix (metric)

$$
\eta = \begin{pmatrix} 1 & 0 & 0 & 0 \\ 0 & -1 & 0 & 0 \\ 0 & 0 & -1 & 0 \\ 0 & 0 & 0 & -1 \end{pmatrix}, \tag{1.82}
$$

we note that we can write the invariant length (1.80) as

$$x^T \eta x = c^2 t^2 - (x^1)^2 - (x^2)^2 - (x^3)^2. \tag{1.83}$$

Under a Lorentz transformation the coordinates transform as

$$x = \begin{pmatrix} x^0 \\ x^1 \\ x^2 \\ x^3 \end{pmatrix} \rightarrow x' = \begin{pmatrix} x'^0 \\ x'^1 \\ x'^2 \\ x'^3 \end{pmatrix} = A \begin{pmatrix} x^0 \\ x^1 \\ x^2 \\ x^3 \end{pmatrix} = Ax, \tag{1.84}$$

where A is a 4×4 matrix implementing the Lorentz transformation. Since the coordinates are real, it follows that the matrix A representing the Lorentz transformation is also real, namely,

$$A^* = A. \tag{1.85}$$

Furthermore, the invariance of the length (1.80) (or (1.83)) now leads to (the constant matrix η does not transform under a Lorentz transformation)

$$x'^T \eta x' = x^T A^T \eta A x = x^T \eta x, \tag{1.86}$$

which is possible only if the matrix A denoting a Lorentz transformation satisfies the relation

$$A^T \eta A = \eta. \tag{1.87}$$

Relation (1.87) is quite similar to (1.67) (see also (1.78)) if we replace the 4×4 constant diagonal matrix η by $E_4 (= \mathbb{1}_4)$. Therefore, we suspect that the Lorentz transformations are related to the orthogonal matrices in some manner.

To understand this better, we note that the diagonal elements of the relation (1.87) lead explicitly to

$$(a_{0\mu})^2 - \sum_{i=1}^{3} (a_{i\mu})^2 = 1, \quad \mu = 0, 1, 2, 3. \tag{1.88}$$

Comparing with (1.71) we see that the matrix elements of a Lorentz transformation satisfy the same relation as those for the orthogonal matrices except for the relative negative signs in the last three terms (which basically arise from the structure of the metric η). Correspondingly, the Lorentz transformations are said to belong to the orthogonal group $O(3, 1)$ which is different from the group $O(4)$ in that the metric in the case of the former has three negative and one positive signature (as opposed to the case of $O(4)$ where the metric is the identity matrix).

We point out here that it is possible to identify

$$x^4 = -ix^0 = -ict, \tag{1.89}$$

and introduce a coordinate four vector as

$$\begin{aligned}
\tilde{x}^\mu &= (\mathbf{x}, x^4) \\
&= (x^1, x^2, x^3, x^4) \\
&= (x^1, x^2, x^3, -ict), \quad \mu = 1, 2, 3, 4,
\end{aligned} \tag{1.90}$$

and denote it as a four component column matrix as

$$\tilde{x} = \begin{pmatrix} \mathbf{x} \\ x^4 \end{pmatrix} = \begin{pmatrix} x^1 \\ x^2 \\ x^3 \\ -ict \end{pmatrix}. \tag{1.91}$$

In this case the invariant length would have the standard form without (the metric) η (see (1.75))

$$\begin{aligned}
-\tilde{x}^T \tilde{x} &= -(\mathbf{x}^2 + (x^4)^2) = -((x^1)^2 + (x^2)^2 + (x^3)^2 - c^2 t^2) \\
&= c^2 t^2 - (x^1)^2 - (x^2)^2 - (x^3)^2.
\end{aligned} \tag{1.92}$$

If we now define a Lorentz transformation through a 4×4 matrix $\tilde{A}$ as

$$\tilde{x} = \begin{pmatrix} \tilde{x}^1 \\ \tilde{x}^2 \\ \tilde{x}^3 \\ \tilde{x}^4 \end{pmatrix} \rightarrow \tilde{x}' = \begin{pmatrix} \tilde{x}'^1 \\ \tilde{x}'^2 \\ \tilde{x}'^3 \\ \tilde{x}'^4 \end{pmatrix} = \tilde{A} \begin{pmatrix} \tilde{x}^1 \\ \tilde{x}^2 \\ \tilde{x}^3 \\ \tilde{x}^4 \end{pmatrix} = \tilde{A}\tilde{x}, \tag{1.93}$$

then the invariance of the length (1.92) under the Lorentz transformation (1.93) would require

$$-\tilde{x}'^T \tilde{x}' = -\tilde{x} \tilde{A}^T \tilde{A} \tilde{x} = -\tilde{x}^T \tilde{x}, \tag{1.94}$$

which is possible only if

$$\tilde{A}^T \tilde{A} = E_4. \tag{1.95}$$

This is precisely the orthogonality condition in (1.67) and this would suggest that the Lorentz transformations belong to the orthogonal group $O(4)$. However, as we have already seen, this is not true and the difference arises from the fact that because the coordinates in (1.90) or (1.91) are now complex (and the Lorentz transformation preserves the nature of the coordinates), the transformation matrix $\tilde{A}$ is no longer real (see (1.66)),

$$\tilde{A}^* \neq \tilde{A}. \tag{1.96}$$

In fact, from the nature of the coordinates in (1.91), it follows that the matrix elements of the transformation matrix would have the structure

$$\tilde{a}_{ij}^* = \tilde{a}_{ij}, \quad \tilde{a}_{44}^* = \tilde{a}_{44}, \quad \tilde{a}_{4i}^* = -\tilde{a}_{4i}, \quad \tilde{a}_{i4}^* = -\tilde{a}_{i4}, \tag{1.97}$$

where $i, j = 1, 2, 3$. As a result, the diagonal elements of the orthogonality relation in (1.87) in this case lead to

$$\sum_{\nu=1}^{4} (\tilde{a}_{\nu\mu})^2 = 1, \tag{1.98}$$

for any fixed value of $\mu = 1, 2, 3, 4$. For example, for $\mu = 4$, relation (1.98), together with (1.97), leads to

$$\sum_{i=1}^{3} (\tilde{a}_{i4})^2 + (\tilde{a}_{44})^2 = |\tilde{a}_{44}|^2 - |\tilde{a}_{i4}|^2 = 1, \tag{1.99}$$

which brings out the $O(3, 1)$ nature of the Lorentz transformations discussed in (1.88). Note that (1.99) implies that in this case we can have $|\tilde{a}_{44}|, |\tilde{a}_{i4}| \geq 1$ in contrast to the $U(N)$ case (see (1.51)).

Let us conclude this section by simply noting that when we combine space-time translations with the Lorentz transformations, we obtain the Poincaré transformations and they form an interesting group. This is an important symmetry group in the study of relativistic systems. In four dimensions, for example, the Poincaré transformations act on the space-time coordinates as

$$x = \begin{pmatrix} x^0 \\ x^1 \\ x^2 \\ x^3 \end{pmatrix} \rightarrow x' = \begin{pmatrix} x'^0 \\ x'^1 \\ x'^2 \\ x'^3 \end{pmatrix} = A \begin{pmatrix} x^0 \\ x^1 \\ x^2 \\ x^3 \end{pmatrix} + \begin{pmatrix} a^0 \\ a^1 \\ a^2 \\ a^3 \end{pmatrix}, \tag{1.100}$$

where the 4×4 matrix A defines a Lorentz transformation (discussed in (1.84)) and $a^\mu = (a^0, a^1, a^2, a^3) = (a^\mu)^*$ denotes the four vector representing space-time translations. We will go into more details of these two important groups in physics in a later chapter.

1.2.6 Symplectic group $Sp(2N)$. Let us introduce the $2N \times 2N$ matrix ϵ of the form

$$\epsilon_{2N} = \begin{pmatrix} \begin{pmatrix} 0 & 1 \\ -1 & 0 \end{pmatrix} & 0 \; 0 & 0 \; 0 & \cdots & 0 \; 0 \\ 0 \; 0 & \begin{pmatrix} 0 & 1 \\ -1 & 0 \end{pmatrix} & 0 \; 0 & \cdots & 0 \; 0 \\ 0 \; 0 & 0 \; 0 & \ddots & \cdots & 0 \; 0 \\ 0 \; 0 & 0 \; 0 & 0 \; 0 & \cdots & \begin{pmatrix} 0 & 1 \\ -1 & 0 \end{pmatrix} \end{pmatrix}, \tag{1.101}$$

which is a block diagonal matrix with the anti-symmetric 2×2 matrices $i\sigma_2 = \begin{pmatrix} 0 & 1 \\ -1 & 0 \end{pmatrix}$ along the diagonal (σ_2 represents the second Pauli matrix). By construction, this matrix is anti-symmetric,

$$\epsilon_{2N}^T = -\epsilon_{2N}, \quad (\epsilon_{2N})^2 = -E_{2N}, \tag{1.102}$$

where $E_{2N} = \mathbb{1}_{2N}$ denotes the $2N \times 2N$ identity (unit) matrix. Such a matrix is known as a symplectic structure (also known as a complex structure) and is by definition even dimensional. Let us next consider a $2N \times 2N$ matrix A which preserves the symplectic structure ϵ_{2N}, namely,

$$A^T \epsilon_{2N} A = \epsilon_{2N} = A \epsilon_{2N} A^T. \tag{1.103}$$

(The relation (1.103) can be compared with (1.67) with $\epsilon_{2N} \to E_{2N} = \mathbb{1}_{2N}$.) Such a matrix is known as a symplectic matrix in $2N$ dimensions and the set of all $2N \times 2N$ matrices satisfying (1.103) can be shown to define a group known as the symplectic group in $2N$ dimensions denoted by $Sp(2N)$ (or Sp_{2N}). Any symplectic matrix satisfying (1.103) is known to have unit determinant ($\det A = 1$) so that it is automatically special. We do not go into the general proof of this here since it is rather involved.

The symplectic groups are of some relevance in the study of quantum mechanics. For example, it is straightforward to show that the real symplectic group $Sp(2N, \mathbb{R})$ (which consists of symplectic matrices which are real) is the invariance group of the N-dimensional Heisenberg algebra. Let us recall that if q_i and p_i with $i = 1, 2, \cdots, N$ denote the canonical coordinate and momentum (Hermitian) operators, they satisfy the canonical (Heisenberg) commutation relations

$$[q_i, q_j] = 0 = [p_i, p_j], \quad [q_i, p_j] = i\hbar\delta_{ij}, \tag{1.104}$$

for $i, j = 1, 2, \cdots, N$. This is commonly known as the Heisenberg algebra. Let us next introduce a $2N$-dimensional vector operator ξ_μ, $\mu = 1, 2, \cdots, 2N$ with the components

$$\xi_\mu = (q_1, p_1, q_2, p_2, \cdots, q_N, p_N). \tag{1.105}$$

In terms of ξ_μ, the Heisenberg algebra (1.104) can be written compactly as

$$[\xi_\mu, \xi_\nu] = i\hbar\epsilon_{\mu\nu}, \quad \mu, \nu = 1, 2, \cdots, 2N, \tag{1.106}$$

where $\epsilon_{\mu\nu}$ corresponds to the (μ, ν) element of the matrix ϵ defined in (1.101).

Let us now make a transformation of the vector ξ_μ as

$$\xi_\mu \rightarrow \xi'_\mu = \sum_{\nu=1}^{2N} \beta_{\mu\nu}\xi_\nu, \tag{1.107}$$

where $\beta_{\mu\nu}$ are constant parameters. It follows now that

$$[\xi'_\mu, \xi'_\nu] = \left[\sum_{\lambda=1}^{2N} \beta_{\mu\lambda}\xi_\lambda, \sum_{\sigma=1}^{2N} \beta_{\nu\sigma}\xi_\sigma \right] = \sum_{\lambda,\sigma=1}^{2N} \beta_{\mu\lambda}\beta_{\nu\sigma}[\xi_\lambda, \xi_\sigma]$$

$$= i\hbar \sum_{\lambda,\sigma=1}^{2N} \beta_{\mu\lambda}\beta_{\nu\sigma}\epsilon_{\lambda\sigma}. \tag{1.108}$$

This implies that the transformed variables ξ'_μ would also satisfy the Heisenberg algebra

$$[\xi'_\mu, \xi'_\nu] = i\hbar\epsilon_{\mu\nu}, \tag{1.109}$$

if the transformation parameters $\beta_{\mu\nu}$ satisfy

$$\sum_{\lambda,\sigma=1}^{2N} \beta_{\mu\lambda}\beta_{\nu\sigma}\epsilon_{\lambda\sigma} = \sum_{\lambda,\sigma=1}^{2N} \beta_{\mu\lambda}\epsilon_{\lambda\sigma}\beta_{\nu\sigma} = \epsilon_{\mu\nu}. \tag{1.110}$$

Introducing the $2N \times 2N$ matrix

$$A = \begin{pmatrix} \beta_{11} & \beta_{12} & \cdots & \beta_{1(2N)} \\ \beta_{21} & \beta_{22} & \cdots & \beta_{2(2N)} \\ \vdots & \vdots & \vdots & \vdots \\ \beta_{(2N)1} & \beta_{(2N)2} & \cdots & \beta_{(2N)(2N)} \end{pmatrix}, \tag{1.111}$$

we note that we can write (1.110) in the matrix form as (compare with (1.103))

$$A \epsilon A^T = \epsilon. \tag{1.112}$$

Namely, we see that the real symplectic group of transformations $Sp(2n, \mathbb{R})$ defines the invariance group of the N-dimensional Heisenberg algebra (the reality condition arises from the fact that the operators are Hermitian). Unfortunately, however, the Hamiltonian, say for the harmonic oscillator system, is not invariant under the $Sp(2n, \mathbb{R})$ group of transformations defined by (1.107) and (1.112). More explicitly,

$$H = \frac{1}{2} \sum_{i=1}^{N} \left(p_i^2 + q_i^2 \right) \neq \frac{1}{2} \sum_{i=1}^{N} \left((p_i')^2 + (q_i')^2 \right), \tag{1.113}$$

where

$$\xi_\mu' = (q_1', p_1', q_2', p_2', \cdots, q_N', p_N'), \tag{1.114}$$

so that it is not a symmetry of the harmonic oscillator system.

The special case of the symplectic group for $N = 1$ is of some interest because it arises in various branches of physics. In this case, we can identify $Sp(2) \equiv SL(2)$ as follows. Let

$$A = \begin{pmatrix} \alpha & \beta \\ \gamma & \delta \end{pmatrix}, \tag{1.115}$$

denote a 2×2 symplectic matrix. Therefore, by definition (see (1.103)) it satisfies

$$A^T \epsilon_2 A = \epsilon_2,$$

or, $\begin{pmatrix} \alpha & \gamma \\ \beta & \delta \end{pmatrix} \begin{pmatrix} 0 & 1 \\ -1 & 0 \end{pmatrix} \begin{pmatrix} \alpha & \beta \\ \gamma & \delta \end{pmatrix} = \begin{pmatrix} 0 & 1 \\ -1 & 0 \end{pmatrix}$,

or, $\begin{pmatrix} 0 & (\alpha\delta - \beta\gamma) \\ -(\alpha\delta - \beta\gamma) & 0 \end{pmatrix} = \begin{pmatrix} 0 & 1 \\ -1 & 0 \end{pmatrix}$,

or, $\det A \begin{pmatrix} 0 & 1 \\ -1 & 0 \end{pmatrix} = \begin{pmatrix} 0 & 1 \\ -1 & 0 \end{pmatrix}$,

or, $\det A \, \epsilon_2 = \epsilon_2$, \tag{1.116}

where we have used the definition of the determinant,

$$\det A = \alpha\delta - \beta\gamma. \tag{1.117}$$

Equation (1.116) determines that

$$\det A = \alpha\delta - \beta\gamma = 1, \tag{1.118}$$

for a mtrix to belong to $Sp(2)$. However, this is also the condition for a matrix to belong to $SL(2)$ and this proves the equivalence (isomorphism) between the two dimensional groups $Sp(2)$ and $SL(2)$. This fact can be used to construct 2-component spinors in physics as we will show in a later chapter.

These two dimensional groups are also quite relevant in the study of Bogoliubov transformations (among other things). For example, let a and $a^\dagger$ denote the annihilation and creation operators for a harmonic oscillator system satisfying (see, for example, (1.57))

$$[a, a] = 0 = [a^\dagger, a^\dagger], \quad [a, a^\dagger] = 1. \tag{1.119}$$

Let us consider a transformation of the operators of the form

$$a \to b = \alpha a + \beta a^\dagger,$$
$$a^\dagger \to b^\dagger = \beta^* a + \alpha^* a^\dagger, \tag{1.120}$$

where we assume that the constant parameters of the transformation α, β satisfy

$$\alpha\alpha^* - \beta\beta^* = 1. \tag{1.121}$$

In this case, it is straightforward to check that the transformed operators satisfy the algebra

$$[b, b] = 0 = [b^\dagger, b^\dagger], \quad [b, b^\dagger] = 1. \tag{1.122}$$

For example, we note that

$$[b, b^\dagger] = [\alpha a + \beta a^\dagger, \beta^* a + \alpha^* a^\dagger]$$
$$= \alpha\alpha^* [a, a^\dagger] + \beta\beta^* [a^\dagger, a] = \alpha\alpha^* - \beta\beta^* = 1. \tag{1.123}$$

We note that if we define the matrices

$$\hat{a} = \begin{pmatrix} a \\ a^\dagger \end{pmatrix}, \quad \hat{b} = \begin{pmatrix} b \\ b^\dagger \end{pmatrix}, \quad A = \begin{pmatrix} \alpha & \beta \\ \beta^* & \alpha^* \end{pmatrix}, \tag{1.124}$$

then we can write the transformation (1.120) in the matrix form as

$$\hat{a} \to \hat{b} = A\hat{a}, \quad A\epsilon_2 A^T = A^T \epsilon_2 A = \det A \, \epsilon_2. \tag{1.125}$$

Here we have used the identification in (1.116). The commutation relations in (1.119) can now be written compactly as

$$[\hat{a}_i, \hat{a}_j] = (\epsilon_2)_{ij}, \quad i, j = 1, 2. \tag{1.126}$$

Calculating the commutation relations for the transformed operators in (1.125) (see (1.108)), it follows now that

$$[\hat{b}_i, \hat{b}_j] = \left(A\epsilon_2 A^T\right)_{ij} = \det A \, (\epsilon_2)_{ij}, \tag{1.127}$$

for $i, j = 1, 2$. The invariance condition for the commutation relations, (1.121), is now seen to correspond to

$$\det A = 1. \tag{1.128}$$

A transformation mixing the annihilation and creation operators is an example of a Bogoliubov transformation and we see that it is implemented through a matrix $A \in SL(2)$. (This is useful in the case of harmonic oscillator, for example, in the construction of coherent states.) We note here that the matrix A is not unitary, namely, $A^\dagger A \neq E_2$.

To conclude this discussion, we note that the groups $SL(N), U(N), O(N), SO(N), Sp(2N)$ are termed classical groups and discussed in detail by Hermann Weyl in his famous book *Classical Groups*. The book also includes a discussion of the symmetric group S_N as well as a more general linear group $GL(N)$. Without going into details, we simply note here that the general linear group $GL(N)$ consists of all $N \times N$ matrices A with only the restriction $\det A \neq 0$ (nonsingular) so that the inverse A^{-1} always exists.

1.3 Group manifold

In this chapter, we have discussed a finite group (with a finite number of elements) as well as several (Lie) groups which are quite important in physics. The (Lie) groups, as we have seen depend on some continuous parameters and correspondingly they are known as continuous groups which have an (nondenumerably) infinite number of elements. Let us introduce the concept of a group manifold for such groups which would help us in classifying infinite dimensional continuous groups (examples of which we have studied in the last section) into compact and non-compact groups. Basically, the parameters on which elements of a continuous group depend define a topological space known as the parameter space. Since the group elements depend on these parameters, every element can be put in one-to-one correspondence with points on the parameters. In this way, a continuous group inherits the topology of the parameter space and correspondingly is also known as a topological group.

We will illustrate these ideas through examples. Let us consider the one dimensional translation group T_1 discussed in section 1.2.2. In this case an element of the group $T(a)$ represents a translation of the coordinate (which can be a space or a time coordinate) by a real continuous parameter "a" with $-\infty < a < \infty$. Therefore, the parameter space, in this case, corresponds to the real line and it is clear that we can represent every group element of T_1 diagrammatically as a point on the real line. For example, a specific element $T(a)$ would have the coordinate "a" on this real line from the origin which can be identified with the identity element $e = T(0)$ (see (1.37)) as shown in Fig. 1.2. Topologically the real line can be identified with a Cartesian axes, say for example, the x-axis. Therefore, this shows that there is a one-to-one correspondence between the elements of the group T_1 and points on the x-axis and we can identify the group manifold of T_1 with the whole (entire) real line.

$$e = T(0) \qquad\qquad T(a)$$

$$0 \qquad\qquad\qquad a$$

Figure 1.2: Diagrammatic representation of one dimensional translations on a real line.

Let us next consider the example of the one dimensional unitary group U_1 discussed in section 1.2.3. A typical group element $U(\theta) = e^{i\theta}, 0 \leq \theta < 2\pi$ corresponds to a simple (one dimensional) phase rotation by an angle θ (see, for example, (1.41) and (1.42)). As a result, the parameter space, in such a case, can be identified with the unit circle (labelled only by an angle) and every element $U(\theta)$ of the group U_1 can be assigned a unique angle θ on the unit circle (angle which a vector from the origin to the unit circle makes with the x-axis) as in Fig. 1.3. Namely, there is a one-to-one correspondence between the elements of the group U_1 and points on the unit circle and we can identify the group manifold of U_1 with the unit circle. It is worth noting here that since $U(\theta + 2\pi n) = U(\theta)$ for any integer N, if the parameter θ were allowed to take any arbitrary value (without the restriction $0 \leq \theta < 2\pi$), there can no longer be a simple one-to-one correspondence between the group elements and the unit circle. In this case, the parameter space coincides with that of the real line and the topological space known as the covering group of $U(1)$ becomes equivalent to the group manifold of T_1.

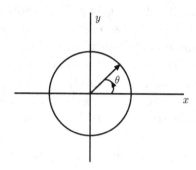

Figure 1.3: Diagrammatic representation of the one dimensional unitary group on a unit circle.

We now turn to the more complicated case of the group U_N discussed in section 1.2.4. Here the group elements are realized as $N \times N$ unitary matrices A whose elements $\alpha_{ij}, i, j = 1, 2, \cdots, N$ are complex parameters. In general, a matrix with N^2 complex parameters (elements) can be equivalently thought of as described by $2N^2$ independent real parameters. However, the unitarity condition (1.48)

$$AA^\dagger = E_N = A^\dagger A, \qquad (1.129)$$

imposes N^2 real conditions on the elements of a $N \times N$ unitary matrix (note that the identity (unit) matrix is real which is the reason for the N^2 real conditions). As a result, the matrix A can be thought of as being parameterized by N^2 independent real parameters which we can take to be, for example, $\mathrm{Re}\,\alpha_{ij}, i, j = 1, 2, \cdots, N$. Therefore, it would appear that the parameter space can be identified with a Euclidean manifold with N^2 independent axes. However, we note from (1.51) that the magnitudes of the elements α_{ij} are bounded by unity so that all the points in this space must satisfy

$$|\mathrm{Re}\,\alpha_{ij}| \leq 1, \qquad (1.130)$$

and the space corresponding to the unit ball in N^2 dimensions is bounded just like the unit circle (in the case of $U(1)$). Since every element of $U(N)$ can be mapped to a point in this unit ball, in this case the group manifold can be identified with the unit ball in N^2 dimensions. The group manifolds for more complicated continuous groups can similarly be constructed and are much more complicated in general. Nonetheless we can always assign a correspondence between the elements of a given group and points on a topological space (defined by the parameters) which is known as the group manifold for the particular group.

A topological space with a given topology (see, for example, the book by J. L. Kelley) is called compact if it has a bounded and closed topology. The space is known as non-compact otherwise. The continuous groups are correspondingly classified as compact or non-compact depending on the nature of the group manifold (topological space) associated with them. It is clear from this that the group of translations T_1 is non-compact since the real line associated with this group is not bounded. On the other hand, the unitary group $U(N)$ is compact since the group manifold is bounded. In a similar manner, it can be shown that the groups $SU(N), SO(N)$ and $O(N)$ are also compact. However, it is worth noting that orthogonal groups of the form $O(m, n)$ with nontrivial m, n are non-compact. For example, in the case of the group $O(3, 1)$ we note from (1.88) that the parameters of the transformation (matrix elements $a_{\mu\nu}$) do not have to be bounded in order to satisfy (1.88) (because it involves a difference). As a consequence, Lorentz transformations are non-compact (the boosts are not bounded). Similarly, the Poincaré group is also non-compact since the space-time translations are not bounded (in addition to the fact that the Lorentz transformations are, in general, not bounded). We note here that the group SL_N is also not compact. This can be easily seen by restricting to $n = 2$ where we can label the matrix as (see (1.115))

$$A = \begin{pmatrix} \alpha & \beta \\ \gamma & \delta \end{pmatrix}, \tag{1.131}$$

satisfying (see also (1.117) and (1.118))

$$\det A = \alpha\delta - \beta\gamma = 1. \tag{1.132}$$

However, without the unitarity condition (1.48) (for example, in the case of $SU(2)$) or the orthogonality condition (1.67), the matrix elements $\alpha, \beta, \gamma, \delta$ satisfying only (1.132) are no longer bounded (namely, they can be arbitrarily large and still satisfy (1.132)). As a result, the group manifold (and, therefore, the group) is non-compact. We note that since the groups $SL(2)$ and $Sp(2)$ are equivalent, $Sp(2)$ is non-compact as well (so is $Sp(2N)$). Finally, we note that any finite group can be thought of as compact since it is bounded and closed (in the sense of a finite set).

1.4 References

H. Weyl, *The Classical Groups*, Princeton University Press, Princeton, NJ (1939).

H. Weyl, *The Theory of Groups and Quantum Mechanics*, (translated from German by H. P. Robertson), Dover Publication, NY (1950).

J. L. Kelley, *General Topology*, Van Nostrand, NY (1957).

CHAPTER 2

Representation of groups

While the structure of the abstract group is quite interesting, in physics the representation of the group itself is more important. The most common representations of (Lie) groups are in terms of finite dimensional $n \times n$ matrices, although sometimes the infinite dimensional (unitary) representations may be more meaningful for non-compact groups. In this chapter we will discuss the matrix representations as well as representations of groups by linear operators in a vector space.

2.1 Matrix representation of a group

The $n \times n$ matrix representation of a group G is defined as follows. For every element of the group $a \in G$, if we can construct an invertible $n \times n$ matrix $U(a)$ such that

(i) we can identify

$$U(e) = E_n, \tag{2.1}$$

where $E_n = \mathbb{1}_n$ denotes the $n \times n$ identity (unit) matrix,

(ii) and for any two elements $a, b \in G$, the matrices $U(a)$ and $U(b)$ satisfy the relation

$$U(a)U(b) = U(ab), \tag{2.2}$$

then, the set of all such $n \times n$ matrices $\{U(a)\}$ forms a group and provides a $(n \times n$ matrix) representation of the group G. We note that in (2.2) the product on the left hand side is the natural associative matrix product which defines the product rule for such a matrix group.

Indeed it is straightforward to check that such a set of matrices forms a group in the following manner.

(i) We note from the defining relation (2.2) that

$$U(a)U(b) = U(ab) = U(c), \qquad (2.3)$$

is well defined with the matrix product rule and is in the set of matrices $\{U(a)\}$. Here we have used the group property (1.1), namely, for any two elements $a, b \in G$ of the group, the group multiplication leads to $ab = c \in G$.

(ii) Since the matrix product is known to be associative, it follows that for any three matrices in the set we have

$$(U(a)U(b))U(c) = U(a)(U(b)U(c)), \qquad (2.4)$$

which should be compared with (1.2). In fact, given the group associative property (1.2) we note that using (2.2) we can write

$$(U(a)U(b))U(c) = U(ab)U(c) = U((ab)c),$$
$$U(a)(U(b)U(c)) = U(a)U(bc) = U(a(bc)), \qquad (2.5)$$

so that we have

$$(U(a)U(b))U(c) = U(a)(U(b)U(c)) = U(abc), \qquad (2.6)$$

since $(ab)c = a(bc) = abc$ (see (1.2)).

(iii) By definition (2.1), the set has a unique identity (unit) element $U(e) = E_n$ which, using the matrix product, satisfies

$$U(e)U(a) = E_nU(a) = U(a) = U(a)E_n = U(a)U(e), \quad (2.7)$$

which should be compared with (1.3). The relation (2.7) can also be viewed from the point of view of group multiplication as

$$U(e)U(a) = U(ea) = U(a) = U(ae) = U(a)U(e), \qquad (2.8)$$

corresponding to $ea = a = ae$.

(iv) Since by definition

$$U(a)U(a^{-1}) = U(aa^{-1}) = U(e) = E_n$$
$$= U(a^{-1})U(a), \qquad (2.9)$$

it follows that

$$\det U(a) \det U(a^{-1}) = \det \left(U(a)U(a^{-1}) \right)$$
$$= \det U(e) = \det E_n = 1. \qquad (2.10)$$

This shows that

$$\det U(a) \neq 0, \qquad (2.11)$$

so that $(U(a))^{-1}$ exists and satisfies

$$U(a)(U(a))^{-1} = E_n = (U(a))^{-1}U(a). \qquad (2.12)$$

Comparing (2.9) and (2.12) we conclude that

$$(U(a))^{-1} = U(a^{-1}). \qquad (2.13)$$

Conversely, we can multiply (2.12) with $U(a^{-1})$ either from left or from right and use the group composition to show that the identification in (2.13) follows.

This shows that the set of $n \times n$ matrices constructed satisfying (2.1) and (2.2) forms a group and provides a matrix representation for the group G.

We note here that if the set of matrices $\{U(a)\}$ defines a $n \times n$ matrix representation of a group G, then the set $\{S^{-1}U(a)S\}$, where S denotes any arbitrary $n \times n$ invertible matrix, also provides a $n \times n$ matrix representation of the group G. This is easily seen by noting that if we identify

$$W(a) = S^{-1}U(a)S, \qquad (2.14)$$

then, using the matrix product

(i) we have

$$W(e) = S^{-1}U(e)S = S^{-1}E_nS = S^{-1}S = E_n, \qquad (2.15)$$

(ii) and, furthermore,

$$W(a)W(b) = S^{-1}U(a)SS^{-1}U(b)S = S^{-1}U(a)E_nU(b)S$$
$$= S^{-1}U(a)U(b)S = S^{-1}U(ab)S = W(ab), \qquad (2.16)$$

which can be compared with (2.1) and (2.2) respectively. Following the same arguments as before, we conclude that the set $\{W(a) = S^{-1}U(a)S\}$ also defines a group and provides a $n \times n$ matrix representation of the group G. This suggests that the matrix representation of a group G may not be unique since we can construct infinitely many sets of matrices similarly related to the set $\{U(a)\}$ for different invertible matrices S. However, we regard all such sets of matrices, related by a similarity transformation, to be equivalent and do not consider $\{W(a)\}$ to provide a distinct representation of G. This can be understood in the following manner.

In the last chapter, we have pointed out that abstract groups can be realized as groups of transformations acting on appropriate linear vector spaces. For example, we have seen that the $SO(N)$ group can be associated with the group of rotations in a N-dimensional coordinate space while the $U(N)$ group rotates the creation and annihilation operators of N harmonic oscillators. Transformations in a linear vector space may be implemented through linear operators acting on vectors. (Only for a very unique discrete transformation is the corresponding operator anti-linear.) Therefore, given a group G, we may identify every element of the group ($a \in G$) with a linear operator A in a finite dimensional vector space $\mathbb{V}_n$,

$$a \rightarrow A. \tag{2.17}$$

Let us consider an orthonormal basis $\{x_i\}$, $i = 1, 2, \cdots, n$ in $\mathbb{V}_n$. A linear operator in $\mathbb{V}_n$ takes one vector in the vector space into another which can be expanded in the basis $|x_i\rangle$ (the Dirac notation for a state vector). In particular, we note that

$$A x_i = x_i' = \sum_{j=1}^{n} \alpha_{ji} x_j, \quad i = 1, 2, \cdots, n. \tag{2.18}$$

The n^2 coefficients of expansion α_{ij} describing the action of the linear operator A on the n basis vectors x_i, $i = 1, 2, \cdots, n$ can be thought of as the components of a $n \times n$ matrix giving a matrix representation U for the linear operator A. In this way, we can construct a $n \times n$ matrix representation for the group G through linear operators acting on $\mathbb{V}_n$, namely,

$$a \rightarrow A \rightarrow U(a), \tag{2.19}$$

with

$$(U(a))_{ij} = \alpha_{ij}, \tag{2.20}$$

which can also be written explicitly as

$$U(a) = \begin{pmatrix} \alpha_{11} & \alpha_{12} & \cdots & \alpha_{1n} \\ \alpha_{21} & \alpha_{22} & \cdots & \alpha_{2n} \\ \vdots & \vdots & \vdots & \vdots \\ \alpha_{n1} & \alpha_{n2} & \cdots & \alpha_{nn} \end{pmatrix}. \tag{2.21}$$

As a result, the relation (2.18) can also be written as

$$Ax_i = \sum_{j=1}^{n} (U(a))_{ji} x_j. \tag{2.22}$$

We can verify that these matrices indeed satisfy the group properties in the following manner.

(i) We note that if we identify (where $\mathbb{1}$ denotes the identity operator)

$$e \to \mathbb{1}, \tag{2.23}$$

then it follows that since the identity operator leaves every vector invariant

$$\mathbb{1}x_i = x_i, \quad i = 1, 2, \cdots, n, \tag{2.24}$$

and we can identify

$$(U(e))_{ij} = \delta_{ij},$$

$$\text{or,} \quad U(e) = \begin{pmatrix} 1 & 0 & \cdots & 0 \\ 0 & 1 & \cdots & 0 \\ \vdots & 0 & \ddots & \vdots \\ 0 & 0 & \cdots & 1 \end{pmatrix} = E_n. \tag{2.25}$$

(ii) Furthermore, for any two other elements of the group

$$b \to B, \quad c \to C, \tag{2.26}$$

where B and C are linear operators acting on $\mathbb{V}_n$, if we have

$$ab = c, \tag{2.27}$$

then we can write

$$AB = C. \tag{2.28}$$

On the other hand, following (2.18)-(2.22), we note that we can write

$$Bx_i = \sum_{j=1}^{n} \beta_{ji} x_j = \sum_{j=1}^{n} (U(b))_{ji} x_j,$$

$$Cx_i = \sum_{j=1}^{n} \gamma_{ji} x_j = \sum_{j=1}^{n} (U(c))_{ji} x_j, \tag{2.29}$$

where we have identified

$$(U(b))_{ij} = \beta_{ij},$$

$$\text{or,} \quad U(b) = \begin{pmatrix} \beta_{11} & \beta_{12} & \cdots & \beta_{1n} \\ \beta_{21} & \beta_{22} & \cdots & \beta_{2n} \\ \vdots & \vdots & \vdots & \vdots \\ \beta_{n1} & \beta_{n2} & \cdots & \beta_{nn} \end{pmatrix},$$

and

$$(U(c))_{ij} = \gamma_{ij},$$

$$\text{or,} \quad U(c) = \begin{pmatrix} \gamma_{11} & \gamma_{12} & \cdots & \gamma_{1n} \\ \gamma_{21} & \gamma_{22} & \cdots & \gamma_{2n} \\ \vdots & \vdots & \vdots & \vdots \\ \gamma_{n1} & \gamma_{n2} & \cdots & \gamma_{nn} \end{pmatrix}. \tag{2.30}$$

It follows now that (by definition, a linear operator acts only on vectors and not constants)

$$ABx_i = A \sum_{j=1}^{n} (U(b))_{ji} x_j = \sum_{j=1}^{n} (U(b))_{ji} \, Ax_j$$

$$= \sum_{j,k=1}^{n} (U(b))_{ji} (U(a))_{kj} x_k$$

$$= \sum_{k=1}^{n} (U(a)U(b))_{ki} x_k. \tag{2.31}$$

Comparing (2.31) with (2.28) and (2.29), we conclude that

$$(U(a)U(b))_{ij} = (U(ab))_{ij} = (U(c))_{ij}. \tag{2.32}$$

It follows now from (2.1) and (2.2) that the set of matrices $\{U(a)\}$ do satisfy the group properties and provide a matrix representation of the group.

However, we note that the matrix elements of a linear operator are basis dependent. For example, as we have seen in (2.20) and (2.22), in the basis x_i, $i = 1, 2, \cdots, n$ we can write

$$Ax_i = \sum_{j=1}^{n} (U(a))_{ji} x_j. \tag{2.33}$$

On the other hand, if we have another basis $|y_i\rangle$, $i = 1, 2, \cdots, n$ in $\mathbb{V}_n$, related to $|x_i\rangle$ through a similarity transformation S ($\det S \neq 0$ by definition of a similarity transformation, namely, it is invertible), we can write

$$y_i = \sum_{j=1}^{n} S_{ji} x_j. \tag{2.34}$$

It follows now that

$$Ay_i = A \sum_{j=1}^{n} S_{ji} x_j = \sum_{j=1}^{n} S_{ji} A x_j$$

$$= \sum_{j,k=1}^{n} S_{ji} (U(a))_{kj} x_k = \sum_{j,k,\ell=1}^{n} S_{ji} (U(a))_{kj} (S^{-1})_{\ell k} y_\ell$$

$$= \sum_{\ell=1}^{n} (S^{-1} U(a) S)_{\ell i} y_\ell, \tag{2.35}$$

where we have used (2.22) as well as (2.34). The coefficients of expansions in the new basis $|y_i\rangle$ are given by the elements of the matrix $S^{-1}U(a)S$ which would also provide a matrix representation of the group, namely, $W(a) = S^{-1}U(a)S$ (see (2.14)). Thus, we see that the same linear operator can lead to different matrix representations of the group depending on the choice of basis. Since the choice of a basis does not alter the physical contents of operators, all such matrix representations are considered to be equivalent to $U(a)$ and not distinct from $U(a)$ unless stated explicitly.

As an aside we note that in (2.34) we have represented the similarity transformation as

$$y_i = \sum_{j=1}^{n} S_{ji} x_j, \tag{2.36}$$

instead of the conventional notation

$$y_i = \sum_{j=1}^{n} S_{ij} x_j. \tag{2.37}$$

The two relations (2.36) and (2.37) are, of course, physically equivalent and can be understood as associated with the active or the passive description of transformations. For example, let us consider a given vector V in the vector space which can be expanded in the basis x_i as

$$V = \sum_{i=1}^{n} v_i x_i. \tag{2.38}$$

Under a transformation

$$V \to V' = SV, \tag{2.39}$$

the change in the vector can be viewed in one of the two following equivalent ways. We may expand the transformed vector in the same basis as

$$V' = \sum_{i=1}^{n} v'_i x_i, \tag{2.40}$$

which can be thought of as implying (see (2.39)) that the basis (of expansion) is unchanged under the transformation, rather the coefficients of expansion change. In other words, the vector transforms, but not the basis (of expansion). This is known as the active picture of transformation. In this case, using the linear nature of operators and (2.18) we determine from (2.38)-(2.39) and (2.40) that

$$V' = SV = \sum_{i=1}^{n} v_i S x_i = \sum_{i,j=1}^{n} v_i S_{ji} x_j$$

$$= \sum_{j=1}^{n} (Sv)_j x_j = \sum_{i=1}^{n} v'_i x_i,$$

$$\text{or,} \quad v'_i = (Sv)_i = \sum_{j=1}^{n} S_{ij} v_j, \tag{2.41}$$

so that we can write

$$V' = \sum_{i=1}^{n} v'_i x_i = \sum_{i,j=1}^{n} S_{ij} v_j x_i. \tag{2.42}$$

In particular, in this picture, if we consider the vector $V = x_i$, for which the expansion (2.38) takes the form (δ_{ij} represents the Kronecker delta)

$$V = x_i = \sum_{j=1}^{n} \delta_{ij} x_j, \tag{2.43}$$

then it follows from (2.42) that under the transformation

$$x_i \to x'_i = y_i = \sum_{j=1}^{n} S_{ij} x_j, \tag{2.44}$$

which coincides with (2.37) and shows that this form is associated with the active view of the transformation.

Alternatively, we can view the transformation in (2.39) as changing the basis of expansion leaving the coefficients of expansion unchanged. This would then correspond to leaving the vector unchanged while changing the basis and is known as the passive view of the coordinate transformation. In this case, we can write

$$V' = \sum_{i=1}^{n} v_i x'_i. \tag{2.45}$$

Using (2.42), we conclude that in this case we have

$$x'_i = y_i = \sum_{j=1}^{n} S_{ji} x_j, \tag{2.46}$$

which coincides with (2.36) and shows that this form is associated with the passive view of the transformation. In our discussion we have been using a passive transformation and (2.36) is correspondingly the appropriate relation.

▶ **Example (Translations in one dimension).** Let us end the discussion of this section with an example of the representation of the translation group in one dimension T_1 through matrices as well as through linear operators. Let us recall that the group of translations in one dimension is defined by (see (1.26))

$$x \xrightarrow{T(a)} x + a, \tag{2.47}$$

where a is the real constant parameter of translation. Let us now introduce a basis in the n-dimensional vector space of coordinates through

$$\psi_m(x) = x^m, \qquad m = 0, 1, 2, \cdots, n, \tag{2.48}$$

and note that the basis functions naturally satisfy

$$\frac{d\psi_m(x)}{dx} = m\psi_{m-1}(x). \tag{2.49}$$

We can arrange the basis functions in the (n-dimensional) column matrix (vector) form

$$\Psi_n(x) = \begin{pmatrix} \psi_0(x) \\ \psi_1(x) \\ \vdots \\ \psi_{n-1}(x) \end{pmatrix}, \tag{2.50}$$

so that the action of translation on this column matrix can be written as

$$\Psi_n(x) \xrightarrow{T(a)} \Psi_n(x+a). \tag{2.51}$$

With this notation, we can now obtain the matrix representations of various dimensions for the translation group. For example, for $n = 2$ we obtain the two dimensional matrix representation

$$\Psi_2(x) = \begin{pmatrix} \psi_0(x) \\ \psi_1(x) \end{pmatrix} = \begin{pmatrix} 1 \\ x \end{pmatrix} \xrightarrow{T(a)} \begin{pmatrix} \psi_0(x+a) \\ \psi_1(x+a) \end{pmatrix} = \begin{pmatrix} 1 \\ x+a \end{pmatrix} = \begin{pmatrix} 1 & 0 \\ a & 1 \end{pmatrix} \begin{pmatrix} 1 \\ x \end{pmatrix}$$

$$= \begin{pmatrix} 1 & 0 \\ a & 1 \end{pmatrix} \begin{pmatrix} \psi_0(x) \\ \psi_1(x) \end{pmatrix} = U_2(a)\Psi_2(x), \tag{2.52}$$

where we have identified

$$U_2(a) = \begin{pmatrix} 1 & 0 \\ a & 1 \end{pmatrix} = \mathbb{1}_2 + a\mathcal{D}_2, \qquad \mathcal{D}_2 = \begin{pmatrix} 0 & 0 \\ 1 & 0 \end{pmatrix}. \tag{2.53}$$

Using (2.49) we note that the matrix $\mathcal{D}_2$ acts like the derivative on the two dimensional matrices, namely,

$$\mathcal{D}_2 \Psi_2(x) = \begin{pmatrix} 0 & 0 \\ 1 & 0 \end{pmatrix} \begin{pmatrix} \psi_0(x) \\ \psi_1(x) \end{pmatrix} = \begin{pmatrix} 0 \\ \psi_0(x) \end{pmatrix} = \Psi_2'(x), \tag{2.54}$$

where the prime denotes a derivative with respect to x. Thus, (2.53) indeed coincides with (1.31). Furthermore, it is easily checked that the matrix $U_2(a)$ satisfies the group property

$$U_2(a)U_2(b) = U_2(a+b), \tag{2.55}$$

which can be compared with (1.34).

For $n = 3$ we can derive the three dimensional matrix representation for the translation group as

$$\Psi_3(x) = \begin{pmatrix} \psi_0(x) \\ \psi_1(x) \\ \psi_2(x) \end{pmatrix} = \begin{pmatrix} 1 \\ x \\ x^2 \end{pmatrix} \xrightarrow{T(a)} \begin{pmatrix} \psi_0(x+a) \\ \psi_1(x+a) \\ \psi_2(x+a) \end{pmatrix} = \begin{pmatrix} 1 \\ x+a \\ (x+a)^2 \end{pmatrix}$$

$$= \begin{pmatrix} 1 & 0 & 0 \\ a & 1 & 0 \\ a^2 & 2a & 1 \end{pmatrix} \begin{pmatrix} 1 \\ x \\ x^2 \end{pmatrix} = \begin{pmatrix} 1 & 0 & 0 \\ a & 1 & 0 \\ a^2 & 2a & 1 \end{pmatrix} \begin{pmatrix} \psi_0(x) \\ \psi_1(x) \\ \psi_2(x) \end{pmatrix}$$

$$= U_3(a)\Psi_3(x), \tag{2.56}$$

where we have identified

$$U_3(a) = \begin{pmatrix} 1 & 0 & 0 \\ a & 1 & 0 \\ a^2 & 2a & 1 \end{pmatrix} = \mathbb{1}_3 + a\mathcal{D}_3 + \frac{a^2}{2!}(\mathcal{D}_3)^2, \tag{2.57}$$

with

$$\mathcal{D}_3 = \begin{pmatrix} 0 & 0 & 0 \\ 1 & 0 & 0 \\ 0 & 2 & 0 \end{pmatrix}. \tag{2.58}$$

Once again, using (2.49) it can be easily checked that

$$\mathcal{D}_3 \Psi_3(x) = \Psi_3'(x), \tag{2.59}$$

and that the matrices $U_3(a)$ satisfy the group property

$$U_3(a)U_3(b) = U_3(a+b). \tag{2.60}$$

In this way, one can obtain the n-dimensional representation for the (one dimensional) translation group to be

$$\Psi_n(x) \xrightarrow{T(a)} \Psi_n(x+a) = U_n(a)\Psi_n(x), \tag{2.61}$$

where we have

$$U_n(a) = \begin{pmatrix} 1 & 0 & 0 & 0 & 0 & \cdots \\ a & 1 & 0 & 0 & 0 & \cdots \\ a^2 & 2a & 1 & 0 & 0 & \\ a^3 & 3a^2 & 3a & 1 & & \\ \vdots & \vdots & \vdots & \ddots & \ddots & \end{pmatrix}$$

$$= \mathbb{1}_n + a\mathcal{D}_n + \frac{a^2}{2!}(\mathcal{D}_n)^2 + \cdots + \frac{a^{n-1}}{(n-1)!}(\mathcal{D}_n)^{n-1}, \tag{2.62}$$

with

$$\mathcal{D}_n = \begin{pmatrix} 0 & & & & \\ 1 & 0 & & & \\ 0 & 2 & 0 & & \\ 0 & 0 & 3 & 0 & \\ \vdots & \vdots & \ddots & \ddots & \ddots \end{pmatrix}, \tag{2.63}$$

denoting the derivative in the n-dimensional space, namely,

$$\mathcal{D}_n \Psi_n(x) = \Psi_n'(x). \tag{2.64}$$

Furthermore, it can be checked that

$$U_n(a)U_n(b) = U_n(a+b). \tag{2.65}$$

It is clear from the structures of the matrices in (2.53), (2.57) and (2.62) that these finite dimensional representations of the translation group are not unitary. Furthermore, taking the limit $n \to \infty$, this leads to an infinite dimensional (reducible) representation of the translation group. (We discuss reducible representations in the next section.)

The effect of a change of basis can also be studied in this example in a straightforward manner. For example, let us consider the two dimensional representation discussed in (2.52)-(2.55). We note that if we change the basis of two dimensional vectors from (2.52) to

$$
\Xi_2(x) = \begin{pmatrix} \xi_0(x) \\ \xi_1(x) \end{pmatrix} = \begin{pmatrix} 1+x \\ 1-x \end{pmatrix}
$$

$$
= \begin{pmatrix} \psi_0(x) + \psi_1(x) \\ \psi_0(x) - \psi_1(x) \end{pmatrix} = \begin{pmatrix} 1 & 1 \\ 1 & -1 \end{pmatrix} \begin{pmatrix} \psi_0(x) \\ \psi_1(x) \end{pmatrix} = S\Psi_2(x), \tag{2.66}
$$

where

$$
S = \begin{pmatrix} 1 & 1 \\ 1 & -1 \end{pmatrix}, \quad S^{-1} = \frac{1}{2} \begin{pmatrix} 1 & 1 \\ 1 & -1 \end{pmatrix}. \tag{2.67}
$$

Under a translation, such a two dimensional vector transforms as

$$
\Xi_2(x) = \begin{pmatrix} \xi_0(x) \\ \xi_1(x) \end{pmatrix} = \begin{pmatrix} 1+x \\ 1-x \end{pmatrix} \xrightarrow{T(a)} \begin{pmatrix} 1+x+a \\ 1-x-a \end{pmatrix}
$$

$$
= \begin{pmatrix} 1+\frac{a}{2} & \frac{a}{2} \\ -\frac{a}{2} & 1-\frac{a}{2} \end{pmatrix} \begin{pmatrix} 1+x \\ 1-x \end{pmatrix}
$$

$$
= W_2(a) \begin{pmatrix} \xi_0(x) \\ \xi_1(x) \end{pmatrix}, \tag{2.68}
$$

where we have identified

$$
W_2(a) = \begin{pmatrix} 1+\frac{a}{2} & \frac{a}{2} \\ -\frac{a}{2} & 1-\frac{a}{2} \end{pmatrix}. \tag{2.69}
$$

It can be checked in a straightforward manner that

$$
W_2(a)W_2(b) = W_2(a+b), \tag{2.70}
$$

so that $W_2(a)$ also provides a two dimensional representation of the translation group. From (2.53), (2.67) and (2.69), we also recognize that we can write

$$
W_2(a) = S^{-1}U_2(a)S, \tag{2.71}
$$

which shows that the two representations are similar and (2.71) can be compared with (2.14).

As we have mentioned, we can also construct a representation of the translation group through linear operators. For example, let us identify

$$
T(a) \to D(a) = \exp\left(a\frac{d}{dx}\right), \tag{2.72}
$$

and note that the linear operator $D(a)$ satisfies

$$
D(a)D(b) = \exp\left(a\frac{d}{dx}\right)\exp\left(b\frac{d}{dx}\right) = \exp\left((a+b)\frac{d}{dx}\right) = D(a+b). \tag{2.73}
$$

Therefore, the operators $D(a)$ satisfy the group properties of translations (see (1.34)). Furthermore, acting on any differentiable function $f(x)$, they lead to

$$
D(a)f(x) = \exp\left(a\frac{d}{dx}\right)f(x) = \sum_{n=0}^{\infty} \frac{1}{n!}\left(a\frac{d}{dx}\right)^n f(x) = f(x+a), \tag{2.74}
$$

so that they truly translate the coordinate of the function and, therefore, provide a representation of the translation group.

We note that we can generalize the construction (2.72) to higher dimensional translation group T_N as follows. Let $\mathbf{x} = (x_1, x_2, \cdots, x_N)$ be the coordinates of a N-dimensional coordinate space and consider the translations

$$\mathbf{x} \xrightarrow{T(\mathbf{a})} \mathbf{x} + \mathbf{a}, \quad \text{or}, \quad x_i \to x_i + a_i, \quad i = 1, 2, \cdots, N. \tag{2.75}$$

where we have identified $\mathbf{a} = (a_1, a_2, \cdots, a_N)$. In this case, we can consider the linear operator

$$D(\mathbf{a}) = \exp\left(\mathbf{a} \cdot \boldsymbol{\nabla}\right) = \exp\left(\sum_{i=1}^{N} a_i \frac{\partial}{\partial x_i}\right), \tag{2.76}$$

which can be easily shown to satisfy

$$D(\mathbf{a})D(\mathbf{b}) = D(\mathbf{a} + \mathbf{b}), \tag{2.77}$$

which coincides with the group composition of the translation group T_N. As in the one dimensional translation group, we can also construct various (matrix) representations of T_N by restricting to appropriate functional spaces.

Finally, let us complete the discussion of this example with a nontrivial one dimensional representation of the translation group T_1. Let us consider an exponential function of the form

$$g_\alpha(x) = \exp(\alpha x), \quad \alpha \neq 0. \tag{2.78}$$

Such a function would transform under a translation as

$$g_\alpha(x) \xrightarrow{T(a)} \exp(\alpha(x + a)) = \exp(\alpha a)\exp(\alpha x)$$
$$= \exp(\alpha a)\, g_\alpha(x) = U_\alpha(a) g_\alpha(x), \tag{2.79}$$

where we have identified

$$U_\alpha(a) = \exp(\alpha a). \tag{2.80}$$

We note that

$$U_\alpha(a)U_\alpha(b) = \exp(\alpha a)\exp(\alpha b) = \exp(\alpha(a + b)) = U_\alpha(a + b), \tag{2.81}$$

so that $U_\alpha(a)$ provides a one dimensional (1×1 matrix) irreducible representation of the translation group. (We will discuss the concept of reducibility and irreducibility in the next section.) We simply note here that $U_\alpha(a)$ and $U_\beta(a)$, for $\alpha \neq \beta$, provide inequivalent representations.

◄

2.2 Unitary and irreducible representations

In physics unitary transformations are quite important. Let $U(a)$ denote a matrix representation of a group G of transformations. The matrix $U(a)$ may or may not be unitary. (We have already seen in the example of the representations of the one dimensional translation, discussed at the end of the last section, that finite dimensional matrix representations are not unitary.) If the matrices are unitary, then the representation of the group is known as a unitary representation. Even when $U(a)$ is not a unitary matrix, if we can find a similarity transformation S leading to an equivalent representation

$$W(a) = S^{-1}U(a)S, \tag{2.82}$$

which satisfies

$$(W(a))^\dagger W(a) = E, \quad \text{for all} \quad a \in G, \tag{2.83}$$

the representation is known as a unitary representation.

Let us give a simple 2×2 matrix example to illustrate this. Let us consider the matrix

$$U = \frac{1}{\sqrt{2}} \begin{pmatrix} 1-i & 0 \\ i & 1+i \end{pmatrix}. \tag{2.84}$$

Clearly, this is not unitary since

$$UU^\dagger = \frac{1}{2} \begin{pmatrix} 1-i & 0 \\ i & 1+i \end{pmatrix} \begin{pmatrix} 1+i & -i \\ 0 & 1-i \end{pmatrix}$$

$$= \frac{1}{2} \begin{pmatrix} 2 & -1-i \\ -1+i & 3 \end{pmatrix} \neq E_2, \tag{2.85}$$

and, similarly,

$$U^\dagger U = \frac{1}{2} \begin{pmatrix} 3 & 1-i \\ 1+i & 2 \end{pmatrix} \neq E_2. \tag{2.86}$$

On the other hand, let us consider a nonsingular matrix

$$S = \begin{pmatrix} 1 & -1 \\ 0 & 1 \end{pmatrix}, \quad S^{-1} = \begin{pmatrix} 1 & 1 \\ 0 & 1 \end{pmatrix}. \tag{2.87}$$

It is a simple calculation to see now that

$$W = S^{-1}US = \frac{1}{\sqrt{2}} \begin{pmatrix} 1 & 1 \\ 0 & 1 \end{pmatrix} \begin{pmatrix} 1-i & 0 \\ i & 1+i \end{pmatrix} \begin{pmatrix} 1 & -1 \\ 0 & 1 \end{pmatrix}$$

$$= \frac{1}{\sqrt{2}} \begin{pmatrix} 1 & i \\ i & 1 \end{pmatrix}, \tag{2.88}$$

which is clearly unitary since

$$WW^\dagger = W^\dagger W = E_2. \tag{2.89}$$

In a similar manner, if we can find a similarity transformation S such that the $n \times n$ matrix $U(a)$ with $n = n_1 + n_2$ and $n_1, n_2 \geq 1$ (providing a representation for the group G) can be brought into either of the forms

$$W(a) = S^{-1}U(a)S = \begin{cases} \begin{pmatrix} W_1(a) & N(a) \\ 0 & W_2(a) \end{pmatrix}, \\ \text{or,} \\ \begin{pmatrix} W_1(a) & 0 \\ \widetilde{N}(a) & W_2(a) \end{pmatrix}, \end{cases} \tag{2.90}$$

the representation is known as reducible. Here $W_1(a)$ and $W_2(a)$ are $n_1 \times n_1$ and $n_2 \times n_2$ square (sub) matrices respectively. On the other hand, $N(a)$ denotes a $n_1 \times n_2$ matrix while $\widetilde{N}(a)$ is a $n_2 \times n_1$ matrix. We note that since $W(a)$ would also provide a matrix representation for the group, we have

$$W(ab) = W(a)W(b), \quad a, b \in G, \tag{2.91}$$

which in turn requires that the diagonal blocks satisfy

$$W_1(ab) = W_1(a)W_1(b), \quad W_2(ab) = W_2(a)W_2(b). \tag{2.92}$$

On the other hand, the off-diagonal blocks in the two cases in (2.90) have to satisfy respectively

$$N(ab) = W_1(a)N(b) + N(a)W_2(b), \tag{2.93}$$

or

$$\widetilde{N}(ab) = \widetilde{N}(a)W_1(b) + W_2(a)\widetilde{N}(b). \tag{2.94}$$

It is clear from (2.92) that both $W_1(a)$ and $W_2(a)$ provide respectively $n_1 \times n_1$ and $n_2 \times n_2$ matrix representations of the group G. In particular, it can be easily verified that

$$W_1(e) = E_{n_1}, \quad W_2(e) = E_{n_2}, \tag{2.95}$$

where E_{n_1} and E_{n_2} denote (unit) identity matrices of dimensions n_1 and n_2 respectively. As we have mentioned earlier, when a similarity transformation S can be found which brings the matrix $U(a)$ to the form (2.90), the (original) representation is known as reducible. In particular, if $N(a) = 0$ (or $\widetilde{N}(a) = 0$) for all $a \in G$, then we have

$$W(a) = S^{-1}U(a)S = \begin{pmatrix} W_1(a) & 0 \\ 0 & W_2(a) \end{pmatrix}. \tag{2.96}$$

If we cannot find a similarity transformation leading to the reducible form of (2.96), we say that the representation $U(a)$ is irreducible. More generally, if we can find a similarity transformation S such that $U(a)$ can be brought to a block diagonal form

$$W(a) = S^{-1}U(a)S = \begin{pmatrix} W_1(a) & & & \\ & W_2(a) & & \\ & & W_3(a) & \\ & & & \ddots \end{pmatrix}, \tag{2.97}$$

where the diagonal (sub) matrices $W_1(a), W_2(a), \cdots$ are irreducible, we call such a representation fully reducible.

The origin of the name "reducible" can be understood as follows. We note that if we identify the linear operator A with the group element $a \in G$ (see (2.17), namely $a \to A$), then following (2.19) as well as (2.33)-(2.35) we have the transformation of a n dimensional vector under the linear transformation A given by

$$Ay_i = \sum_{j=1}^{n} \left(S^{-1}U(a)S\right)_{ji} y_j = \sum_{j=1}^{n} \left(W(a)\right)_{ji} y_j. \tag{2.98}$$

For simplicity, let us consider the first structure for $W(a)$ in (2.90). It is clear, in this case, that if the vector y_i has the special form

$$y_i = (y_1, y_2, \cdots, y_{n_1}, 0, 0, \cdots), \tag{2.99}$$

then, under the transformation (2.98), the transformed vector will have the form

$$Ay_i = \begin{cases} \displaystyle\sum_{j=1}^{n_1} (W_2(a))_{ji}\, y_j, & \text{for} \quad i = 1, 2, \cdots, n_1, \\ \\ 0, & \text{for} \quad i = n_1 + 1, n_1 + 2, \cdots, n. \end{cases} \tag{2.100}$$

Namely, the vector remains form invariant under the transformation signifying that the linear operator A has an invariant subspace. (Namely, the operator A does not take a vector out of this n_1 dimensional space.) It is easy to see that the second form in (2.90) has a n_2 dimensional invariant space of the form

$$y_i = (0, 0, \cdots, 0, y_{n_1+1}, y_{n_1+2}, \cdots, y_n). \qquad (2.101)$$

When there are no invariant subspaces for a linear operator representing a group, the representation is known as irreducible while it is called reducible otherwise. Using these ideas we can prove an important result in group theory known as Schur's lemma (named after Issai Schur) which is stated as follows.

Schur's lemma:

Let $U^{(1)}(a)$ and $U^{(2)}(a)$ denote respectively two irreducible $n_1 \times n_1$ and $n_2 \times n_2$ matrix representations of the group G. If there exists a $n_1 \times n_2$ matrix T such that

$$U^{(1)}(a)T = TU^{(2)}(a), \qquad (2.102)$$

with each product representing a $n_1 \times n_2$ matrix, then

1. If $n_1 \neq n_2$, then the matrix $T = \mathbf{0}$ (corresponds to the null matrix).

2. If $n_1 = n_2$ and $U^{(1)}(a), U^{(2)}(a)$ represent two inequivalent irreducible representations of G, then also $T = \mathbf{0}$ corresponds to the null matrix.

3. If $n_1 = n_2$ and $U^{(1)}(a) = U^{(2)}(a)$, then the matrix T must be a multiple of the identity matrix, namely,

$$T = \lambda E_{n_1}, \qquad (2.103)$$

for some constant λ. Here E_{n_1} denotes the n_1-dimensional (unit) identity matrix.

We will not go into the proof of Schur's lemma which is very nicely discussed in the book by L. Jansen and M. Boon.

▶ **Example (Neither fully reducible nor unitary representation).** As a simple example of a representation which is not fully reducible, let us consider the one dimensional translation group T_1 which, as we have discussed, is a non-compact group. As we have seen in (2.53) it has a two dimensional representation given by

$$U_2(a) = \begin{pmatrix} 1 & 0 \\ a & 1 \end{pmatrix}. \qquad (2.104)$$

We note that

$$U_2(a = 0) = E_2,$$

$$U_2(a)U_2(b) = \begin{pmatrix} 1 & 0 \\ a & 1 \end{pmatrix} \begin{pmatrix} 1 & 0 \\ b & 1 \end{pmatrix} = \begin{pmatrix} 1 & 0 \\ a+b & 1 \end{pmatrix} = U_2(a+b), \tag{2.105}$$

as required. However, this representation is not fully reducible. For if it were, this would require that there exists a (2×2) nonsingular matrix S such that (see (2.90) and (2.96))

$$S^{-1}U_2(a)S = \begin{pmatrix} 1 & 0 \\ 0 & 1 \end{pmatrix} = E_2,$$

$$\text{or,} \quad U_2(a) = SE_2S^{-1} = E_2 = \begin{pmatrix} 1 & 0 \\ 0 & 1 \end{pmatrix}, \tag{2.106}$$

which is not true.

Similarly, we can show that this cannot be a unitary representation as follows. Let us assume that $U_2(a)$ is a unitary representation. Then, there exists a (2×2) nonsingular matrix S such that (see (2.82)-(2.83))

$$W_2(a) = S^{-1}U_2(a)S, \tag{2.107}$$

defines a unitary matrix, namely,

$$(W_2(a))^{\dagger}W_2(a) = E_2. \tag{2.108}$$

Furthermore, since any unitary matrix is diagonalizable, there exists a nonsingular matrix T such that

$$T^{-1}W_2(a)T = \begin{pmatrix} \lambda_1 & 0 \\ 0 & \lambda_2 \end{pmatrix}. \tag{2.109}$$

It now follows from eqs. (2.107) and (2.109) that

$$(ST)^{-1}U_2(a)(ST) = \begin{pmatrix} \lambda_1 & 0 \\ 0 & \lambda_2 \end{pmatrix},$$

$$\text{or,} \quad (ST)^{-1}(U_2(a) - \lambda E_2)(ST) = \begin{pmatrix} \lambda_1 - \lambda & 0 \\ 0 & \lambda_2 - \lambda \end{pmatrix},$$

$$\text{or,} \quad \det\left((ST)^{-1}(U_2(a) - \lambda E_2)(ST)\right) = \det \begin{pmatrix} \lambda_1 - \lambda & 0 \\ 0 & \lambda_2 - \lambda \end{pmatrix},$$

$$\text{or,} \quad \det(U_2(a) - \lambda E_2) = \begin{vmatrix} \lambda_1 - \lambda & 0 \\ 0 & \lambda_2 - \lambda \end{vmatrix},$$

$$\text{or,} \quad \begin{vmatrix} 1 - \lambda & 0 \\ a & 1 - \lambda \end{vmatrix} = \begin{vmatrix} \lambda_1 - \lambda & 0 \\ 0 & \lambda_2 - \lambda \end{vmatrix},$$

$$\text{or,} \quad (1 - \lambda)^2 = (\lambda_1 - \lambda)(\lambda_2 - \lambda),$$

$$\text{or,} \quad \lambda_1 = \lambda_2 = 1, \tag{2.110}$$

This, in turn, implies that

$$(ST)^{-1}U_2(a)(ST) = \begin{pmatrix} \lambda_1 & 0 \\ 0 & \lambda_2 \end{pmatrix} = E_2,$$

$$\text{or,} \quad U_2(a) = (ST)E_2(ST)^{-1} = E_2 = \begin{pmatrix} 1 & 0 \\ 0 & 1 \end{pmatrix}, \tag{2.111}$$

which is not true so that $U_2(a)$ cannot define a unitary representation. Therefore, $U_2(a)$ is neither fully reducible nor a unitary representation of the one dimensional translation group. These facts are, of course, expected since T_1 is a non-compact group.

◀

The fact that a non-compact group does not have in general any finite dimensional unitary representation is expressed by the following theorem.

Theorem. *Any non-compact group cannot have a faithful finite dimensional unitary representation. (A faithful representation denotes a one-to-one correspondence between the group elements and its representation, namely, for two distinct elements $g, g' \in G$ with $g \neq g'$, the corresponding representations $U(g), U(g')$ must satisfy $U(g) \neq U(g')$ in order to provide a faithful representation.)*

Proof. Suppose that G is a non-compact group and that it possesses a finite dimensional nontrivial unitary matrix representation $U(G)$ in a vector space $\mathbb{V}$. As we already know (see (1.51)) any matrix element of a finite dimensional unitary matrix $U(g)$ satisfies $|(U(g))_{ij}| \leq 1$. Therefore, the group manifold of finite dimensional unitary matrices is bounded (and also closed), i.e., it corresponds to a compact space. Since for a faithful representation, G and $U(G)$ are one-to-one, if $U(G)$ denote finite dimensional unitary matrices, this would imply that a non-compact group G can have a continuous one-to-one map on to a compact space which is impossible.

■

▶ **Example (Infinite dimensional unitary representation of T_1).** In section 2.1, we constructed (see the example) finite dimensional representations of the non-compact translation group in one dimension which were not unitary. We also note here that although the (infinite dimensional) representation in (2.72) seems to be formally unitary, questions such as Hermiticity, unitarity etc. cannot be discussed until a suitable space of functions has been specified. Let us construct here an infinite dimensional unitary representation of the one dimensional translation group T_1.

Let $|f\rangle, |g\rangle$ denote states in a Hilbert space with the inner product defined by

$$\langle f|g \rangle = \int_{-\infty}^{\infty} dx \, \langle f|x \rangle \langle x|g \rangle = \int_{-\infty}^{\infty} dx \, f^*(x) g(x), \tag{2.112}$$

where $f(x), g(x)$ denote (in general, complex) functions of the one dimensional coordinate x spanning the real axis $(-\infty \leq x \leq \infty)$. The Hilbert space is defined by the set of all L_2 (Lebesgue square integrable) functions $f(x)$ satisfying

$$\|f\|^2 = \langle f|f \rangle = \int_{-\infty}^{\infty} dx \, f^*(x) f(x) < \infty. \tag{2.113}$$

In such a space, the action of the one dimensional translation T_1 on any function $f(x) \in L_2$ can be defined by

$$f(x) \xrightarrow{U(a)} f(x + a) \in L_2, \tag{2.114}$$

for any real constant a $(-\infty < a < \infty)$. It follows now that

$$\langle U(a)f | U(a)g \rangle = \int_{-\infty}^{\infty} dx\, f^*(x + a)g(x + a) = \int_{-\infty}^{\infty} dx\, f^*(x)g(x),$$

or, $\quad \langle U(a)f | U(a)g \rangle = \langle f | g \rangle. \tag{2.115}$

Since this holds for any two arbitrary L_2 functions $f(x), g(x)$, it follows that

$$U^{\dagger}(a)U(a) = \mathbb{1}, \tag{2.116}$$

so that $U(a)$ provides a unitary representation of the translation group T_1.

Let us note that the set of functions (see (2.48))

$$\xi_n(x) = x^n, \quad n = 0, 1, 2, \cdots, \tag{2.117}$$

do not define elements of the L_2 space since

$$\langle \xi_n | \xi_n \rangle = \int_{-\infty}^{\infty} dx\, x^{2n} \to \infty. \tag{2.118}$$

In order to find a suitable L_2 basis functions (not normalized), let us consider the functions defined in (2.118) with an appropriate weight (damping factor), namely, let us define

$$\psi_n(x) = x^n\, e^{-\frac{1}{2}x^2} = \xi_n\, e^{-\frac{1}{2}x^2}, \quad n = 0, 1, 2, \cdots. \tag{2.119}$$

It is clear now that

$$\int_{-\infty}^{\infty} dx\, \psi_n^*(x)\psi_n(x) = \int_{-\infty}^{\infty} dx\, x^{2n}\, e^{-x^2} = \Gamma\left(n + \frac{1}{2}\right) < \infty, \tag{2.120}$$

always exists so that $\psi_n(x) \in L_2$. Furthermore, under a finite translation $x \to x+a$ (see (2.47))

$$\psi_n(x) \xrightarrow{U(a)} \psi_n(x + a) = (x + a)^n\, e^{-\frac{1}{2}(x+a)^2} = (x + a)^n e^{-\frac{1}{2}a^2 - ax}\, e^{-\frac{1}{2}x^2}$$

$$= \sum_{m=0}^{\infty} C_{mn}(a)x^m\, e^{-\frac{1}{2}x^2} = \sum_{m=0}^{\infty} C_{mn}(a)\psi_m(x), \tag{2.121}$$

which shows that $\psi_n(x + a) \in L_2$. The constants C_{mn} can, in principle, be detrmined by Taylor expanding the function and give an infinite dimensional representation of T_1. In this case, it is straightforward to check that the operator

$$D = \frac{d}{dx}, \tag{2.122}$$

is anti-Hermitian in this space, namely, for any two functions $f(x), g(x) \in L_2$,

$$\int_{-\infty}^{\infty} dx\, (Df(x))^*\, g(x) = -\int_{-\infty}^{\infty} dx\, f^*(x)(Dg(x)), \tag{2.123}$$

which follows from the boundary conditions $f(x), g(x) \to 0$ as $|x| \to \infty$ (see (2.119)). As a result, the representation of the translation group in (2.72)

$$U(a) = e^{aD} = e^{a\frac{d}{dx}}, \tag{2.124}$$

is unitary in this space of functions.

However, in order to calculate explicitly the unitary matrix, we need to change the basis functions. Let us consider the normalized infinite dimensional basis functions

$$\phi_n(x) = \left(\frac{1}{\sqrt{\pi}2^n n!}\right)^{\frac{1}{2}} H_n(x) e^{-\frac{1}{2}x^2}, \quad n = 0, 1, 2, \cdots, \tag{2.125}$$

where $H_n(x)$ denotes the nth order Hermite polynomial and

$$\langle \phi_n | \phi_m \rangle = \int_{-\infty}^{\infty} dx\, \phi_n^*(x)\phi_m(x) = \delta_{nm}. \tag{2.126}$$

As we have seen in (2.121), under a translation, we can write

$$\phi_n(x) \xrightarrow{U(a)} \phi_n(x+a) = \sum_{m=0}^{\infty} U_{mn}(a)\phi_m(x), \tag{2.127}$$

where $U_{mn}(a)$ denote the matrix elements of the infinite dimensional matrix $U(a)$ providing a unitary representation for T_1.

To calculate these matrix elements, let us recall that the basis functions $\phi_n(x)$ correspond to the energy eigenfunctions of the one dimensional harmonic oscillator ($\hbar = m = \omega = 1$), namely,

$$\frac{1}{2}\left(-\frac{d^2}{dx^2} + x^2\right)\phi_n(x) = \left(n + \frac{1}{2}\right)\phi_n(x),$$

or, $\alpha^\dagger \alpha \phi_n(x) = n\phi_n(x), \tag{2.128}$

where we have introduced the annihilation and creation operators (we denote them as $\alpha, \alpha^\dagger$ in order to avoid confusion with the parameter of translation a)

$$\alpha = \frac{1}{\sqrt{2}}\left(x + \frac{d}{dx}\right), \quad \alpha^\dagger = \frac{1}{\sqrt{2}}\left(x - \frac{d}{dx}\right), \tag{2.129}$$

satisfying

$$[\alpha, \alpha^\dagger] = \mathbb{1}, \quad [\alpha, \alpha] = 0 = [\alpha^\dagger, \alpha^\dagger]. \tag{2.130}$$

The one dimensional harmonic oscillator enters into the discussion of the unitary representation of T_1 in the following way. We note that we can invert the relations in (2.129) o write

$$x = \frac{1}{\sqrt{2}}\left(\alpha + \alpha^\dagger\right), \quad \frac{d}{dx} = \frac{1}{\sqrt{2}}\left(\alpha - \alpha^\dagger\right), \tag{2.131}$$

so that we can write the operator implementing translations as (see (2.124))

$$U(a) = e^{a\frac{d}{dx}} = e^{\frac{a}{\sqrt{2}}\left(\alpha - \alpha^\dagger\right)}. \tag{2.132}$$

Using the Baker-Cambell-Hausdorff formula (which will be discussed later in section 4.3)

$$e^A e^B = e^{A+B+\frac{1}{2}[A,B]}, \quad \text{if} \quad [A,[A,B]] = 0 = [B,[A,B]], \tag{2.133}$$

we can rewrite

$$U(a) = e^{\frac{a}{\sqrt{2}}(\alpha - \alpha^\dagger)} = e^{-\frac{a}{\sqrt{2}}\alpha^\dagger} e^{\frac{a}{\sqrt{2}}\alpha} e^{-\frac{a^2}{4}}, \tag{2.134}$$

where we have used (2.130).

Let us recall that the energy eigenstates of the harmonic oscillator can be written in terms of the creation operators as

$$|n\rangle = \frac{(\alpha^\dagger)^n}{\sqrt{n!}} |0\rangle, \quad n = 0, 1, 2, \cdots, \tag{2.135}$$

where $|0\rangle$ denotes the ground state satisfying $\alpha|0\rangle = 0$. The corresponding wave functions (eigenfunctions) are obtained from the inner product with the coordinate basis as

$$\phi_n(x) = \langle x|n\rangle. \tag{2.136}$$

It follows that the translated wave function can be written as

$$\phi_n(x + a) = \langle x + a|n\rangle = \langle x|U^\dagger(a)|n\rangle = \langle x|U(-a)|n\rangle, \tag{2.137}$$

where we have used the fact that $U(a)$ is unitary so that (see (2.134))

$$U^\dagger(a) = U(-a) = e^{-\frac{a^2}{4}} e^{\frac{a}{\sqrt{2}}\alpha^\dagger} e^{-\frac{a}{\sqrt{2}}\alpha}. \tag{2.138}$$

Using the standard properties of the annihilation and creation operators, it follows that

$$e^{-\frac{a}{\sqrt{2}}\alpha}|n\rangle = \sum_{m=0}^{\infty} \frac{1}{m!} \left(-\frac{a}{\sqrt{2}}\right)^m \alpha^m |n\rangle = \sum_{m=0}^{n} \frac{1}{m!} \left(-\frac{a}{\sqrt{2}}\right)^m \alpha^m |n\rangle$$

$$= \sum_{m=0}^{n} \frac{1}{m!} \left(-\frac{a}{\sqrt{2}}\right)^m \sqrt{\frac{n!}{(n-m)!}} |n-m\rangle$$

$$= \sum_{m=0}^{n} \frac{1}{(n-m)!} \left(-\frac{a}{\sqrt{2}}\right)^{n-m} \sqrt{\frac{n!}{m!}} |m\rangle, \tag{2.139}$$

where we have let $m \to n - m$ in the last step. It follows now that

$$U(-a)|n\rangle = e^{-\frac{a^2}{4}} \sum_{m=0}^{n} \frac{1}{(n-m)!} \left(-\frac{a}{\sqrt{2}}\right)^{n-m} \sqrt{\frac{n!}{m!}} e^{\frac{a}{\sqrt{2}}\alpha^\dagger} |m\rangle$$

$$= e^{-\frac{a^2}{4}} \sum_{m=0}^{n} \frac{1}{(n-m)!} \left(-\frac{a}{\sqrt{2}}\right)^{n-m} \sqrt{\frac{n!}{m!}} \sum_{k=0}^{\infty} \frac{1}{k!} \left(\frac{a}{\sqrt{2}}\right)^k (\alpha^\dagger)^k |m\rangle$$

$$= e^{-\frac{a^2}{4}} \sum_{m=0}^{n} \frac{1}{(n-m)!} \left(-\frac{a}{\sqrt{2}}\right)^{n-m} \sqrt{\frac{n!}{m!}} \sum_{k=0}^{\infty} \frac{1}{k!} \left(\frac{a}{\sqrt{2}}\right)^k \sqrt{\frac{(k+m)!}{m!}} |k+m\rangle$$

$$= e^{-\frac{a^2}{4}} \sum_{m=0}^{n} \sum_{k=0}^{\infty} (-1)^{n-m} \frac{\sqrt{(n!)(k+m)!}}{(n-m)!(k!)(m!)} \left(\frac{a}{\sqrt{2}}\right)^{n-m+k} |k+m\rangle$$

$$= e^{-\frac{a^2}{4}} \sum_{\ell=0}^{\infty} \sum_{m=0}^{\min(n,\ell)} (-1)^{n-m} \left(\frac{a}{\sqrt{2}}\right)^{n-2m+\ell} \frac{\sqrt{(n!)(\ell!)}}{(n-m)!(\ell-m)!(m!)} |\ell\rangle$$

$$= \sum_{\ell=0}^{\infty} U_{\ell n}(a)|\ell\rangle. \tag{2.140}$$

Here we have redefined $k = \ell - m$ in the final step and have used the fact that

$$\frac{1}{(\ell - m)!} = 0, \quad \text{for } m \geq \ell + 1. \tag{2.141}$$

We have also identified

$$U_{\ell n}(a) = e^{-\frac{a^2}{4}} \sum_{m=0}^{\min(n,\ell)} (-1)^{n-m} \left(\frac{a}{\sqrt{2}}\right)^{n+\ell-2m} \frac{\sqrt{(n!)(\ell!)}}{(n-m)!(\ell-m)!(m!)}, \tag{2.142}$$

where "$\min(n, \ell)$" stands for the smaller of n and ℓ. Taking the inner product with $\langle x|$, it follows now from (2.127), (2.137) and (2.140) that

$$\phi_n(x + a) = \sum_{\ell=0}^{\infty} U_{\ell n}(a)\phi_\ell(x), \tag{2.143}$$

where the constants $U_{\ell n}(a)$ representing the matrix elements of the translation operator are given in (2.142). We note that the annihilation and creation operators of the harmonic oscillator also manifest in the ray representation of the two dimensional translation group T_2 as we will discuss later in section 4.4.

◀

2.3 Group integration

In section 1.3 we introduced the concept of a group manifold which allowed us to classify continuous (Lie) groups into compact and non-compact groups. For compact groups, we can also define the notion of a volume integral on this manifold which is invariant under the group action and satisfies the following relations.

1. The volume of the group manifold is finite

$$\int_G dg = \Omega < \infty, \tag{2.144}$$

where Ω denotes the volume of the group manifold.

2. For any fixed element $g_0 \in G$, we can make the change of variables

$$g \to g_0 g, \quad \text{or,} \quad g \to g g_0, \tag{2.145}$$

such that

$$d(g_0 g) = d(g g_0) = dg, \tag{2.146}$$

namely, the integration measure is invariant under the action of the group.

3. The volume integration also satisfies

$$d(g^{-1}) = dg. \tag{2.147}$$

An integration measure satisfying (2.144)-(2.147) is known as the Haar measure in the theory of Lie groups. For non-compact groups such a measure may not exist. However, if a non-compact group is locally compact (in the sense of a topological space) we may define a group integration as defined above, but the volume of the group manifold would no longer be finite, namely,

$$\int_G dg = \Omega = \infty, \tag{2.148}$$

in contrast to the compact case. For example, for translations in one dimension, T_1, whose group manifold is locally compact, has an infinite volume as we will show in (2.165).

▶ **Example (Unitary groups).** As we have discussed in section 1.3, unitary groups are compact groups. Let us begin with the example of the one dimensional unitary group $U(1)$ (also denoted by U_1). As we have discussed earlier, the group manifold in this case corresponds to the unit circle and every group element can be represented on this manifold as

$$g \to U(\theta) = e^{i\theta}, \quad 0 \le \theta \le 2\pi. \tag{2.149}$$

Therefore, every group element corresponds to a unique angle θ on this manifold (the unit circle) and we can identify (we emphasize here that dg is an abstract notation for the integration measure and does not represent the differential of g, more rigorously we should denote the measure as $\mu(dg)$ which we do not for simplicity)

$$dg \to d\theta, \tag{2.150}$$

which leads to the following.

1. It follows from (2.150) that

$$\int_G dg = \int_0^{2\pi} d\theta = 2\pi < \infty. \tag{2.151}$$

2. Any fixed group element g_0 corresponds to a fixed angle θ_0 on this manifold and we have

$$gg_0 \rightarrow e^{i\theta}e^{i\theta_0} = e^{i(\theta+\theta_0)}, \tag{2.152}$$

so that the product $gg_0 = g_0 g$ corresponds to the unique angle $\theta + \theta_0$ on the group manifold and the group action corresponds to a translation of the angle θ by a fixed θ_0. As a result, we have

$$d(gg_0) = d(g_0 g) = d(\theta + \theta_0) = d\theta = dg. \tag{2.153}$$

3. We note from (2.149) that

$$g^{-1} \rightarrow (U(\theta))^{-1} = U(-\theta) = e^{-i\theta}, \tag{2.154}$$

so that the inverse element corresponds to the point $(-\theta)$ on the unit circle and the integration measure remains invariant, namely,

$$dg^{-1} = dg. \tag{2.155}$$

We have also noted in section 1.3 that the the group manifold for $U(N)$, the unitary group in N-dimensions, can be identified with the subspace of the unit ball in N^2 dimensions which can be parameterized by N^2 complex parameters $\alpha_{ij}, i, j = 1, 2, \cdots, N$ satisfying (see (1.130))

$$|\alpha_{ij}| \leq 1, \tag{2.156}$$

such that the unitarity condition is satisfied. (Basically, we can parameterize the unit ball by N^2 real parameters, say, Re α_{ij}.) In this case, therefore, we can identify

$$dg = \prod_{i,j=1}^{n} d(\text{Re}\,\alpha_{ij}) d(\text{Im}\,\alpha_{ij})\, \delta(U^\dagger(g)U(g) - \mathbb{1}). \tag{2.157}$$

Since the coordinates of the unit ball are bounded as in (2.156), it follows that

$$\int_G dg < \infty. \tag{2.158}$$

◀

▶ **Example ($SO(3)$ group).** The group manifold of 3-dimensional rotation group $SO(3)$ or $R(3)$ may be of some interest to the readers. As discussed in section 1.2.5 (see, for example, (1.74)-(1.78) and the subsequent discussion) we may identify the group action with a rotation of coordinates in a 3-dimensional Euclidean space by an angle θ around a fixed 3-dimensional unit vector $\mathbf{n}$

$$\mathbf{n} = (n_1, n_2, n_3), \quad \mathbf{n}^2 = n_1^2 + n_2^2 + n_3^2 = 1, \tag{2.159}$$

which can be represented as $(\mathbf{n}, \theta)$ as shown in Fig. 2.1. However, since $(\mathbf{n}, \theta)$ and $(-\mathbf{n}, 2\pi - \theta)$ correspond to the same rotation, we need to restrict the range of the angle of rotation θ to $0 \leq \theta \leq \pi$. Therefore, we see that the group manifold of $R(3)$ can be identified with a three dimensional space consisting of all vectors of the form $\mathbf{r} = \theta\mathbf{n}$ with $0 \leq \theta \leq \pi$. Since $\mathbf{n}$ represents a unit vector, it follows that

$$r^2 = \mathbf{r}^2 = \theta^2, \tag{2.160}$$

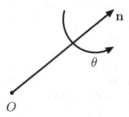

Figure 2.1: A rotation by an angle θ around a unit vector $\mathbf{n}$ in the 3-dimensional Euclidean space with O denoting the origin of the coordinate system.

so that the group manifold corresponds to the interior of a three dimensional sphere of radius π. In spherical coordinates, we can parameterize the components of the unit vector $\mathbf{n}$ as

$$\mathbf{n} = (n_1, n_2, n_3) = (\sin\alpha\cos\beta, \sin\alpha\sin\beta, \cos\alpha), \qquad (2.161)$$

where $0 \le \alpha \le \pi$ and $0 \le \beta \le 2\pi$. With this the group invariant volume element can be written as (recall that in spherical coordinates $d^3x = r^2 dr \sin\theta d\theta d\phi$)

$$dg = \theta^2 d\theta \sin\alpha d\alpha \, d\beta, \qquad (2.162)$$

where, as we have noted earlier, $0 \le \theta \le \pi$. (In this case, $\int dg = \frac{4\pi^4}{3}$.)

However, the group manifold for $R(3)$ has the following subtlety. We note that $(\mathbf{n}, \pi)$ and $(-\mathbf{n}, \pi)$ represent the same rotation. Therefore, we must identify the diametrically opposite points on the surface of the sphere (namely, the two points representing the unit vectors $\mathbf{n}$ and $-\mathbf{n}$ on the surface of the sphere). This leads to the fact that the group manifold for rotations is not simply connected. This is the origin of the spinor representations which are double-valued representations of $R(3)$ (or $SO(3)$). In contrast to $R(N)$, the group manifolds of $U(N)$ and $SL(N)$ are known to be simply connected (note that $SL(N)$ is a non-compact group, see the discussion after (1.132)).   ◀

Let us also note here that any finite group, such as the permutation group S_N (see section 1.2.1), can always be regarded as compact with the group integration

$$\int_G dg = \sum_{g \in G}. \qquad (2.163)$$

The group invariant volume, in this case, basically counts the total number of independent elements in the group and since the finite groups have a finite number of elements, the group invariant volume is finite.

In contrast to these examples, we note that the group manifold for the one dimensional translation group T_1 is the real line. Every group element can be uniquely specified on this manifold by a point "a" denoting the parameter of translation. Consequently, we can write

$$dg = da, \tag{2.164}$$

so that the group volume follows to be

$$\int_G dg = \int_{-\infty}^{\infty} da = \infty, \tag{2.165}$$

in spite of the fact that the real line is a locally compact manifold (as we have mentioned earlier).

2.4 Peter-Weyl theorem

One important difference between the compact and non-compact groups is given by the Peter-Weyl theorem (named after Hermann Weyl and his student Fritz Peter) which essentially implies the following. First, for a compact group, any of its representations is always finite dimensional, fully reducible and equivalent to a unitary representation. In contrast, for a non-compact group, its representations may not be fully reducible and the only two unitary representations of the group are either the one dimensional trivial representation or an infinite dimensional representation as we have already noted in connection with the example of the one dimensional translation group T_1 earlier. Let us prove some of these assertions below.

2.4.1 Fully reducible representation. Let us change notation (for a reason to be clear soon) and denote the finite dimensional $(n \times n)$ matrix representation of a (compact) group element by $U(g)$ instead of $U(a)$ as we have done earlier. Let us assume that $U(g)$ is reducible and write it in the form (see (2.90))

$$U(g) = \begin{pmatrix} U_1(g) & N(g) \\ 0 & U_2(g) \end{pmatrix}. \tag{2.166}$$

Let us identify

$$W(g) = \begin{pmatrix} U_1(g) & 0 \\ 0 & U_2(g) \end{pmatrix}. \tag{2.167}$$

Since $U_1(g)$ and $U_2(g)$ correspond to representations of G, as we have already seen in (2.92), it follows that

$$W(g_1)W(g_2) = W(g_1g_2). \tag{2.168}$$

Let us next define

$$S = \int_G dg\, W(g^{-1})U(g), \tag{2.169}$$

where the integral exists since $W(g^{-1})$ and $U(g)$ are finite dimensional matrices. For any fixed element $g_0 \in G$, it follows from (2.169) that (the reason for changing notation for the group element is to take advantage of the group integration property below)

$$
\begin{aligned}
W(g_0^{-1})SU(g_0) &= \int_G dg\, W(g_0^{-1})W(g^{-1})U(g)U(g_0) \\
&= \int_G dg\, W(g_0^{-1}g^{-1})U(gg_0) \\
&= \int_G d(gg_0)\, W((gg_0)^{-1})U(gg_0) \\
&= \int_G dg'\, W((g')^{-1})U(g') = S, \tag{2.170}
\end{aligned}
$$

which leads to

$$W(g_0^{-1})SU(g_0) = S, \quad \text{or,} \quad SU(g_0) = W(g_0)S. \tag{2.171}$$

Since g_0 is any arbitrary element of the group, (2.171) can be written in general as

$$SU(g) = W(g)S. \tag{2.172}$$

Let us note that

$$
\begin{aligned}
W(g^{-1})U(g) &= \begin{pmatrix} U_1(g^{-1}) & 0 \\ 0 & U_2(g^{-1}) \end{pmatrix} \begin{pmatrix} U_1(g) & N(g) \\ 0 & U_2(g) \end{pmatrix} \\
&= \begin{pmatrix} U_1(g^{-1})U_1(g) & U_1(g^{-1})N(g) \\ 0 & U_2(g^{-1})U_2(g) \end{pmatrix} \\
&= \begin{pmatrix} E_{n_1} & U_1(g^{-1})N(g) \\ 0 & E_{n_2} \end{pmatrix}, \tag{2.173}
\end{aligned}
$$

where E_{n_1} and E_{n_2} are respectively $n_1 \times n_1$ and $n_2 \times n_2$ unit matrices (with $n_1 + n_2 = n$). Therefore, we can write

$$S = \int_G dg\, W(g^{-1})U(g) = \begin{pmatrix} \Omega E^{(1)} & K \\ 0 & \Omega E^{(2)} \end{pmatrix}, \qquad (2.174)$$

where we have identified

$$\Omega = \int_G dg, \quad K = \int_G dg\, U_1(g^{-1})N(g). \qquad (2.175)$$

It follows from (2.174) that

$$\det S = \Omega^{n_1+n_2} = \Omega^n \neq 0, \qquad (2.176)$$

so that S^{-1} exists and we can rewrite (2.172) as (see also (2.167))

$$SU(g)S^{-1} = W(g) = \begin{pmatrix} U_1(g) & 0 \\ 0 & U_2(g) \end{pmatrix}. \qquad (2.177)$$

This proves that $U(g)$ is fully reducible since if $U_1(g)$ and/or $U_2(g)$ are reducible, we can follow the same procedure for $U_1(g)$ and $U_2(g)$ as well.

2.4.2 Unitary representation. Let us next show that any (finite dimensional, $n \times n$) matrix representation $U(g)$ of such a (compact) group is equivalent to a fully reducible unitary representation. To prove this, let us define

$$T = \int_G dg\, U^\dagger(g)U(g). \qquad (2.178)$$

Since $U^\dagger(g)U(g)$ is a Hermitian and positive matrix (namely, all the eigenvalues are positive), so is T. Let us next note that for any fixed element of the group, $g_0 \in G$,

$$
\begin{aligned}
U^\dagger(g_0)TU(g_0) &= \int_G dg\, U^\dagger(g_0)U^\dagger(g)U(g)U(g_0) \\
&= \int_G dg\, (U(g)U(g_0))^\dagger U(gg_0) \\
&= \int_G d(gg_0)\, U^\dagger(gg_0)U(gg_0) = T. \qquad (2.179)
\end{aligned}
$$

Since g_0 is arbitrary, we can write the relation (2.179) in general for any element of the group as

$$U^\dagger(g)TU(g) = T. \tag{2.180}$$

As we have seen, T is a Hermitian and positive $n \times n$ matrix so that we can write it as

$$T = S^\dagger S, \tag{2.181}$$

for some nonsingular (namely, S^{-1} exists) $n \times n$ matrix S. Therefore, it follows from (2.180) that

$$U^\dagger(g)S^\dagger SU(g) = S^\dagger S,$$

$$\text{or,} \quad \left(SU(g)S^{-1}\right)^\dagger \left(SU(g)S^{-1}\right) = E_n. \tag{2.182}$$

Therefore, if we define

$$W(g) = SU(g)S^{-1}, \tag{2.183}$$

we can write (2.182) as

$$W^\dagger(g)W(g) = E_n, \tag{2.184}$$

namely, $W(g)$ is a $n \times n$ unitary matrix. This proves that the representation is equivalent to a unitary representation, and furthermore as we have shown earlier this is also fully reducible.

2.5 Orthogonality relations

Let $U^{(1)}(g)$ and $U^{(2)}(g)$ denote two $n_1 \times n_1$ and $n_2 \times n_2$ unitary irreducible matrix representations of a compact group G. Then we can prove the following relations in a straightforward manner.

1. If $U^{(1)}(g)$ and $U^{(2)}(g)$ are inequivalent representations, then

$$\int_G \mathrm{d}g \left(U^{(1)}_{ij}(g)\right)^* U^{(2)}_{\mu\nu}(g) = 0, \tag{2.185}$$

for all $i, j = 1, 2, \cdots, n_1$ and $\mu, \nu = 1, 2, \cdots, n_2$.

2. For $n_1 = n_2$ and $U^{(1)}(g) = U^{(2)}(g)$ (equivalent representations), we have

$$\int_G \mathrm{d}g \left(U^{(1)}_{ij}(g)\right)^* U^{(1)}_{k\ell}(g) = \lambda \delta_{ik}\delta_{jl}, \tag{2.186}$$

for $i, j, k, \ell = 1, 2, \cdots, n_1$ and the constant λ is given by

$$\lambda = \frac{1}{n_1} \Omega, \tag{2.187}$$

where Ω denotes the volume of the group manifold defined in (2.175). We note that for a finite group G such as the symmetric group S_n, we replace the integration over the group manifold $\int_G dg$ by the summation (over all group elements) $\sum_{g \in G}$ as in (2.163) so that the volume Ω corresponds to the total number of independent elements of G. Thus, for example, we have $\Omega = N!$ for the group S_N.

To prove these statements, let M denote an arbitrary $n_1 \times n_2$ constant matrix and define another $n_1 \times n_2$ (constant) matrix T by

$$T = \int_G dg\, U^{(1)}(g^{-1}) M U^{(2)}(g). \tag{2.188}$$

Then, for any fixed element $g_0 \in G$, we have

$$U^{(1)}(g_0^{-1}) T U^{(2)}(g_0) = \int_G dg\, U^{(1)}(g_0^{-1}) U^{(1)}(g^{-1}) M U^{(2)}(g) U^{(2)}(g_0)$$

$$= \int_G dg\, U^{(1)}(g_0^{-1} g^{-1}) M U^{(2)}(g g_0)$$

$$= \int_G d(g g_0)\, U^{(1)}((g g_0)^{-1}) M U^{(2)}(g g_0)$$

$$= \int_G dg'\, U^{(1)}((g')^{-1}) M U^{(2)}(g') = T. \tag{2.189}$$

Since g_0 is an arbitrary element of the group, we can also write (2.189) in general as

$$U^{(1)}(g^{-1}) T U^{(2)}(g) = T,$$

$$\text{or,} \quad U^{(1)}(g) T = T U^{(2)}(g). \tag{2.190}$$

It is clear now from Schur's lemma (see (2.102)) that if $U^{(1)}(g)$ and $U^{(2)}(g)$ denote two inequivalent and irreducible representations, then

$$T = 0, \tag{2.191}$$

which leads to

$$\int_G dg\, U^{(1)}(g^{-1})MU^{(2)}(g) = \int_G dg\, M_{i\mu}U^{(1)}_{ji}(g^{-1})U^{(2)}_{\mu\nu}(g) = 0,$$

$$(2.192)$$

where summation over repeated indices is assumed and $i, j = 1, 2, \cdots, n_1$ while $\mu, \nu = 1, 2, \cdots, n_2$. On the other hand, the matrix M is an arbitrary $n_1 \times n_2$ matrix so that for (2.192) to hold, we must have

$$\int_G dg\, U^{(1)}_{ji}(g^{-1})U^{(2)}_{\mu\nu}(g) = 0, \qquad (2.193)$$

for any $i, j = 1, 2, \cdots, n_1$ and $\mu, \nu = 1, 2, \cdots, n_2$. Since $U^{(1)}(g)$ is unitary, it follows that

$$U^{(1)}(g^{-1}) = \left(U^{(1)}(g)\right)^\dagger,$$

$$\text{or,} \quad U^{(1)}_{ji}(g^{-1}) = \left(U^{(1)}_{ij}(g)\right)^*. \qquad (2.194)$$

Therefore, (2.193) can be written as

$$\int_G dg\, \left(U^{(1)}_{ij}(g)\right)^* U^{(2)}_{\mu\nu}(g) = 0, \qquad (2.195)$$

which proves (2.185).

On the other hand, when $n_1 = n_2$ and $U^{(1)}(g) = U^{(2)}(g)$ (equivalent representations), we note again from Schur's lemma (2.103) that for

$$U^{(1)}(g)T = TU^{(1)}(g) \qquad (2.196)$$

to hold we must have $T = \lambda E_{n_1}$ so that we have from (2.188)

$$\int_G dg\, U^{(1)}(g^{-1})MU^{(1)}(g) = \lambda E_{n_1}. \qquad (2.197)$$

Taking the trace of both sides of (2.197) and using $\text{Tr}\, E_{n_1} = n_1$ as well as

$$\text{Tr}\, U^{(1)}(g^{-1})MU^{(1)}(g) = \text{Tr}\, U^{(1)}(g)U^{(1)}(g^{-1})M = \text{Tr}\, U(e)M$$

$$= \text{Tr}\, E_{n_1}M = \text{Tr}\, M, \qquad (2.198)$$

we determine

$$\lambda = \frac{1}{n_1} \operatorname{Tr} M \int_G dg = \frac{1}{n_1} \Omega \operatorname{Tr} M. \tag{2.199}$$

Substituting this into (2.197) we obtain

$$\int_G dg\, U^{(1)}(g^{-1}) M U^{(1)} = \frac{1}{n_1} \Omega E_{n_1} \operatorname{Tr} M,$$

$$\text{or,} \quad \int_G dg\, M_{ik} U_{ji}^{(1)}(g^{-1}) U_{k\ell}^{(1)} = \frac{1}{n_1} \Omega \delta_{j\ell} \operatorname{Tr} M. \tag{2.200}$$

Since M is an arbitrary $n_1 \times n_1$ constant matrix, using (2.194) we conclude that

$$\int_G dg\, \left(U_{ij}^{(1)}(g) \right)^* U_{k\ell}^{(1)}(g) = \frac{1}{n_1} \Omega \delta_{ik} \delta_{j\ell}, \tag{2.201}$$

which proves (2.186).

2.6 Character of a representation

Let $U(g)$ be a $n \times n$ matrix representation (not necessarily irreducible) of a compact group G. We call

$$\chi(g) = \operatorname{Tr} U(g), \tag{2.202}$$

to be the character of the representation. Since $U(e) = E_N$, we obtain from (2.202) that

$$\chi(e) = \operatorname{Tr} U(e) = \operatorname{Tr} E_n = n, \tag{2.203}$$

where n denotes the dimension of the representation space.

If $U(g)$ is fully reducible and is a direct sum of irreducible representation matrices $U^{(1)}(g), U^{(2)}(g), \cdots, U^{(n)}(g)$, namely, (see (2.97))

$$S^{-1}U(g)S = \begin{pmatrix} U^{(1)}(g) & & & \\ & U^{(2)}(g) & & \\ & & \ddots & \\ & & & U^{(k)}(g) \end{pmatrix}, \tag{2.204}$$

then taking the trace of both sides of (2.204) and noting that

$$\text{Tr}\left(S^{-1}U(g)S\right) = \text{Tr}\,U(g) = \chi(g), \tag{2.205}$$

we obtain

$$\chi(g) = \sum_{a=1}^{n_k} \chi_a(g), \tag{2.206}$$

where $\chi_a(g)$ denotes the character of the a-th irreducible representation $U^{(a)}(g)$.

Since $\text{Tr}\left(S^{-1}U(g)S\right) = \text{Tr}\,U(g)$, the character of a representation is independent of the choice of a particular equivalent representation. Moreover, the number of all possible irreducible representations of a compact group is known to be at most denumerable. Hence we may denote the characters of the (irreducible) submatrices as $\chi_1(g), \chi_2(g), \chi_3(g), \cdots$ in such a way that $\chi_a(g) \neq \chi_b(g)$ if $U^{(a)}(g)$ and $U^{(b)}(g)$ are inequivalent representations. In the orthogonality relations for $U^{(a)}(g)$ and $U^{(b)}(g)$ in (2.185)-(2.187), if we set $i = j$ and $\mu = \nu$ and sum over i as well as μ, we obtain the orthogonality relation for the characters to be

$$\int_G \mathrm{d}g\, \chi_a^*(g)\chi_b(g) = \Omega\,\delta_{ab}. \tag{2.207}$$

where, for irreducible representations $U^{(a)}(g)$ and $U^{(b)}(g)$,

$$\delta_{ab} = \begin{cases} 1, & \text{if } U^{(a)}(g) \text{ and } U^{(b)}(g) \text{ are equivalent,} \\ 0, & \text{if } U^{(a)}(g) \text{ and } U^{(b)}(g) \text{ are inequivalent.} \end{cases} \tag{2.208}$$

The characters of irreducible representations of classical groups such as $U(N), SU(N), SO(N)$ as well as the symmetric group have been computed by H. Weyl.

2.7 References

H. Weyl, *The Classical Groups*, Princeton University Press, Princeton, NJ (1939).

M. Naimark, *Normed Rings*, Noordhoff, Groningen (1960).

L. Jansen and M. Boon, *Theory of finite groups*, John Wiley and sons, N. Y. (1964). This book has a nice proof of Schur's lemma as well as detailed discussions on orthogonality relations.

L. Nachbin, *Haar Integral*, Van Nostrand, N. J. (1965).

CHAPTER 3
Lie algebras

3.1 Definition of a Lie algebra

Let us start with a the formal definition of a Lie algebra L as follows.

$(L1)$: The Lie algebra L is a vector space over a field F which, in our discussions, will be assumed to be the field of either real or complex numbers. In other words, for any $x, y \in L$ and for any real or complex numbers α, β, λ, the product $\alpha x \in L$ and the sum $x + y \in L$ are defined and satisfy the conditions

$$\alpha(\beta x) = (\alpha\beta)x, \quad \lambda(\alpha x + \beta y) = (\lambda\alpha)x + (\lambda\beta)y. \tag{3.1}$$

$(L2)$: For any two elements of the Lie algebra, a bilinear product $[x, y] \in L$ is defined in L satisfying the anti-symmetry condition

$$[x, y] = -[y, x]. \tag{3.2}$$

Here by *bilinear product*, we imply the validity of the condition

$$[\alpha x + \beta y, z] = \alpha[x, z] + \beta[y, z], \tag{3.3}$$

for any $\alpha, \beta \in F$ and for any $x, y, z \in L$.

$(L3)$: Finally, for any three elements $x, y, z \in L$, the bilinear product is assumed to satisfy the Jacobi identity

$$[[x, y], z] + [[y, z], x] + [[z, x], y] = 0. \tag{3.4}$$

Any vector space L satisfying $(L1)$-$(L3)$ is known as a Lie algebra. If the field F is real or complex, then L is correspondingly called a real or a complex Lie algebra. Unless otherwise stated, we will assume that L is a complex Lie algebra in all our discussions.

$(L3')$: We note here that the condition $(L3)$ can also be written in the equivalent form

$$[x, [y, z]] + [y, [z, x]] + [z, [x, y]] = 0, \tag{3.5}$$

which can be seen as follows. First, using the anti-symmetry property of the bilinear product (3.2), we can write

$$[[x, y], z] = -[z, [x, y]], \quad [y, [z, x]] = -[[z, x], y],$$
$$[z, [x, y]] = -[[x, y], z], \tag{3.6}$$

and substituting this into (3.4) we obtain

$$[[x, y], z] + [[y, z], x] + [[z, x], y]$$
$$= -[z, [x, y]] - [x, [y, z]] - [y, [z, x]]$$
$$= -\left([x, [y, z]] + [y, [z, x]] + [z, [x, y]]\right) = 0, \tag{3.7}$$

which coincides with (3.5).

We note that if

$$[x, y] = 0, \quad \text{for all } x, y \in L, \tag{3.8}$$

it corresponds to a Lie algebra which is called Abelian. A Lie algebra is called non-Abelian otherwise. It is worth emphasizing here that $[x, y]$ is an abstract bilinear product and need not necessarily correspond to a commutator

$$[x, y] = xy - yx, \tag{3.9}$$

since the product xy may not itself be defined in this context. However, if A defines an associative algebra such that for $x, y, z \in A$

$$(xy)z = x(yz) = xyz, \tag{3.10}$$

then the commutator algebra $L = [A, A]$ is a Lie algebra defined by the bilinear product $[x, y] = xy - yx$. We note that in this case we have $[x, y] = z \in L$ and we can write

$$[[x, y], z] = [x, y]z - z[x, y] = (xy - yx)z - z(xy - yx)$$
$$= xyz - yxz - zxy + zyx, \tag{3.11}$$

which leads to

$$[[x, y], z] + [[y, z], x] + [[z, x], y]$$
$$= xyz - yxz - zxy + zyx + yzx - zyx - xyz + xzy$$
$$+ zxy - xzy - yzx + yxz = 0, \tag{3.12}$$

so that the Jacobi identity is satisfied.

For a Lie algebra L, let us suppose that we can find $n \times n$ matrices (or more generally, linear operators in a Hilbert space) $\rho(x)$ for all $x \in L$ (recall that matrices define a vector space) satisfying

(i) $\rho(x)$ is linear in x, namely,

$$\rho(\alpha x + \beta y) = \alpha \rho(x) + \beta \rho(y), \tag{3.13}$$

for any $x, y \in L$ and for any real or complex numbers α, β.

(ii) $\rho(x)$ also satisfies

$$\rho([x, y]) = [\rho(x), \rho(y)] = \rho(x)\rho(y) - \rho(y)\rho(x). \tag{3.14}$$

We note that since

$$[\rho(x), \rho(y)] = \rho(x)\rho(y) - \rho(y)\rho(x), \tag{3.15}$$

$\rho(x)$ generates a Lie algebra because the matrix product $\rho(x)\rho(y)$ is associative. In this case, we say that $\rho(x)$ provides a (linear) $n \times n$ matrix representation of the Lie algebra L. (It is worth pointing out here that even though the matrix product is associative, in general, there is no guarantee that $[\rho(x), \rho(y)] = \rho(z)$, for some $z \in L$. However, by our assumption (3.14), if $[x, y] = z \in L$, it follows that $\rho([x, y]) = \rho(z) = [\rho(x), \rho(y)] \in L$ and $\rho(x)$ satisfying (3.13) and (3.14) defines a Lie algebra.)

A familiar example of a matrix representation of a Lie algebra in quantum mechanics is given by the spin matrices

$$s_i = \frac{1}{2}\sigma_i, \quad i = 1, 2, 3, \tag{3.16}$$

where σ_i denote the three Pauli matrices

$$\sigma_1 = \begin{pmatrix} 0 & 1 \\ 1 & 0 \end{pmatrix}, \quad \sigma_2 = \begin{pmatrix} 0 & -i \\ i & 0 \end{pmatrix}, \quad \sigma_3 = \begin{pmatrix} 1 & 0 \\ 0 & -1 \end{pmatrix}, \tag{3.17}$$

which satisfy the angular momentum (Lie) algebra

$$[s_1, s_2] = is_3, \quad [s_2, s_3] = is_1, \quad [s_3, s_1] = is_2. \tag{3.18}$$

Here $i = \sqrt{-1}$ and the algebra (3.18) can be written in the more compact form

$$[s_i, s_j] = i\epsilon_{ijk}s_k, \tag{3.19}$$

where the subscripts $i, j, k = 1, 2, 3$ and ϵ_{ijk} represents the three dimensional Levi-Civita tensor which is completely anti-symmetric with $\epsilon_{123} = 1$.

3.2 Examples of commonly used Lie algebras in physics

In this section, we describe some of the more common examples of Lie algebras in physics as well as their representations.

3.2.1 Lie algebra of $gl(N)$ and $sl(N)$.

Let x, y, z denote any three $N \times N$ matrices and we define the bilinear product of any pair of them $[x, y]$ to correspond to their matrix commutator, namely,

$$[x, y] = xy - yx, \tag{3.20}$$

so that we have a Lie algebra. Such a Lie algebra without any restrictions on the matrices is known as $gl(N)$ (gl stands for general linear) since it is intrinsically related to the general linear group $GL(N)$ as we will explore later. Furthermore, (3.20) gives a N-dimensional matrix representation of $gl(N)$ known as the defining representation. (There can be several possible representations of a Lie algebra, not just the defining representation, corresponding to other possible $n \times n$ matrix realizations as we will see and as we know from our studies on the representations of the angular momentum algebra.)

If we restrict the matrices x, y, z further to be traceless $N \times N$ matrices,

$$\text{Tr}\, x = \text{Tr}\, y = \text{Tr}\, z = 0, \tag{3.21}$$

then, all such matrices also define a Lie algebra. We note that for any two finite dimensional matrices x, y, the trace of their commutator vanishes, namely,

$$\text{Tr}\, [x, y] = 0, \tag{3.22}$$

because of the cyclicity of trace, so that $[x, y]$ also belongs to the Lie algebra (it is a traceless $N \times N$ matrix). The Lie algebra of traceless general $N \times N$ matrices defines the special linear Lie algebra $sl(N)$ which is related to the special linear Lie group $SL(N)$ (see section 1.2.4).

3.2.2 Lie algebra of $so(N)$. Let us define a set of second rank anti-symmetric tensor operators in a Hilbert space

$$X_{\mu\nu} = -X_{\nu\mu}, \quad \mu, \nu = 1, 2, \cdots, N, \tag{3.23}$$

which satisfy the commutation relations ($\mu, \nu, \alpha, \beta = 1, 2, \cdots, N$ and $i = \sqrt{-1}$)

$$\begin{aligned}[X_{\mu\nu}, X_{\alpha\beta}] &= X_{\mu\nu}X_{\alpha\beta} - X_{\alpha\beta}X_{\mu\nu} \\ &= i\left(\delta_{\mu\alpha}X_{\nu\beta} + \delta_{\nu\beta}X_{\mu\alpha} - \delta_{\mu\beta}X_{\nu\alpha} - \delta_{\nu\alpha}X_{\mu\beta}\right).\end{aligned} \tag{3.24}$$

This defines a Lie algebra known as the $so(N)$ Lie algebra and is related to the Lie group $SO(N)$ and the operators $X_{\mu\nu}$ are known as the generators of the Lie algebra $so(N)$. Because of anti-symmetry, there are only $\frac{1}{2}N(N-1)$ such operators. (We note here that the Lie algebra is defined up to a scaling of the operators $X_{\mu\nu}$ and here we have chosen the operators to be Hermitian.)

A familiar example of $so(N)$ can be given as follows. Let $(\hat{x}_\mu, \hat{p}_\mu)$ with $\mu, \nu = 1, 2, \cdots, N$ denote the canonical coordinate and momentum operators in N-dimensional quantum mechanics so that we have the canonical commutation relations (we assume $\hbar = 1$)

$$\begin{aligned}[\hat{x}_\mu, \hat{p}_\nu] &= -[\hat{p}_\nu, \hat{x}_\mu] = i\delta_{\mu\nu}, \\ [\hat{x}_\mu, \hat{x}_\nu] &= 0 = [\hat{p}_\mu, \hat{p}_\nu],\end{aligned} \tag{3.25}$$

which can be realized in the coordinate representation as $\hat{p}_\mu = -i\frac{\partial}{\partial x_\mu}$. If we now define

$$L_{\mu\nu} \equiv \hat{x}_\mu \hat{p}_\nu - \hat{x}_\nu \hat{p}_\mu = -L_{\nu\mu}, \tag{3.26}$$

then it can be checked, using (3.25), in a straightforward manner that

$$[L_{\mu\nu}, L_{\alpha\beta}] = i\left(\delta_{\mu\alpha}L_{\nu\beta} + \delta_{\nu\beta}L_{\mu\alpha} - \delta_{\mu\beta}L_{\nu\alpha} - \delta_{\nu\alpha}L_{\mu\beta}\right), \tag{3.27}$$

so that it satisfies the $so(N)$ Lie algebra (see (3.24)). $L_{\mu\nu}$ can be thought of as the angular momentum operators generating rotations in the μ-ν plane.

In quantum mechanics, $(\hat{x}_\mu, \hat{p}_\mu)$ are Hermitian operators, namely, $\hat{x}_\mu^\dagger = \hat{x}_\mu, \hat{p}_\mu^\dagger = \hat{p}_\mu$, so that

$$L_{\mu\nu}^\dagger = (\hat{p}_\nu)^\dagger(\hat{x}_\mu)^\dagger - (\hat{p}_\mu)^\dagger(\hat{x}_\nu)^\dagger = \hat{p}_\nu\hat{x}_\mu - \hat{p}_\mu\hat{x}_\nu$$

$$= \hat{x}_\mu\hat{p}_\nu - \hat{x}_\nu\hat{p}_\mu = L_{\mu\nu}, \tag{3.28}$$

namely, they are Hermitian.

For $N = 3$, for example, we recognize that there are only three independent angular momentum $(L_{\mu\nu})$ operators which we can label as

$$L_1 = L_{23} = -L_{32}, \quad L_2 = L_{31} = -L_{13}, \quad L_3 = L_{12} = -L_{21}. \tag{3.29}$$

It now follows from (3.27) that

$$[L_1, L_2] = iL_3, \quad [L_2, L_3] = iL_1, \quad [L_3, L_1] = iL_2, \tag{3.30}$$

which we recognize to be the familiar angular momentum algebra (see also (3.19))

$$[L_i, L_j] = i\epsilon_{ijk}L_k, \quad i, j, k = 1, 2, 3, \tag{3.31}$$

with $L_i^\dagger = L_i$, $i = 1, 2, 3$.

We can now construct some finite dimensional non-spinorial representations for the $so(N)$ Lie algebra as follows. Let us first consider the case when the operators $L_{\mu\nu}$ act on constants C. From the fact that $\hat{p}_\mu C = -i\frac{\partial C}{\partial x_\mu} = 0$, we conclude from the definition (3.26) that

$$L_{\mu\nu}C = 0, \tag{3.32}$$

which furnishes the trivial one dimensional representation of $so(N)$ corresponding to the zero angular momentum state $\ell = 0$ in the case $N = 3$.

When $L_{\mu\nu}$ acts on a linear monomial x_λ, then we have $\hat{p}_\mu x_\lambda = -i\frac{\partial x_\lambda}{\partial x_\mu} = -i\delta_{\mu\lambda}$ so that

$$L_{\mu\nu}x_\lambda = (\hat{x}_\mu\hat{p}_\nu - \hat{x}_\nu\hat{p}_\mu)\hat{x}_\lambda = -i(\delta_{\nu\lambda}\,x_\mu - \delta_{\mu\lambda}\,x_\nu), \tag{3.33}$$

and x_λ lead to the N-dimensional representation for $L_{\mu\nu}$. In general, finding a vector space on which the operators act leads to a representation for them in the following way. In this particular case, we note that $x_\mu = (x_1, x_2, \cdots, x_N)$ represents the coordinate vector of the N-dimensional vector space $\mathbb{V}_N$. Since the coordinates x_λ define a complete basis in this space, we can expand

$$L_{\mu\nu}x_\lambda = \sum_{\beta=1}^{N} x_\beta \left(\widetilde{L}_{\mu\nu}\right)_{\beta\lambda}, \tag{3.34}$$

and comparing with (3.33) we obtain the $N \times N$ matrix representation for the matrix $\widetilde{L}_{\mu\nu} = \rho(L_{\mu\nu})$ as

$$\left(\widetilde{L}_{\mu\nu}\right)_{\alpha\beta} = -i\left(\delta_{\mu\alpha}\delta_{\nu\beta} - \delta_{\nu\alpha}\delta_{\mu\beta}\right), \tag{3.35}$$

for $\mu, \nu, \alpha, \beta = 1, 2, \cdots, N$ and it can be checked that it satisfies (3.27) with the standard matrix products, namely,

$$[\widetilde{L}_{\mu\nu}, \widetilde{L}_{\lambda\rho}] = i\left(\delta_{\mu\lambda}\widetilde{L}_{\nu\rho} + \delta_{\nu\rho}\widetilde{L}_{\mu\lambda} - \delta_{\mu\rho}\widetilde{L}_{\nu\lambda} - \delta_{\nu\lambda}\widetilde{L}_{\mu\rho}\right). \tag{3.36}$$

This gives the N dimensional ($N \times N$ matrix) representation of $so(N)$, namely, the defining representation. For $N = 3$, for example, we can determine (using the identification in (3.29))

$$\widetilde{L}_1 = -i\begin{pmatrix} 0 & 0 & 0 \\ 0 & 0 & 1 \\ 0 & -1 & 0 \end{pmatrix}, \quad \widetilde{L}_2 = -i\begin{pmatrix} 0 & 0 & -1 \\ 0 & 0 & 0 \\ 1 & 0 & 0 \end{pmatrix},$$

$$\widetilde{L}_3 = -i\begin{pmatrix} 0 & 1 & 0 \\ -1 & 0 & 0 \\ 0 & 0 & 0 \end{pmatrix}, \tag{3.37}$$

which are manifestly Hermitian and purely imaginary and, as we know from quantum mechanics, provide the 3×3 matrix reprsentation corresponding to the angular momentum value $\ell = 1$.

Similarly, let us consider a symmetric second rank tensor of the form

$$T_{\lambda\rho} = x_\lambda x_\rho = T_{\rho\lambda}. \tag{3.38}$$

From the fact that $\hat{p}_\mu T_{\lambda\rho} = -i\frac{\partial x_\lambda x_\rho}{\partial x_\mu} = -i(\delta_{\mu\lambda}x_\rho + \delta_{\mu\rho}x_\lambda)$, we can now directly evaluate and check that

$$L_{\mu\nu}T_{\lambda\rho} = i\left(\delta_{\mu\lambda}T_{\nu\rho} + \delta_{\nu\rho}T_{\mu\lambda} - \delta_{\mu\rho}T_{\nu\lambda} - \delta_{\nu\lambda}T_{\mu\rho}\right), \tag{3.39}$$

so that $T_{\mu\nu} = T_{\nu\mu}$ in (3.38) defines a (matrix) vector space on which the operators $L_{\mu\nu}$ act and leads to a matrix representation of dimension $\frac{1}{2}N(N+1)$ for the operators (as in (3.34)-(3.35)). However, such a representation is not irreducible which can be seen as follows. If we set

$$T = T_{\lambda\lambda},\qquad(3.40)$$

where repeated indices are summed, it follows from (3.39) that

$$L_{\mu\nu}T = 0,\qquad(3.41)$$

and T corresponds to the trivial one dimensional representation (see also (3.32)). As a result, we can construct the traceless symmetric tensor (repeated indices are summed)

$$S_{\lambda\rho} = T_{\lambda\rho} - \frac{1}{N}\delta_{\lambda\rho}T = S_{\rho\lambda}, \quad S_{\lambda\lambda} = 0\qquad(3.42)$$

and it follows from (3.39) and (3.41) that it also satisfies

$$L_{\mu\nu}S_{\lambda\rho} = i\left(\delta_{\mu\lambda}S_{\nu\rho} + \delta_{\nu\rho}S_{\mu\lambda} - \delta_{\mu\rho}S_{\nu\lambda} - \delta_{\nu\lambda}S_{\mu\rho}\right).\qquad(3.43)$$

Th action of the operators $L_{\mu\nu}$ does not take a vector out of the vector space defined by $S_{\lambda\rho}$. Therefore, $S_{\lambda\rho}$ also leads to a representation of $so(N)$ which is $\frac{1}{2}N(N+1) - 1$ dimensional. Noting from (3.42) that $T_{\lambda\rho} = S_{\lambda\rho} + \frac{1}{N}\delta_{\lambda\rho}T$, we conclude that the $\frac{1}{2}N(N+1)$ dimensional representation of $so(N)$ generated by $T_{\lambda\rho}$ is fully reducible since the vector space T_λ is a direct sum of $S_{\lambda\rho}$ and T (see also the discussion around (2.98)-(2.101)). For the case of $N = 3$, the traceless symmetric tensor $S_{\lambda\rho}$ corresponds to the angular momentum state of $\ell = 2$. This is easily seen in the spherical coordinates (r, θ, ϕ) instead of Cartesian coordinates (x_1, x_2, x_3). For example, in the spherical coordinates, we can write

$$x_1 = r\sin\theta\cos\phi,$$
$$x_2 = r\sin\theta\sin\phi,$$
$$x_3 = r\cos\theta,\qquad(3.44)$$

leading to

$$x_\pm \equiv x_1 \pm ix_2 = r\sin\theta\, e^{\pm i\phi}.\qquad(3.45)$$

We know that for $\ell = 2$, the azimuthal quantum number takes the values $m = \pm 2, \pm 1, 0$. The spherical harmonics which are known to provide a representation for the angular momentum states define a 5-dimensional space in this case and can be expressed as (with a suitable normalization)

$$Y_{2,\pm2}(\theta,\phi) = \left(\frac{15}{32\pi}\right)^{\frac{1}{2}} \sin^2\theta e^{\pm 2i\phi} = \frac{1}{r^2}\left(\frac{15}{32\pi}\right)^{\frac{1}{2}} x_\pm x_\pm,$$

$$Y_{2,\pm1}(\theta,\phi) = \mp\left(\frac{15}{8\pi}\right)^{\frac{1}{2}} \sin\theta\cos\theta e^{\pm i\phi} = \frac{1}{r^2}\left(\frac{15}{32\pi}\right)^{\frac{1}{2}} (\mp x_\pm x_3),$$

$$Y_{2,0}(\theta,\phi) = \left(\frac{5}{16\pi}\right)^{\frac{1}{2}} (3\cos^2\theta - 1)$$

$$= \frac{1}{r^2}\left(\frac{15}{32\pi}\right)^{\frac{1}{2}}\sqrt{\frac{2}{3}} (2x_3^2 - x_+x_-). \tag{3.46}$$

3.2.3 Lie algebras of $u(N)$ and $su(N)$. Let a_μ and $a_\mu^\dagger$ denote respectively the annihilation and creation operators of a N-dimensional harmonic oscillator satisfying

$$[a_\mu, a_\nu] = 0 = [a_\mu^\dagger, a_\nu^\dagger], \quad [a_\mu, a_\nu^\dagger] = \delta_\mu^\nu, \tag{3.47}$$

for $\mu,\nu = 1,2,\cdots,N$. Here and in the following discussions we use the notation that a "super" index corresponds to that of a creation operator and define,

$$X^\mu_{\ \nu} = a_\mu^\dagger a_\nu. \tag{3.48}$$

It can be checked easily using (3.47) that

$$[X^\mu_{\ \nu}, a_\alpha^\dagger] = [a_\mu^\dagger a_\nu, a_\alpha^\dagger] = a_\mu^\dagger[a_\nu, a_\alpha^\dagger] = \delta_\nu^\alpha a_\mu^\dagger,$$

$$[X^\mu_{\ \nu}, a_\alpha] = [a_\mu^\dagger a_\nu, a_\alpha] = [a_\mu^\dagger, a_\alpha]a_\nu = -\delta_\alpha^\mu a_\nu. \tag{3.49}$$

Using this, it follows that

$$[X^\mu_{\ \nu}, X^\alpha_{\ \beta}] = [X^\mu_{\ \nu}, a_\alpha^\dagger a_\beta]$$

$$= [X^\mu_{\ \nu}, a_\alpha^\dagger]a_\beta + a_\alpha^\dagger[X^\mu_{\ \nu}, a_\beta]$$

$$= \delta_\nu^\alpha a_\mu^\dagger a_\beta - \delta_\beta^\mu a_\alpha^\dagger a_\nu$$

$$= \delta_\nu^\alpha X^\mu_{\ \beta} - \delta_\beta^\mu X^\alpha_{\ \nu}, \tag{3.50}$$

for $\mu,\nu,\alpha,\beta = 1,2,\cdots,N$. Any set of linear operators (or matrices) $X^\mu_{\ \nu}$ satisfying the commutation relations (3.50) defines the Lie algebra $u(N)$ which is connected to the Lie group $U(N)$. The operators

X^μ_ν are known as the generators of the Lie algebra $u(N)$. From the definition in (3.48) we note the Hermiticity condition

$$(X^\mu_\nu)^\dagger = X^\nu_\mu. \tag{3.51}$$

Let us note that the operator (the repeated index α is summed over $\alpha = 1, 2, \cdots, N$)

$$X = X^\alpha_\alpha, \tag{3.52}$$

commutes with X^μ_ν for all $\mu, \nu = 1, 2, \cdots, N$. Namely, using (3.50) we have

$$\begin{aligned}
[X^\mu_\nu, X] &= [X^\mu_\nu, X^\alpha_\alpha] \\
&= \delta^\alpha_\nu X^\mu_\alpha - \delta^\mu_\alpha X^\alpha_\nu = X^\mu_\nu - X^\mu_\nu = 0.
\end{aligned} \tag{3.53}$$

The operator X is an example of a Casimir invariant (named after the Dutch physicist Hendrik Casimir) of $u(N)$ Lie algebra which will be explained in more detail later in this chapter (as well as in chapter 10).

Let us next define an operator

$$A^\mu_\nu = X^\mu_\nu - \frac{1}{N} \delta^\mu_\nu X. \tag{3.54}$$

Then it follows that

(i) the trace of A^μ_ν vanishes

$$A^\mu_\mu = X^\mu_\mu - \frac{1}{N} N X = X - X = 0. \tag{3.55}$$

(ii) the operator A^μ_ν defined in (3.54) also satisfies the $u(N)$ Lie algebra, namely,

$$\begin{aligned}
[A^\mu_\nu, A^\alpha_\beta] &= [X^\mu_\nu - \frac{1}{N} \delta^\mu_\nu X, X^\alpha_\beta - \frac{1}{N} \delta^\alpha_\beta X] \\
&= [X^\mu_\nu, X^\alpha_\beta] = \delta^\alpha_\nu X^\mu_\beta - \delta^\mu_\beta X^\alpha_\nu \\
&= \delta^\alpha_\nu A^\mu_\beta - \delta^\mu_\beta A^\alpha_\nu,
\end{aligned} \tag{3.56}$$

where we have used (3.53) as well as the definition in (3.54).

Any set of linear operators satisfying (3.55) and (3.56) defines the $su(N)$ Lie algebra which is related to the Lie group $SU(N)$.

The special case of $N = 2$ is of some interest in physics. If we identify (recall that, in this case, we have $A^1{}_1 + A^2{}_2 = 0$ by definition (3.55))

$$J_3 = \frac{1}{2}\left(A^1{}_1 - A^2{}_2\right) = A^1{}_1 = -A^2{}_2,$$

$$J_+ = A^1{}_2,$$

$$J_- = A^2{}_1 = \left(A^1{}_2\right)^\dagger, \tag{3.57}$$

where we have used (3.51) as well as (3.55), then it is easy to check using the algebra in (3.56) that

$$[J_3, J_\pm] = \pm J_\pm, \quad [J_+, J_-] = 2J_3. \tag{3.58}$$

Redefining

$$J_\pm = J_1 \pm i J_2, \tag{3.59}$$

we can rewrite the commutation relations in (3.58) in the compact form

$$[J_i, J_j] = i\epsilon_{ijk} J_k, \quad i, j, k = 1, 2, 3, \tag{3.60}$$

which we recognize to be the angular momentum algebra in 3-dimensions (see also (3.31)) and it follows from the definitions in (3.57) and (3.59) as well as the condition (3.51) that

$$(J_i)^\dagger = J_i, \quad i = 1, 2, 3, \tag{3.61}$$

so that the generators of the algebra are Hermitian. Equations (3.31) and (3.60) show the well known equivalence of the Lie algebras $so(3)$ and $su(2)$.

We can construct matrix representations of $u(N)$ as follows. Let us denote by $|0\rangle$ the vacuum state which satisfies

$$a_\mu|0\rangle = 0, \quad \mu = 1, 2, \cdots, N, \tag{3.62}$$

and construct a completely symmetric tensor of rank n as

$$T^{\mu_1\mu_2\cdots\mu_n} = a_{\mu_1}^\dagger a_{\mu_2}^\dagger \cdots a_{\mu_n}^\dagger |0\rangle, \tag{3.63}$$

where $\mu_j \leq N$, $j = 1, 2, \cdots, n$. (Namely, the totally symmetric tensor operator $a_{\mu_1}^\dagger a_{\mu_2}^\dagger \cdots a_{\mu_n}^\dagger$ acting on vacuum creates a symmetric tensor

state $T^{\mu_1\mu_2\cdots\mu_n}$.) Using (3.50) as well as (3.62) we have $X^\alpha_\beta|0\rangle = 0$
and hence it follows now that

$$X^\alpha_\beta T^{\mu_1\mu_2\cdots\mu_n} = \sum_{j=1}^n \delta^{\mu_j}_\beta T^{\mu_1\cdots\mu_{j-1}\alpha\mu_{j+1}\cdots\mu_n},$$

$$A^\alpha_\beta T^{\mu_1\mu_2\cdots\mu_n} = \left(X^\alpha_\beta - \frac{1}{N}\delta^\alpha_\beta X\right) T^{\mu_1\mu_2\cdots\mu_n}$$

$$= \sum_{j=1}^n \delta^{\mu_j}_\beta T^{\mu_1\cdots\mu_{j-1}\alpha\mu_{j+1}\cdots\mu_n} - \frac{n}{N}\delta^\alpha_\beta T^{\mu_1\mu_2\cdots\mu_n}. \tag{3.64}$$

Therefore, the symmetric tensor $T^{\mu_1\mu_2\cdots\mu_n}$ provides an irreducible
matrix representation, for both $u(N)$ and $su(N)$, of dimensionality

$$d = \frac{1}{n!} N(N+1)(N+2)\cdots(N+n-1)$$

$$= \frac{(N+n-1)!}{(N-1)!n!} = \binom{N+n-1}{n}. \tag{3.65}$$

For $N = 2$, we are dealing with the Lie algebra $su(2) \simeq so(3)$ and
(3.65) yields

$$d = \frac{(2+n-1)!}{(1!)(n!)} = \frac{(n+1)!}{n!} = (n+1). \tag{3.66}$$

In this case, with the conventional identification $d = 2j + 1$ where j
denotes the angular momentum quantum number, the representation
(3.63) describes the angular momentum states of $j = \frac{(d-1)}{2} = \frac{n}{2}$ so
that $j = \frac{1}{2}, 1, \frac{3}{2}, 2, \cdots$ corresponding to $n = 1, 2, 3, \cdots$.

Returning to the $u(N)$ algebra, we note that the tensor

$$T^{\mu_1\mu_2\cdots\mu_n} = a^\dagger_{\mu_1} a^\dagger_{\mu_2} \cdots a^\dagger_{\mu_n}|0\rangle, \tag{3.67}$$

with $\mu_1, \cdots, \mu_n = 1, 2, \cdots, N$, is totally symmetric in its indices.
Therefore, we cannot construct any more complicated representations from this tensor through the use of the Young tableaux symbol
$[f_1, f_2, \cdots, f_n]$ (Young tableau will be discussed in chapter 5). Therefore, in order to construct such representations, let us introduce N^2
independent annihilation (and creation) operators $a^{(j)}_\mu, (a^{(j)}_\mu)^\dagger$ with
$\mu, \nu, j, k = 1, 2, \cdots, N$ satisfying

$$[a^{(j)}_\mu, \left(a^{(k)}_\nu\right)^\dagger] = \delta_{jk}\delta^\nu_\mu,$$

$$[a^{(j)}_\mu, a^{(k)}_\nu] = 0 = [\left(a^{(j)}_\mu\right)^\dagger, \left(a^{(k)}_\nu\right)^\dagger]. \tag{3.68}$$

With these let us construct the mixed tensor

$$X^\mu{}_\nu = \left(a^{(j)}_\mu\right)^\dagger a^{(j)}_\nu,$$ (3.69)

where the repeated index (j) is being summed. It is straightforward to check using (3.68) that these operators also satisfy the $u(N)$ Lie algebra (3.50), namely,

$$[X^\mu{}_\nu, X^\alpha{}_\beta] = \delta^\alpha_\nu X^\mu{}_\beta - \delta^\mu_\beta X^\alpha{}_\nu.$$ (3.70)

We can now form the more general tensor (state)

$$T^{\mu_1\mu_2\cdots\mu_n} = \sum_{j_1,\cdots j_n=1}^n C_{j_1,j_2,\cdots,j_n} \left(a^{(j_1)}_{\mu_1}\right)^\dagger \cdots \left(a^{(j_n)}_{\mu_n}\right)^\dagger |0\rangle,$$ (3.71)

for constant tensor coefficients $C_{j_1,j_2,\cdots,j_n}$ which satisfy the Young tableau symmetry $[f_1, f_2, \cdots, f_n]$. This automatically leads to the tensor $T^{\mu_1\cdots\mu_n}$ satisfying the Young tableau $[f_1, f_2, \cdots, f_n]$. Such a representation theory of $u(N)$ has been extensively studied by Biedenharn and Louck. An alternative method is to use fermionic annihilation and creation operators a_μ and $a^\dagger_\mu$ satisfying the anti-commutation relations (corresponding to a fermion oscillator system)

$$a_\mu a^\dagger_\nu + a^\dagger_\nu a_\mu = \delta^\nu_\mu,$$
$$a_\mu a_\nu + a_\nu a_\mu = 0 = a^\dagger_\mu a^\dagger_\nu + a^\dagger_\nu a^\dagger_\mu, \quad \mu, \nu = 1, 2, \cdots, N,$$ (3.72)

and define

$$X^\mu{}_\nu = a^\dagger_\mu a_\nu.$$ (3.73)

In this case, the state

$$T^{\mu_1\mu_2\cdots\mu_n} = a^\dagger_{\mu_1} a^\dagger_{\mu_2} \cdots a^\dagger_{\mu_n} |0\rangle, \quad n \le N,$$ (3.74)

would provide the completely anti-symmetric (tensor) representation of the Lie algebra. However, we would not go into details of such a representation here.

We remark here that if we define an operator (in the the coordinate representation) as

$$Y_\mu{}^\nu = -x_\mu \frac{\partial}{\partial x_\nu} - \frac{1}{2}\delta_\mu^\nu = -\frac{1}{2}\left(x_\mu \frac{\partial}{\partial x_\nu} + \frac{\partial}{\partial x_\nu} x_\mu\right),$$ (3.75)

which is anti-Hermitian (remember our convention of upper index corresponding to the creation operator)

$$\left(Y_\mu{}^\nu\right)^\dagger = -Y_\mu{}^\nu (\neq Y_\nu{}^\mu), \quad \mu, \nu = 1, 2, \cdots, N, \tag{3.76}$$

also satisfies

$$\left[Y_\mu{}^\nu, Y_\alpha{}^\beta\right] = \delta_\mu^\beta Y_\alpha{}^\nu - \delta_\alpha^\nu Y_\mu{}^\beta, \tag{3.77}$$

which gives the same $u(N)$ Lie algebra relation as in (3.50) with the identification $Y_\mu{}^\nu = -X^\mu{}_\nu$. However, this form of the generator corresponds to a rather non-compact realization of the $g\ell(N)$ Lie algebra. For example, if we express x_μ and $\frac{\partial}{\partial x_\mu}$ in terms of annihilation and creation operators (of a harmonic oscillator system) as

$$a_\mu = \frac{1}{\sqrt{2}}\left(x_\mu + \frac{\partial}{\partial x_\mu}\right), \quad a_\mu^\dagger = \frac{1}{\sqrt{2}}\left(x_\mu - \frac{\partial}{\partial x_\mu}\right), \tag{3.78}$$

then we obtain from (3.75) that

$$\begin{aligned}
Y_\mu{}^\nu &= -\frac{1}{2}\left(a_\mu + a_\mu^\dagger\right)\left(a_\nu - a_\nu^\dagger\right) - \frac{1}{2}\delta_\mu^\nu \\
&= -\frac{1}{2}\left(a_\mu a_\nu + a_\mu^\dagger a_\nu - a_\nu^\dagger a_\mu - a_\mu^\dagger a_\nu^\dagger\right),
\end{aligned} \tag{3.79}$$

where we have used the commutation relations in (3.47). It follows now that $Y_\mu{}^\nu$ acting on the vacuum state generates an infinite number of states leading to an infinite dimensional representation which is unitary for the corresponding Lie group $GL(N)$ (we will discuss the connection between the Lie algebra and the Lie group in the next chapter). On the other hand, we note that Y_μ^ν in (3.75) operating on a finite dimensional tensor

$$T_{\mu_1\cdots\mu_n} = x_{\mu_1} x_{\mu_2} \cdots x_{\mu_n}, \tag{3.80}$$

gives a finite dimensional representation of the corresponding Lie group $GL(N)$ which is not unitary.

Finally, returning to the Lie algebra $so(3) \simeq su(2)$ (discussed in the last section and see also (3.57)-(3.60)), we introduce two sets of creation and annihilation operators as

$$a = a_1, \quad b = a_2, \tag{3.81}$$

and construct the operators

$$J_3 = \frac{1}{2}\left(a^\dagger a - b^\dagger b\right), \quad J_+ = a^\dagger b, \quad J_- = b^\dagger a. \tag{3.82}$$

It can be checked in a straightforward manner that the operators satisfy the $so(3) \simeq su(2)$ Lie algebra. In this case, the standard quantum mechanical state vector (angular momentum eigenstates) $|j, m\rangle, m = j, j - 1, \cdots, -j$ with j integer or half integer can be realized as

$$|j, m\rangle = \frac{1}{\sqrt{(j + m)!(j - m)!}} \left(a^\dagger\right)^{j+m} \left(b^\dagger\right)^{j-m} |0\rangle, \qquad (3.83)$$

which satisfies the relations

$$J_3 |j, m\rangle = m |j, m\rangle,$$
$$J_\pm |j, m\rangle = \sqrt{(j \mp m)(j \pm m + 1)} |j, m \pm 1\rangle, \qquad (3.84)$$

and can be checked to satisfy the orthonormality relation

$$\langle j, m | j', m' \rangle = \delta_{jj'} \delta_{mm'}. \qquad (3.85)$$

This construction of the representations of the angular momentum algebra is due to J. Schwinger. We note that with respect to the inner product (3.85) the generators satisfy the Hermiticity properties

$$(J_3)^\dagger = J_3, \quad (J_\pm)^\dagger = J_\mp. \qquad (3.86)$$

In terms of the tensor $T^{\mu_1 \cdots \mu_n}$ defined in (3.71) for the $su(2)$ algebra, we note from (3.81) as well as (3.83) that this construction corresponds to the choice of the constant coefficients in (3.71) to correspond to

$$C_{j_1, \cdots, j_n} = C_{1,1,\cdots,1,2,2,\cdots,2}. \qquad (3.87)$$

We also note that if we define J_3 and $J_\pm$ by (as opposed to (3.82))

$$J_3 = \frac{1}{2} \left(a^\dagger a + b^\dagger b + 1\right), \quad J_- = iab, \quad J_+ = ib^\dagger a^\dagger, \qquad (3.88)$$

these generators can be checked to also define the angular momentum algebra (see, for example, (3.58))

$$[J_3, J_\pm] = \pm J_\pm, \quad [J_+, J_-] = 2J_3, \qquad (3.89)$$

However, in the present case we note that

$$(J_3)^\dagger = J_3, \quad (J_\pm)^\dagger = -J_\mp, \qquad (3.90)$$

This change in the Hermiticity of the generators is reflected in the fact that the corresponding Lie group, in this case, corresponds to the non-compact group $SO(1,2)$ for which the invariant length is given by $x_3^2 - (x_1^2 + x_2^2)$. The associated Lie algebra under consideration is $so(1,2)$ (which is isomorphic to $su(1,1)$). In this case, the quadratic Casimir operator (which commutes with all the generators as in (3.53)) can be calculated to have the form

$$I_2 = \mathbf{J}^2 = J_3^2 + \frac{1}{2}\left(J_+ J_- + J_- J_+\right) = \frac{1}{4}\left(a^\dagger a - b^\dagger b\right)^2 - \frac{1}{2}, \quad (3.91)$$

which leads to an infinite dimensional unitary representation of the Lie group $SO(1,2)$.

Before concluding this section, we note that the validity of (3.64) would depend on the definitions of $X^\alpha{}_\beta$ in (3.48) (and the corresponding $A^\alpha{}_\beta$) and $T^{\mu_1\mu_2\cdots\mu_n}$ in (3.63). This necessarily brings in the symmetry properties associated with the tensor $T^{\mu_1\mu_2\cdots\mu_n}$. The construction, in fact, has a much more general validity independent of the identifications in (3.48) and (3.63). Let us consider a more general mixed tensor (state) of the form $T^{\mu_1\mu_2\cdots\mu_n}_{\nu_1\nu_2\cdots\nu_m}$ which may or may not have any symmetry such as total symmetry or anti-symmetry associated with its indices. If $X^\alpha{}_\beta$ and X operate on this tensor yielding

$$X^\alpha{}_\beta T^{\mu_1\mu_2\cdots\mu_n}_{\nu_1\nu_2\cdots\nu_m} = \sum_{j=1}^{n} \delta^{\mu_j}_\beta T^{\mu_1\mu_2\cdots\mu_{j-1}\alpha\mu_{j+1}\cdots\mu_n}_{\nu_1\nu_2\cdots\nu_m}$$

$$- \sum_{k=1}^{m} \delta^\alpha_{\nu_k} T^{\mu_1\mu_2\cdots\mu_n}_{\nu_1\nu_2\cdots\nu_{k-1}\beta\nu_{k+1}\cdots\nu_m},$$

$$X T^{\mu_1\mu_2\cdots\mu_n}_{\nu_1\nu_2\cdots\nu_m} = (n-m) T^{\mu_1\mu_2\cdots\mu_n}_{\nu_1\nu_2\cdots\nu_m}, \quad (3.92)$$

then, it can be verified easily that the operators $X^\alpha{}_\beta$ satisfy the $u(N)$ algebra (3.50) without assuming any identification such as (3.48). The verification proceeds as in the calculation of the eigenvalues of the second order Casimir invariant I_2 described in the next section.

3.3 Structure constants and the Killing form

Let L denote a Lie algebra. Since L defines a vector space (see $(L1)$), we can find a linearly independent basis $e_1, e_2, \cdots, e_d$ in this space where $d = \dim L$. Here we assume L to be finite dimensional, namely, we are considering a finite dimensional Lie algebra L. We note that even though L is finite dimensional, the associated Lie group is infinite dimensional. For example, let us consider the Lie algebra $L = so(3)$ which is three dimensional (there are only three elements or generators in the algebra), but the Lie group $G = SO(3)$ is clearly infinite dimensional (the number of elements of rotation corresponding to different rotation parameters is infinite). This is one of the reasons why Lie algebras are easier to deal with than Lie groups.

Returning to our discussion, let us note that if $e_1, e_2, \cdots, e_d$ define a basis of L, then the bilinear product $[e_\mu, e_\nu] \in L$ where $\mu, \nu = 1, 2, \cdots, d$ (see (3.2)) so that the product can be expanded in the basis e_λ as

$$[e_\mu, e_\nu] = \sum_{\lambda=1}^{d} C_{\mu\nu}^{\lambda} e_\lambda, \tag{3.93}$$

for some constants $C_{\mu\nu}^{\lambda}$ which satisfy the conditions

$$C_{\mu\nu}^{\lambda} = -C_{\nu\mu}^{\lambda}, \tag{3.94}$$

$$C_{\mu\nu}^{\alpha} C_{\alpha\lambda}^{\beta} + C_{\nu\lambda}^{\alpha} C_{\alpha\mu}^{\beta} + C_{\lambda\mu}^{\alpha} C_{\alpha\nu}^{\beta} = 0, \tag{3.95}$$

where summation over the repeated index α is understood in (3.95). Here $\mu, \nu, \lambda, \alpha, \beta = 1, 2, \cdots, d$. The anti-symmetry in (3.94) is obvious from the definition (3.2) since

$$[e_\mu, e_\nu] = -[e_\nu, e_\mu]. \tag{3.96}$$

The nonlinear relation in (3.95) can be seen to follow from the Jacobi identity in (3.4) in the following way. Using (3.93), we note that

$$[[e_\mu, e_\nu], e_\lambda] = \left[\sum_{\alpha=1}^{d} C_{\mu\nu}^{\alpha} e_\alpha, e_\lambda \right]$$

$$= \sum_{\alpha=1}^{d} C_{\mu\nu}^{\alpha} [e_\alpha, e_\lambda] \qquad \text{(by (3.3))}$$

$$= \sum_{\alpha=1}^{d} C_{\mu\nu}^{\alpha} \sum_{\beta=1}^{d} C_{\alpha\lambda}^{\beta} e_\beta \qquad \text{(by (3.93))}$$

$$= \sum_{\alpha,\beta=1}^{d} C_{\mu\nu}^{\alpha} C_{\alpha\lambda}^{\beta} e_\beta, \tag{3.97}$$

so that the Jacobi identity (3.4) leads to

$$0 = [[e_\mu, e_\nu], e_\lambda] + [[e_\nu, e_\lambda], e_\mu] + [[e_\lambda, e_\mu], e_\nu]$$

$$= \sum_{\alpha,\beta=1}^{d} \left(C_{\mu\nu}^{\alpha} C_{\alpha\lambda}^{\beta} + C_{\nu\lambda}^{\alpha} C_{\alpha\mu}^{\beta} + C_{\lambda\mu}^{\alpha} C_{\alpha\nu}^{\beta} \right) e_\beta, \tag{3.98}$$

which proves (3.95) since the basis is assumed to be linearly independent.

Conversely, if we have a set of constants $C_{\mu\nu}^{\lambda}$ satisfying the conditions (3.94) and (3.95) and if we define the bilinear product as in (3.93), then we can readily show that it satisfies the Jacobi identity (3.4) and completely determines the Lie algebra. The constants $C_{\mu\nu}^{\lambda}$ are known as the structure constants of the Lie algebra.

For any Lie algebra L, we can find at least one nontrivial $d \times d$ matrix representation (corresponding to the dimension of L) as follows. Let $e_1, e_2, \cdots, e_d$ be a basis of L with the structure constants $C_{\mu\nu}^{\lambda}$. Let us introduce the $d \times d$ matrix $\rho_{ad}(e_\mu)$ by

$$(\rho_{ad}(e_\mu))_{\nu\lambda} \equiv -C_{\mu\nu}^{\lambda},$$

$$(\rho_{ad}(x))_{\mu\nu} = \rho_{ad}(\sum_{\lambda=1}^{d} \xi_\lambda e_\lambda) = \sum_{\lambda=1}^{d} \xi_\lambda \left(\rho_{ad}(e_\lambda)\right)_{\mu\nu}, \tag{3.99}$$

where $x = \sum_{\lambda=1}^{d} \xi_\lambda e_\lambda$ with constant coefficients ξ_λ is an arbitrary element of L. We note that the second relation in (3.99) is due to the linearity of $\rho_{ad}(x)$. Using the standard matrix multiplication, we now obtain

$$[\rho_{ad}(e_\mu), \rho_{ad}(e_\nu)]_{\alpha\beta} = (\rho_{ad}(e_\mu)\rho_{ad}(e_\nu) - \rho_{ad}(e_\nu)\rho_{ad}(e_\mu))_{\alpha\beta}$$

$$= \sum_{\lambda=1}^{d} \left((\rho_{ad}(e_\mu))_{\alpha\lambda} (\rho_{ad}(e_\nu))_{\lambda\beta} - (\rho_{ad}(e_\nu))_{\alpha\lambda} (\rho_{ad}(e_\mu))_{\lambda\beta} \right)$$

$$= \sum_{\lambda=1}^{d} \left(C_{\mu\alpha}^{\lambda} C_{\nu\lambda}^{\beta} - C_{\nu\alpha}^{\lambda} C_{\mu\lambda}^{\beta} \right)$$

$$= \sum_{\lambda=1}^{d} \left(-C_{\mu\alpha}^{\lambda} C_{\lambda\nu}^{\beta} - C_{\alpha\nu}^{\lambda} C_{\lambda\mu}^{\beta} \right) = \sum_{\lambda=1}^{d} C_{\nu\mu}^{\lambda} C_{\lambda\alpha}^{\beta}$$

$$= \sum_{\lambda=1}^{d} C_{\mu\nu}^{\lambda} \left(-C_{\lambda\alpha}^{\beta} \right) = \sum_{\lambda=1}^{d} C_{\mu\nu}^{\lambda} \left(\rho_{ad}(e_\lambda)\right)_{\alpha\beta}, \tag{3.100}$$

where we have used (3.94) and (3.95) in the intermediate steps. Therefore, we see that the $d \times d$ matrices $\rho_{ad}(e_\mu)$ satisfy the relation

$$[\rho_{ad}(e_\mu), \rho_{ad}(e_\nu)] = \sum_{\lambda=1}^{d} C_{\mu\nu}^\lambda \rho_{ad}(e_\lambda)$$

$$= \rho_{ad}(\sum_{\lambda=1}^{d} C_{\mu\nu}^\lambda e_\lambda) = \rho_{ad}([e_\mu, e_\nu]), \qquad (3.101)$$

where we have used (3.93) as well as (3.99). Consequently, it follows that $\rho_{ad}(x)$ for

$$x = \sum_{\mu=1}^{d} \xi_\mu e_\mu, \qquad (3.102)$$

where ξ_μ denote real or complex constants, provides a $d \times d$ matrix representation of L known as the adjoint representation of the Lie algebra.

The adjoint representation can actually be derived more easily in the following manner. Let us recall the relation between linear operators and matrix representations in a vector space (see, for example, (2.17)-(2.22)). We can define a linear operator called the adjoint operator $ad\, x$ with respect to any element of the Lie algebra $x \in L$ acting on vectors in the vector space L by

$$(ad\, x)\, y = [x, y]. \qquad (3.103)$$

Equivalently, we can also define the linear operator $A_\mu = ad\, e_\mu, \mu = 1, 2, \cdots, d$ which corresponds to the adjoint operator with respect to a set of basis vectors in this space and we see, from (3.103), that it satisfies (see also (3.97))

$$A_\mu e_\nu = [e_\mu, e_\nu] = \sum_{\lambda=1}^{d} C_{\mu\nu}^\lambda e_\lambda. \qquad (3.104)$$

Using (3.103) we note that we can write

$$[[x, y], z] = (ad\, [x, y])\, z,$$

$$[[y, z], x] = -[x, [y, z]] = -[x, (ad\, y)\, z] = -(ad\, x)\, (ad\, y)\, z,$$

$$[[z, x], y] = [y, [x, z]] = [y, (ad\, x)\, z] = (ad\, y)\, (ad\, x)\, z. \qquad (3.105)$$

It follows now that the Jacobi identity (3.4) can be written as

$$[[x,y],z] + [[y,z],x] + [[z,x],y] = 0,$$

$$\text{or,} \quad ((ad\,[x,y]) - (ad\,x)\,(ad\,y) + (ad\,y)\,(ad\,x))\,z = 0, \qquad (3.106)$$

which must hold for any three arbitrary elements $x, y, z \in L$ and, therefore, determines

$$(ad\,[x,y]) = (ad\,x)\,(ad\,y) - (ad\,y)\,(ad\,x)$$

$$= [(ad\,x), (ad\,y)]. \qquad (3.107)$$

Consequently, we can identify

$$(ad\,e_\mu) = \rho_{ad}(e_\mu), \quad (ad\,x) = \rho_{ad}(x), \qquad (3.108)$$

as a representation, the adjoint representation, of the Lie algebra L. We also note here that the adjoint representation of the $so(N)$ Lie algebra is isomorphic to that obtained from any second rank antisymmetric tensor $T_{\mu\nu} = -T_{\nu\mu}$ operator which is left as an exercise.

Let us next introduce a symmetric inner product for any two elements $x, y \in L$ by

$$\langle x|y \rangle = \text{Tr}\,((ad\,x)\,(ad\,y)) = \langle y|x \rangle, \qquad (3.109)$$

which is clearly symmetric because of the cyclicity of trace ("Tr" denotes matrix trace). Furthermore, we note that

$$\langle [x,y]|z \rangle = \text{Tr}\,((ad\,[x,y])\,(ad\,z)) = \text{Tr}\,(([[ad\,x),(ad\,y)])\,(ad\,z))$$

$$= \text{Tr}\,((ad\,x)(ad\,y)(ad\,z)) - \text{Tr}\,((ad\,y)(ad\,x)(ad\,z))$$

$$= \text{Tr}\,((ad\,x)(ad\,y)(ad\,z)) - \text{Tr}\,((ad\,x)(ad\,z)(ad\,y))$$

$$= \text{Tr}\,((ad\,x)\,([ad\,y, ad\,z]))$$

$$= \langle x|[y,z] \rangle, \qquad (3.110)$$

and this is easily seen to be completely anti-symmetric under the exchange of any pair of elements $x, y, z \in L$. Here we have used the cyclicity of trace as well as (3.107) in the intermediate steps.

For a set of basis vectors $e_1, e_2, \cdots, e_d$ of the Lie algebra, let us define the symmetric tensor

$$g_{\mu\nu} = \langle e_\mu|e_\nu \rangle = \text{Tr}\,((ad\,e_\mu)\,(ad\,e_\nu)) = g_{\nu\mu}, \qquad (3.111)$$

which is known as the Killing form (named after the German mathematician Wilhelm Killing) or the metric tensor of the Lie algebra.

Using the identification in (3.99) (see also (3.108)), we can write explicitly

$$g_{\mu\nu} = \sum_{\alpha,\beta=1}^{d} (ad\,e_\mu)_{\alpha\beta} (ad\,e_\nu)_{\beta\alpha} = \sum_{\alpha,\beta=1}^{d} C_{\mu\alpha}^{\beta} C_{\nu\beta}^{\alpha}. \qquad (3.112)$$

Similarly, following (3.110) we can introduce a completely anti-symmetric third rank tensor as

$$f_{\mu\nu\lambda} = \langle [e_\mu, e_\nu] | e_\lambda \rangle = \langle e_\mu | [e_\nu, e_\lambda] \rangle. \qquad (3.113)$$

Using (3.93) as well as (3.111) we see from the definition (3.113) that

$$f_{\mu\nu\lambda} = \langle [e_\mu, e_\nu] | e_\lambda \rangle = \sum_{\alpha=1}^{d} C_{\mu\nu}^{\alpha} \langle e_\alpha | e_\lambda \rangle = \sum_{\alpha=1}^{d} C_{\mu\nu}^{\alpha} g_{\alpha\lambda},$$

$$f_{\mu\nu\lambda} = \langle e_\mu | [e_\nu, e_\lambda] \rangle = \sum_{\alpha=1}^{d} C_{\nu\lambda}^{\alpha} \langle e_\mu | e_\alpha \rangle = \sum_{\alpha=1}^{d} C_{\nu\lambda}^{\alpha} g_{\mu\alpha}, \qquad (3.114)$$

so that we can write this completely anti-symmetric tensor explicitly as

$$f_{\mu\nu\lambda} = \sum_{\alpha=1}^{d} C_{\mu\nu}^{\alpha} g_{\alpha\lambda} = \sum_{\alpha=1}^{d} C_{\nu\lambda}^{\alpha} g_{\alpha\mu} = \sum_{\alpha=1}^{d} C_{\lambda\mu}^{\alpha} g_{\alpha\nu}, \qquad (3.115)$$

where we have used the symmetry of the metric tensor and the last relation comes from the total anti-symmetry (under the interchange of any pair of indices) in the definition in (3.113) as well as the anti-symmetry of the structure constants $C_{\mu\nu}^{\lambda}$ (see (3.94)). Such a completely anti-symmetric third rank tensor $f_{\mu\nu\lambda}$ is often used in physics in connection with gauge theories, as we will see in chapter 8.

As we will see in the next section $g_{\mu\nu}$ will be non-degenerate for the class of Lie algebras known as semi-simple Lie algebras. Namely, for such Lie algebras, there exists an inverse metric tensor $g^{\mu\nu}$ satisfying

$$\sum_{\lambda=1}^{d} g_{\mu\lambda} g^{\lambda\nu} = \delta_\mu^\nu, \quad \mu, \nu = 1, 2, \cdots, d. \qquad (3.116)$$

In this case, we can define the second order (quadratic) Casimir invariant as

$$I_2 \equiv \sum_{\mu,\nu=1}^{d} g^{\mu\nu} \rho(e_\mu) \rho(e_\nu), \qquad (3.117)$$

for any representation $\rho(e_\mu)$ of L, not necessarily the adjoint representation $\rho_{ad}(e_\mu)$. (Namely, the metric tensor is clearly defined from the basis matrices in the adjoint representation, but once it is defined, it becomes a fixed tensor in the vector space and can be used for raising and lowering indices in any representation.) It now follows that

$$
\begin{aligned}
[I_2, \rho(e_\lambda)] &= \sum_{\mu,\nu=1}^{d} g^{\mu\nu}[\rho(e_\mu)\rho(e_\nu), \rho(e_\lambda)] \\
&= \sum_{\mu,\nu=1}^{d} g^{\mu\nu}\left(\rho(e_\mu)[\rho(e_\nu), \rho(e_\lambda)] + [\rho(e_\mu), \rho(e_\lambda)]\rho(e_\nu)\right) \\
&= \sum_{\mu,\nu=1}^{d}\sum_{\alpha=1}^{d} g^{\mu\nu}\left(C_{\nu\lambda}^{\alpha}\rho(e_\mu)\rho(e_\alpha) + C_{\mu\lambda}^{\alpha}\rho(e_\alpha)\rho(e_\nu)\right) \\
&= \sum_{\alpha,\mu,\nu=1}^{d}\left(g^{\mu\alpha}C_{\alpha\lambda}^{\nu} + g^{\nu\alpha}C_{\alpha\lambda}^{\mu}\right)\rho(e_\mu)\rho(e_\nu),
\end{aligned}
\tag{3.118}
$$

where we have interchanged $\alpha \leftrightarrow \nu$ in the first term and $\alpha \leftrightarrow \mu$ in the second term in the last line (together with the symmetry of the inverse metric). We can simplify the right hand side by using the definition (3.116) as well as the anti-symmetry (3.94) of the structure constants as follows,

$$
\begin{aligned}
\sum_{\alpha=1}^{d}\left(g^{\mu\alpha}C_{\alpha\lambda}^{\nu} + g^{\nu\alpha}C_{\alpha\lambda}^{\mu}\right) &= \sum_{\alpha=1}^{d} g^{\mu\alpha}C_{\alpha\lambda}^{\nu} + \sum_{\beta=1}^{d} C_{\beta\lambda}^{\mu}g^{\nu\beta} \\
&= \sum_{\alpha,\beta,\sigma=1}^{d} g^{\mu\alpha}\left(C_{\alpha\lambda}^{\sigma}g_{\beta\sigma} + g_{\alpha\sigma}C_{\beta\lambda}^{\sigma}\right)g^{\nu\beta} \\
&= \sum_{\alpha,\beta,\sigma=1}^{d} g^{\mu\alpha}\left(C_{\alpha\lambda}^{\sigma}g_{\sigma\beta} - C_{\lambda\beta}^{\sigma}g_{\sigma\alpha}\right)g^{\nu\beta} \\
&= 0,
\end{aligned}
\tag{3.119}
$$

where we have used (3.115) in the last step. This shows that

$$
[I_2, \rho(e_\lambda)] = 0,
\tag{3.120}
$$

namely, the quadratic Casimir operator commutes with every basis element and, therefore, with any arbitrary element $\rho(x), x \in L$

of the Lie algebra. (clearly, the Casimir operator is defined upto a multiplicative (normalization) constant.) If $\rho(x)$ is irreducible, then Schur's Lemma (2.103) implies that I_2 is proportional to the $n \times n$ unit matrix E_n (n denotes the dimensionality of the matrix representation), namely,

$$I_2 = C_2 E_n. \tag{3.121}$$

The constant C_2 corresponds to the eigenvalue of the Casimir invariant I_2 for the particular representation under consideration. In the case of $su(2)$, we have

$$C_2 = j(j+1), \quad j = 0, \frac{1}{2}, 1, \frac{3}{2}, \cdots, \tag{3.122}$$

if we choose a suitable normalization.

Let us consider the examples of $u(N)$ or $su(N)$ Lie algebras and denote the matrix representation $\rho(X^\mu{}_\nu)$ as $X^\mu{}_\nu$, for simplicity (see (3.50)). Then, for $u(N)$ the second order Casimir can be written as

$$I_2 = X^\mu{}_\nu X^\nu{}_\mu, \tag{3.123}$$

while for $su(N)$ (see (3.54) and (3.56)) it is given by

$$I_2' = A^\mu{}_\nu A^\nu{}_\mu = X^\mu{}_\nu X^\nu{}_\mu - \frac{1}{N} X^2 = I_2 - \frac{1}{N} X^2, \tag{3.124}$$

where we have used the definitions in (3.52) and (3.54) (and repeated indices are summed). If we operate I_2 on the completely symmetric tensor (state) in (3.64), namely,

$$T^{\mu_1\mu_2\cdots\mu_n} = a^\dagger_{\mu_1} a^\dagger_{\mu_2} \cdots a^\dagger_{\mu_n} |0\rangle, \tag{3.125}$$

and calculate using (3.65), we obtain (recall that $\alpha, \beta = 1, 2, \cdots, N$ for $u(N)$ or $su(N)$)

$$I_2 T^{\mu_1\mu_2\cdots\mu_n} = X^\alpha{}_\beta X^\beta{}_\alpha T^{\mu_1\mu_2\cdots\mu_n}$$

$$= X^\alpha{}_\beta \sum_{j=1}^{n} \delta^{\mu_j}_\alpha T^{\mu_1\mu_2\cdots\mu_{j-1}\beta\mu_{j+1}\cdots\mu_n}$$

$$= \sum_{j=1}^{n} X^{\mu_j}{}_\beta T^{\mu_1\mu_2\cdots\mu_{j-1}\beta\mu_{j+1}\cdots\mu_n}$$

$$= \sum_{j=1}^{n} \left[\sum_{k\neq j} \delta^{\mu_k}_\beta T^{\mu_1\cdots\mu_{k-1}\mu_j\mu_{k+1}\cdots\mu_{j-1}\beta\mu_{j+1}\cdots\mu_n} \right.$$

$$+ \delta_\beta^\beta T^{\mu_1 \mu_2 \cdots \mu_{j-1} \mu_j \mu_{j+1} \cdots \mu_n} \Big]$$

$$= \sum_{j \neq k} T^{\mu_1 \mu_2 \cdots \mu_{k-1} \mu_j \mu_{k+1} \cdots \mu_{j-1} \mu_k \mu_{j+1} \cdots \mu_N}$$

$$+ Nn \, T^{\mu_1 \mu_2 \cdots \mu_n}$$

$$= (n(n-1) + Nn) \, T^{\mu_1 \mu_2 \cdots \mu_n}$$

$$= n(N + n - 1) T^{\mu_1 \mu_2 \cdots \mu_n}, \tag{3.126}$$

so that the eigenvalues of I_2 (the quadratic Casimir for $u(N)$) for such a representation is given by $n(N + n - 1)$. Here we have used the symmetric nature of $T^{\mu_1 \mu_2 \cdots \mu_n}$ in the intermediate step.

Similarly, it follows that

$$X T^{\mu_1 \mu_2 \cdots \mu_n} = X^\alpha_{\ \alpha} T^{\mu_1 \mu_2 \cdots \mu_n} = \sum_{j=1}^n \delta_\alpha^{\mu_j} T^{\mu_1 \mu_2 \cdots \mu_{j-1} \alpha \mu_{j+1} \cdots \mu_n}$$

$$= \sum_{j=1}^n T^{\mu_1 \mu_2 \cdots \mu_n} = n T^{\mu_1 \mu_2 \cdots \mu_n}, \tag{3.127}$$

which leads to

$$I_2' T^{\mu_1 \mu_2 \cdots \mu_n} = \left(I_2 - \frac{1}{N} X^2 \right) T^{\mu_1 \mu_2 \cdots \mu_n}$$

$$= \left(n(N + n - 1) - \frac{n^2}{N} \right) T^{\mu_1 \mu_2 \cdots \mu_n}$$

$$= \frac{(N-1)n(N+n)}{N} T^{\mu_1 \mu_2 \cdots \mu_n}. \tag{3.128}$$

Therefore, the eigenvalues of the quadratic Casimir operator for $su(N)$ for such a representation is given by $\frac{(N-1)n(N+n)}{N}$. For $su(2)$, this leads to

$$I_2' T^{\mu_1 \mu_2 \cdots \mu_n} = \frac{1}{2} n(n+2) T^{\mu_1 \mu_2 \cdots \mu_n} = 2j(j+1) T^{\mu_1 \mu_2 \cdots \mu_n}, \tag{3.129}$$

for $j = \frac{n}{2}$. This reproduces the well known result that

$$\mathbf{J}^2 = \frac{1}{2} A^\mu_{\ \nu} A^\nu_{\ \mu}, \tag{3.130}$$

has eigenvalues $j(j + 1)$ with $j = 0, \frac{1}{2}, 1, \frac{3}{2}, \cdots$.

3.4 Simple and semi-simple Lie algebras

Let L be a Lie algebra so that we have

$$[L, L] \subseteq L. \tag{3.131}$$

However, L can also possess its sub-Lie algebra $L_0 \subseteq L$ such that

$$[L_0, L_0] \subseteq L_0. \tag{3.132}$$

Moreover, it can also happen that the subalgebra satisfies a stronger relation

$$[L_0, L] \subseteq L_0, \tag{3.133}$$

and in this case, we say that the subalgebra L_0 is an ideal of L. When the solution to (3.133) excludes the rather (obvious) trivial cases of $L_0 = L$ and $L_0 = 0$ (null), we say that L_0 defines a proper ideal of the Lie algebra L. A Lie algebra L which is not one-dimensional and does not contain any proper ideal is known as a simple Lie algebra (or simply a "simple" algebra). On the other hand, if L contains proper ideals, but none of the proper ideals is Abelian, then it is known as a semi-simple Lie algebra. We can further classify Lie algebras into categories such as solvable, nilpotent, *etc.* However, we will not go into details of these since we will not use them in any essential way in our discussions. As noted earlier, we will consider L to denote a complex Lie algebra. Let us next note some important theorems within the context of Lie algebras without going into proofs which would be beyond the scope of such a book.

Theorem. *Any semi-simple Lie algebra is a direct sum of simple Lie algebras*, i.e. *a semi-simple Lie algebra L can be written as*

$$L = L_1 \oplus L_2 \oplus \cdots \oplus L_m, \tag{3.134}$$

where L_i, $i = 1, 2, \cdots, m$ denote simple Lie algebras with

$$[L_i, L_j] = 0, \quad \text{for} \quad i \neq j. \tag{3.135}$$

Theorem. (**Cartan's criterion**)

A necessary and sufficient condition for a complex Lie algebra
L to be semi-simple is that its killing form

$$\langle x|y \rangle = \text{Tr} \, (ad\,x)\,(ad\,y), \tag{3.136}$$

is non-degenerate. In other words, if $e_1, e_2, \cdots, e_d$ denote a basis of
L, then (see also (3.111) as well as (3.116))

$$g_{\mu\nu} = \text{Tr} \, (ad\,e_\mu)\,(ad\,e_\nu), \tag{3.137}$$

has an inverse $g^{\mu\nu}$ satisfying $\sum_{\lambda=1}^{d} g_{\mu\lambda} g^{\lambda\nu} = \delta_\mu^\nu$ for $\mu, \nu = 1, 2, \cdots, d$.

Theorem. *Any finite dimensional matrix representation of a semi-simple*
Lie algebra is fully reducible.

▶ **Example ($u(N)$ algebra).** As we have seen in section 3.2.3, we can identify the
basis elements of $u(N)$ as

$$X^\mu{}_\nu = a_\mu^\dagger a_\nu, \tag{3.138}$$

where $a_\mu, a_\mu^\dagger$, $\mu = 1, 2, \cdots, N$ denote respectively the annihilation and creation
operators of a N-dimensional harmonic oscillator and satisfy the $u(N)$ algebra
(see (3.50))

$$[X^\mu{}_\nu, X^\alpha{}_\beta] = \delta_\nu^\alpha X^\mu{}_\beta - \delta_\beta^\mu X^\alpha{}_\nu, \tag{3.139}$$

which defines a non-Abelian algebra. As we have seen in (3.52), in this case, we
can identify the (linear) invariant operator (linear Casimir operator)

$$I_1 = X = X^\alpha{}_\alpha, \tag{3.140}$$

which commutes with all the basis elements (generators) of L (and, therefore, with
itself)

$$[I_1, X^\mu{}_\nu] = [X, X^\mu{}_\nu] = 0, \quad [I_1, I_1] = [X, X] = 0. \tag{3.141}$$

As a result, the set of elements $L_0 = \{\lambda I_1\}$ for any complex number λ defines a
proper ideal of $L = u(N)$, since

$$[L_0, L] = 0 \subseteq L_0. \tag{3.142}$$

Therefore, $u(N)$ cannot be a simple algebra. But, we also note that $u(N)$ cannot
be a semi-simple algebra because L_0 is an Abelian ideal since

$$[L_0, L_0] = 0. \tag{3.143}$$

This shows that the $u(N)$ algebra is neither simple nor semi-simple. In fact, let us note that, since $su(N)$ is known to be a simple Lie algebra, if we write (of course, dimensionally this is permissible)

$$u(N) = su(N) \oplus \{\lambda I_1\}, \tag{3.144}$$

where $\{\lambda I_1\}$ defines a one-dimensional Abelian algebra, then this would have the general structure of a direct sum of a semi-simple and an Abelian Lie algebras. Such a Lie algebra is known to be a reductive Lie algebra and any (real) reductive Lie algebra is known to be derived from a (real) compact Lie algebra and *vice versa* in some sense. This is true for $u(N)$ and, therefore, $u(N)$ is a reductive Lie algebra. It is worth pointing out here that the Poincaré Lie algebra is neither simple nor semi-simple as well, but is not reductive.

◀

▶ **Example** ($so(2,1)$ **Lie algebra).** We have already studied the angular momentum algebra

$$[J_i, J_j] = i\epsilon_{ijk} J_k, \tag{3.145}$$

corresponding to the equivalent Lie algebras $so(3) \simeq su(2)$ (see (3.31) and (3.60)) which are equivalent. More explicitly, we can write the algebra in (3.145) as

$$[J_1, J_2] = iJ_3, \quad [J_2, J_3] = iJ_1, \quad [J_3, J_1] = iJ_2. \tag{3.146}$$

As we have discussed earlier, the groups $SO(3)$ and $SU(2)$ are compact groups describing rotations. Therefore, the finite dimensional representations of these Lie algebras are unitary and describe the angular momentum states which are essential in studying rotation invariant quantum mechanical systems.

There are two other equivalent Lie algebras, $so(2,1) \simeq su(1,1)$, which are also very useful in studying some quantum mechanical systems and we will discuss here the unitary representations of these algebras. (This can be thought of as the Lorentz algebra in three dimensions.) Like angular momentum, this algebra also consists of three generators $J_i, i = 1, 2, 3$ satisfying the commutation relations

$$[J_1, J_2] = -iJ_3, \quad [J_2, J_3] = iJ_1, \quad [J_3, J_1] = iJ_2, \tag{3.147}$$

which is similar to (3.146) except for the sign in the first relation. It is clear that the algebra (3.147) can be mapped to the angular momentum algebra with the redefinition

$$J_1 \to iJ_1, \quad J_2 \to iJ_2. \tag{3.148}$$

However, under this redefinition, J_1, J_2 cease to be Hermitian. Of course, the finite dimensional representations of $so(2,1) \simeq su(1,1)$ can be obtained in this manner from those of the angular momentum algebra, but these representations are not unitary. This is simply a reflection of the fact that the groups $SO(2,1)$ and $SU(1,1)$ are non-compact (like the Lorentz group) for which there is no finite dimensional unitary representation. (The generator J_3 can still be thought of as generating compact rotations around the z-axis, but J_1, J_2 generate non-compact transformations along the x, y-axes.) In quantum mechanics, on the other hand, we are interested in unitary representations. For a non-compact symmetry these are necessarily infinite dimensional.

To construct the infinite dimensional unitary representations, let us continue with (3.147) where all the generators are Hermitian

$$(J_i)^\dagger = J_i, \quad i = 1, 2, 3. \tag{3.149}$$

The quadratic Casimir invariant of the algebra (3.147) can be checked to correspond to

$$I_2 = (J_1)^2 + (J_2)^2 - (J_3)^2, \quad [I_2, J_i] = 0, \quad i = 1, 2, 3. \tag{3.150}$$

This can be compared with the Casimir invariant of the angular momentum algebra given by $I_2 = \mathbf{J}^2 = (J_1)^2 + (J_2)^2 + (J_3)^2$. We can, of course, diagonalize only one of the generators of the algebra along with I_2 and we choose to diagonalize the compact generator J_3. In the usual manner we can conclude that the eigenvalues of J_3 take only integer or half integer values (allowing for double valued representations), namely,

$$J_3 : \; m = 0, \pm 1, \pm 2, \cdots, \quad \text{or,} \quad m = \pm\frac{1}{2}, \pm\frac{3}{2}, \cdots. \tag{3.151}$$

As in the study of angular momentum algebra, let us define the operators

$$J_\pm = J_1 \pm i J_2, \quad (J_\pm)^\dagger = J_\mp, \tag{3.152}$$

which can be checked (using (3.147)) to satisfy the commutation relations

$$[J_\pm, J_3] = [J_1 \pm i J_2, J_3] = -i J_2 \pm i(i J_1)$$
$$= \mp (J_1 \pm i J_2) = \mp J_\pm. \tag{3.153}$$

Therefore, $J_\pm$ correspond to the raising and lowering operators which raise/lower the eigenvalues of J_3 by one unit. Namely, acting on an eigenstate of J_3 with eigenvalue m, they take us to another eigenstate of J_3 with eigenvalue $(m \pm 1)$.

We note that

$$J_\pm J_\mp = (J_1 \pm i J_2)(J_1 \mp i J_2) = (J_1)^2 + (J_2)^2 \mp i[J_1, J_2]$$
$$= (J_1)^2 + (J_2)^2 \mp i(-i J_3) = \left((J_1)^2 + (J_2)^2 - (J_3)^2\right) + (J_3)^2 \mp J_3$$
$$= I_2 + J_3(J_3 - 1). \tag{3.154}$$

Equivalently, we can write

$$I_2 = J_\pm J_\mp - J_3(J_3 - 1). \tag{3.155}$$

From (3.152) we see that $J_\pm J_\mp = J_\mp^\dagger J_\mp$ is a positive semi-definite operator and, therefore, we conclude from (3.155) that the quadratic Casimir operator for $so(2, 1)$ is not positive semi-definite, unlike in the case of the angular momentum algebra. The eigenvalues of I_2, in this case, can be positive or negative or zero. Let us denote the simultaneous eigenstates of I_2 and J_3 as satisfying

$$J_3|j, m\rangle = m|j, m\rangle, \quad m = 0, \pm 1, \pm 2, \cdots, \quad \text{or,} \quad m = \pm\frac{1}{2}, \pm\frac{3}{2}, \cdots,$$

$$I_2|j, m\rangle = -j(j + 1)|j, m\rangle. \tag{3.156}$$

We point out here that in some literature the eigenvalues of I_2 are denoted by $k(1-k)$ with the identification $k = -j$. Furthermore, since the second equation is invariant under $j \leftrightarrow -(j+1)$, without any loss of generality, we will restrict ourselves to $j < 0$. Depending on the values of (j, m), the unitary representations of the algebra are classified into three categories which we discuss below.

(i) **Discrete/analytic series:** Here we assume that the infinite dimensional representations are bounded from below or from above. In this case, taking the expectation value of (3.154) in the state $|_J, m\rangle$ and recognizing that the operator $J_\pm J_\mp$ is positive semi-definite, we obtain

$$\langle j, m | J_\pm J_\mp | j, m \rangle = \langle j, m | I_2 + J_3 (J_3 \mp 1) | j, m \rangle$$
$$= -j(j+1) + m(m \mp 1) \geq 0, \tag{3.157}$$

where we have assumed that the eigenstates are normalized. For a fixed value of j, let us assume that there exists a state $|j, m_{\min}\rangle$ such that

$$J_- |j, m_{\min}\rangle = 0. \tag{3.158}$$

In this state, (3.157) leads to

$$- j(j+1) + m_{\min}(m_{\min} - 1) = 0,$$

$$\text{or,} \quad m_{\min} = -j > 0. \tag{3.159}$$

Such a representation is bounded from below and is denoted by D_j^+. It is labelled by

$$m = -j, -j+1, -j+2, \cdots, \tag{3.160}$$

and since m can take only integer or half integer values, it follows that

$$j = -\frac{1}{2}, -1, -\frac{3}{2}, -2, \cdots. \tag{3.161}$$

It is clear that given a value of j, the representation is uniquely determined and is infinite dimensional (since m takes an infinite number of values). We note here that for $j = -\frac{1}{2}$, the eigenvalue of the quadratic Casimir operator is positive while for $j = -1$ it vanishes. For all other values, it is negative. The trivial one dimensional representation corresponding to $j = -1$ will not be considered further.

In a similar manner, if there is a state $|j, m_{\max}\rangle$ such that

$$J_+ |j, m_{\max}\rangle = 0, \tag{3.162}$$

then (3.157) in this state determines

$$- j(j+1) + m_{\max}(m_{\max} + 1) = 0,$$

$$\text{or,} \quad m_{\max} = j < 0. \tag{3.163}$$

Such a representation is bounded from above and is denoted by D_j^-. It is labelled by

$$m = j, j-1, j-2, \cdots, \quad j = -1, -\frac{3}{2}, \cdots. \tag{3.164}$$

Principal series: Besides the discrete representations discussed above, there are also continuous unitary representations of the Lie algebra and the principal series is one of them. They are known as continuous representations because j takes a continuum of values. Furthermore, these are representations where there does not exist any state satisfying

$$J_-|j,m\rangle = 0, \quad \text{or}, \quad J_+|j,m\rangle = 0. \tag{3.165}$$

Correspondingly, the value of m is not restricted either from above or below.

In the continuous representations, we assume that $j(j+1) < 0$ (so that the eigenvalue of the Casimir operator I_2 is positive). In the case of the principal series we allow the quantum number j to be complex. Therefore, writing

$$j = \alpha + i\lambda, \tag{3.166}$$

where α, λ are real, we obtain

$$j(j+1) = (\alpha + i\lambda)(\alpha + 1 + i\lambda) = \text{real},$$

$$\text{or}, \quad \left(\alpha(\alpha + 1) - \lambda^2\right) + i\lambda(2\alpha + 1) = \text{real},$$

$$\text{or}, \quad \alpha = -\frac{1}{2}. \tag{3.167}$$

Furthermore, using this value of α in (3.166), we can check that

$$j(j+1) = (-\frac{1}{2} + i\lambda)(\frac{1}{2} + i\lambda) < 0,$$

$$\text{or}, \quad \frac{1}{4} + \lambda^2 > 0. \tag{3.168}$$

In this case, the eigenvalues of the Casimir operator takes values $\frac{1}{4} < -j(j+1) < \infty$ (which is known as a non-exceptional interval). In this case, we note from (3.154) that

$$\langle j,m|J_\pm J_\mp|j,m\rangle = -j(j+1) + m(m \mp 1) = \lambda^2 + (m \mp \frac{1}{2})^2 > 0, \tag{3.169}$$

which holds for any finite value of λ. As a result, there is no state in the representation for which

$$J_-|j,m\rangle = 0, \quad \text{or}, \quad J_+|j,m\rangle = 0, \tag{3.170}$$

so that the m quantum number in these representations is unbounded from below as well as from above. Further investigation shows that, in this case,

$$m = \pm\frac{1}{2}, \pm\frac{3}{2}, \cdots. \tag{3.171}$$

Complementary series: The complementary series (sometimes also called the supplementtary series) also describes a continuous unitary representation where m is not bounded either from below or from above. Here we assume, as in the case of the principal series, that $j(j+1) < 0$ (so that the eigenvalues of the quadratic Casimir operator is positive). However, we require j to be real unlike the earlier case. Thus, we have

$$j(j+1) < 0,$$

$$\text{or}, \quad (j + \frac{1}{2})^2 < (\frac{1}{2})^2,$$

$$\text{or}, \quad |j + \frac{1}{2}| < \frac{1}{2}. \tag{3.172}$$

This determines (recall that $j < 0$)

$$-1 < j < 0. \tag{3.173}$$

Global properties of the group further restrict this finite interval to

$$-1 < j < -\frac{1}{2}. \tag{3.174}$$

Equivalently, we can identify

$$j = -\frac{1}{2} - \lambda, \tag{3.175}$$

where λ is real and satisfies

$$0 < \lambda < \frac{1}{2}, \tag{3.176}$$

and j can take any continuous value in the finite interval (3.174).

We note that in this interval $-j(j+1) = \frac{1}{4} - \lambda^2$ so that using (3.176) we have $0 < -j(j+1) < \frac{1}{4}$ (which is known as an exceptional interval). From (3.154) we then note that

$$\langle j, m | J_\pm J_\mp | j, m \rangle = -j(j+1) + m(m \mp 1) = (m \mp \frac{1}{2})^2 - \lambda^2 > 0. \tag{3.177}$$

As a result, there is no state in the representation for which

$$J_- | j, m \rangle = 0, \quad \text{or,} \quad J_+ | j, m \rangle = 0, \tag{3.178}$$

so that the m quantum number is not bounded from below or from above. Further investigaten shows that in this case

$$m = 0, \pm 1, \pm 2, \cdots. \tag{3.179}$$

The Lie algebra $so(2,1) \simeq su(1,1)$ is useful in studying scattering states of potentials such as the Pöschl-Teller potential. It also manifests in the study of two dimensional (anti) de Sitter space.

◀

▶ **Example ($so(4)$ Lie algebra).** The $so(N)$ Lie algebras are known to be simple for $N = 3$ and $N \geq 5$. ($so(2)$ is an Abelian algebra.) However, $so(4)$ is not simple, rather it is a semi-simple Lie algebra which can be seen as follows. Let us recall the discussion in section 3.2.2 and note that the basis elements $J_{\mu\nu} = -J_{\nu\mu}$ with (in section 3.2.2 we denoted the basis elements as the angular momentum operators $L_{\mu\nu}$)

$$[J_{\mu\nu}, J_{\alpha\beta}] = i \left(\delta_{\mu\alpha} J_{\nu\beta} + \delta_{\nu\beta} J_{\mu\alpha} - \delta_{\mu\beta} J_{\nu\alpha} - \delta_{\nu\alpha} J_{\mu\beta} \right), \tag{3.180}$$

where $\mu, \nu, \alpha, \beta = 1, 2, 3, 4$. If we now identify

$$J_1 = \frac{1}{2} \left(J_{23} + J_{14} \right), \quad J_2 = \frac{1}{2} \left(J_{31} + J_{24} \right), \quad J_3 = \frac{1}{2} \left(J_{12} + J_{34} \right),$$

$$K_1 = \frac{1}{2} \left(J_{23} - J_{14} \right), \quad K_2 = \frac{1}{2} \left(J_{31} - J_{24} \right), \quad K_3 = \frac{1}{2} \left(J_{12} - J_{34} \right), \tag{3.181}$$

or equivalently

$$J_i = \frac{1}{2} \left(\frac{1}{2} \epsilon_{ijk} J_{jk} + J_{i4} \right), K_i = \frac{1}{2} \left(\frac{1}{2} \epsilon_{ijk} J_{jk} - J_{i4} \right), \tag{3.182}$$

where $i, j, k = 1, 2, 3$, the algebra (3.180) can be written as

$$[J_i, J_j] = i\epsilon_{ijk} J_k, \tag{3.183}$$

$$[K_i, K_j] = i\epsilon_{ijk} K_k, \tag{3.184}$$

$$[J_i, K_j] = 0. \tag{3.185}$$

We recognize from (3.183) and (3.184) that J_i and K_i individually satisfy the $so(3)$ algebra (see (3.31) as well as (3.60) and note that the two algebras $so(3)$ and $su(2)$ are equivalent). Furthermore, from (3.185) we note that J_i and K_i commute so that we can write

$$so(4) = so(3) \oplus so(3), \tag{3.186}$$

which shows that the $so(4)$ Lie algebra is a direct sum of two $so(3)$ Lie algebras which are simple Lie algebras. This proves that $so(4)$ is a semi-simple Lie algebra.

We note here that given the operators $J_{\mu\nu}$, we can define their conjugates (in physics also known as dual) as

$$^*J_{\mu\nu} = \frac{1}{2} \epsilon_{\mu\nu\lambda\rho} J_{\lambda\rho}, \tag{3.187}$$

where $\mu, \nu, \lambda, \rho = 1, 2, 3, 4$ and $\epsilon_{\mu\nu\lambda\rho}$ is the completely anti-symmetric Levi-Civita tensor in four dimensions with $\epsilon_{1234} = 1$. It is clear from this definition as well as from (3.181) that we can write

$$J_1 = \frac{1}{2} \left(J_{23} + {}^*J_{23} \right), \qquad J_2 = \frac{1}{2} \left(J_{31} + {}^*J_{31} \right), \qquad J_3 = \frac{1}{2} \left(J_{12} + {}^*J_{12} \right),$$

$$K_1 = \frac{1}{2} \left(J_{23} - {}^*J_{31} \right), \qquad K_2 = \frac{1}{2} \left(J_{31} - {}^*J_{31} \right), \qquad K_3 = \frac{1}{2} \left(J_{12} - {}^*J_{12} \right), \tag{3.188}$$

which shows that J_i and K_i with $i = 1, 2, 3$ correspond to the self-conjugate and anti-self-conjugate components of $J_{\mu\nu}$. Therefore, finding a representation of $J_{\mu\nu}$ in terms of the representations of J_i and K_i would correspond to finding self-conjugate and anti-self-conjugate representations of $so(4)$.

As a simple application of the $so(4)$ Lie algebra, let us determine the energy levels of a Hydrogenic atom described by the Hamiltonian

$$H = \frac{\mathbf{p}^2}{2\mu} - \frac{Ze^2}{r}, \tag{3.189}$$

where Z denotes the nuclear charge, μ the reduced mass of the system and $r = |\mathbf{r}|$ corresponds to the length of the reduced coordinate vector ($Z = 1$ corresponds to the Hydrogen atom). Since three dimensional rotations define a symmetry of this Hamiltonian, the three angular momentum operators $L_i, i = 1, 2, 3$ defined by

$$\mathbf{L} = \mathbf{x} \times \mathbf{p}, \quad L_i = \epsilon_{ijk} x_j p_k, \tag{3.190}$$

commute with the Hamiltonian

$$[L_i, H] = 0, \quad i = 1, 2, 3, \tag{3.191}$$

As we have seen in (3.31), the angular momentum operators L_i satisfy the $so(3)$ Lie algebra (we restore here the factor of $\hbar$ that arises in quantum mechanics from the identification $\mathbf{p} = -i\hbar\boldsymbol{\nabla}$)

$$[L_i, L_j] = i\hbar\epsilon_{ijk}L_k. \tag{3.192}$$

The Hamiltonian in (3.189) has a second (hidden) $so(3)$ symmetry conventionally known as the *dynamical symmetry* or the *accidental symmetry* of the system. This can be seen by noting that if we define a set of three operators M_i through the relation

$$\mathbf{M} = \frac{\mathbf{p} \times \mathbf{L}}{\mu} - \frac{Ze^2}{r}\mathbf{r}, \quad \mathbf{L} \cdot \mathbf{M} = 0, \tag{3.193}$$

then one can verify using the canonical commutation relations as well as (3.191) that

$$[M_i, H] = 0, \tag{3.194}$$

and the three operators M_i generate transformations which leave the Hamiltonian (3.189) invariant. Furthermore, using (3.192) (as well as the canonical commutation relations) it can also be checked that

$$[M_i, M_j] = -\frac{2i\hbar}{\mu}\epsilon_{ijk}HL_k,$$

$$[M_i, L_j] = i\hbar\epsilon_{ijk}M_k. \tag{3.195}$$

We note here for later use that we can write

$$\mathbf{M}^2 = \frac{2H}{\mu}\left(\mathbf{L}^2 + \hbar^2\right) + e^4, \tag{3.196}$$

where the factor $\hbar^2$ arises because of non-commutativity of quantum operators.

If we work in the subspace of the Hilbert space corresponding to a particular energy value E, then the first relation in (3.195) has the form

$$[M_i, M_j] = -\frac{2i\hbar E}{\mu}\epsilon_{ijk}L_k, \tag{3.197}$$

and a simple rescaling of the operators M_i by

$$M_i \to \left(-\frac{\mu}{2E}\right)^{\frac{1}{2}} M_i, \tag{3.198}$$

allows us to write the relations (3.192) and (3.195) as

$$[L_i, L_j] = i\hbar\epsilon_{ijk}L_k,$$
$$[M_i, M_j] = i\hbar\epsilon_{ijk}L_k,$$
$$[M_i, L_j] = i\hbar\epsilon_{ijk}M_k. \tag{3.199}$$

This is a set of coupled commutation relations which defines an algebra. However, to understand its meaning, let us change the basis and define (compare with (3.181))

$$J_i = \frac{1}{2}\left(L_i + M_i\right), \quad K_i = \frac{1}{2}\left(L_i - M_i\right). \tag{3.200}$$

Using (3.199), the new operators can be checked to satisfy the algebra

$$[J_i, J_j] = i\hbar\epsilon_{ijk}J_k,$$
$$[K_i, K_j] = i\hbar\epsilon_{ijk}K_k,$$
$$[J_i, K_j] = 0, \tag{3.201}$$

which coincides with (3.185). Therefore, we see that there is a larger symmetry

$$so(4) = so(3) \oplus so(3), \tag{3.202}$$

operative in the Hamiltonian for the Hydrogenic atom.

To determine the energy spectrum of the system, we note that each of the two $so(3)$ algebras has an associated quadratic Casimir operator which we denote by $\mathbf{J}^2$ and $\mathbf{K}^2$ respectively. The representations of the two angular momentum algebras in (3.201) would be labelled by the eigenvalues of these operators which have the forms

$$\mathbf{J}^2 : \hbar^2 j(j+1), \quad \mathbf{K}^2 : \hbar^2 k(k+1), \tag{3.203}$$

where $j, k = 0, \frac{1}{2}, 1, \cdots$. Alternatively, we can define

$$C = \mathbf{J}^2 + \mathbf{K}^2 = \frac{1}{2}\left(\mathbf{L}^2 + \mathbf{M}^2\right),$$
$$C' = \mathbf{J}^2 - \mathbf{K}^2 = \mathbf{L} \cdot \mathbf{M} = 0. \tag{3.204}$$

The second of the relations in (3.204) implies that for any allowed representation we must have $j = k$, which in turn implies that in any representation the quadratic Casimir C will have eigenvalues $2\hbar^2 j(j+1)$. It follows now that we can write (recall (3.196), (3.198) as well as the fact that we are in a subspace of the Hamiltonian with energy value E)

$$C = \frac{1}{2}\left(\mathbf{L}^2 + \mathbf{M}^2\right)$$

$$= \frac{1}{2}\left(\mathbf{L}^2 - \mathbf{L}^2 - \hbar^2 - \frac{\mu(Ze^2)^2}{2E}\right)$$

$$= \left(-\frac{\mu(Ze^2)^2}{4E} - \frac{1}{2}\hbar^2\right)\mathbb{1} = 2\hbar^2 j(j+1)\mathbb{1}, \tag{3.205}$$

so that we determine

$$E_n = -\frac{\mu(Ze^2)^2}{2\hbar^2(2j+1)^2} = -\frac{\mu(Ze^2)^2}{2\hbar^2 n^2}, \tag{3.206}$$

where we have identified

$$n = 2j + 1 = 1, 2, 3, \cdots . \tag{3.207}$$

We recognize (3.207) as the correct energy levels for the Hydrogenic atom. However, the symmetry analysis can even determine the degeneracy associated with each of these levels. We note from (3.200) that we can invert these relations to obtain

$$\mathbf{L} = \mathbf{J} + \mathbf{K}, \tag{3.208}$$

and from the composition (addition) of angular momenta, we recognize that the eigenvalues of $\mathbf{L}^2$ denoted by $\hbar^2\ell(\ell+1)$ can take values

$$\ell = j+k, j+k-1, \cdots, |j-k|-1, |j-k| = 2j, 2j-1, \cdots, 1, 0$$

$$= n-1, n-2, \cdots, 1, 0. \tag{3.209}$$

Namely, the energy level (3.207) would have a n-fold degeneracy associated with the quantum number of the orbital angular momentum ℓ which is a consequence of the larger $so(4)$ symmetry of the system. (This is in addition to the degeneracy associated with the azimuthal quantum number m which results from three dimensional rotational symmetry.)

◀

3.5 Universal enveloping Lie algebra

We end this discussion of Lie algebras with only a brief comment on the enveloping algebras. Let L denote a Lie algebra. Then, by definition (see (3.2) and discussion there) if $x, y \in L$, then $[x, y] \in L$ is defined, but xy is not well defined in L in general. However, we can formally introduce an abstract associative product $xy, xyz, \cdots$ for any $x, y, z \in L$ satisfying the Jacobi identity. The resulting associative algebra is known as the universal enveloping algebra of the Lie algebra L. It is infinite dimensional even if L is finite dimensional. In this case, the original Lie product $[x, y]$ is identified with the commutator of L, namely, $[x, y] = xy - yx$. For example, let us consider the Lie algebra of $so(3)$. The generators in the coordinate representation have the forms $L_i = -i\epsilon_{ijk} \left(x_j \frac{\partial}{\partial x_k} - x_k \frac{\partial}{\partial x_j} \right), i, j, k = 1, 2, 3$. In this case we note that the product of two generators $L_i L_j$ is a quadratic differential operator not in the Lie algebra of $so(3)$ (the commutator, however, is linear and belongs to the Lie algebra). However, if we formally, introduce this operator product, the commutator can be identified with the product in $so(3)$. The operator products define an associative algebra in the sense that $L_1(L_2 L_3) = (L_1 L_2)L_3 = L_1 L_2 L_3$ which is infinite dimensional and defines the universal enveloping algebra of $so(3)$. Moreover, we can also introduce the concept of a Casimir invariant in this universal enveloping algebra. For example, for the $u(N)$ Lie algebra, we can define the quadratic Casimir invariant to be the product

$$I_2 = X_\mu^{\ \nu} X_\nu^{\ \mu}, \tag{3.210}$$

without appealing to the matrix product of a matrix representation of the Lie algebra L.

3.6 References

V. Bargmann, Ann. of Math. 48, 568 (1947).

L. C. Biedenharn and H. Van Dam (ed), *Quantum Theory of Angular Momentum; a Collection of Reprints and Original Papers*, Academic Press (1965).

L. C. Biedenharn and J. D. Louck, *Angular Momentum in Quantum Physics*, Cambridge University Press (1981).

Y. Alhassid, F. Gursey and F. Iachello, Phys. Rev. Lett. 50, 873 (1983).

J. F. Humphreys, *A course in group theory*, Oxford University Press (2008).

CHAPTER 4

Relationship between Lie algebras and Lie groups

In the last two chapters we discussed the concepts of Lie groups and Lie algebras. In this chapter we will explore systematically the connection between the two. We start in the next section with the concept of an infinitesimal group and its generators and see how it helps derive the Lie algebra from a Lie group.

4.1 Infinitesimal group and the Lie algebra

Let G be a Lie group which we assume to be $G = GL(N), U(N), SU(N)$ for simplicity. Let A denote its defining ($N \times N$ dimensional) matrix representation (sometimes also called the fundamental representation) corresponding to the group element $a \in G$. (As we have discussed in chapter 2, a representation of the group can be obtained either as matrices or as linear operators acting on a Hilbert space. However, here we will restrict ourselves only to matrices.) Let us next consider a group element $a \in G$ which differs only infinitesimally from the identity element so that, in this case, we can write

$$a \to A = E_N + \epsilon \beta + O(\epsilon^2), \tag{4.1}$$

for a real, infinitesimal constant parameter ϵ. As matrices (4.1) takes the explicit form ($\mu, \nu = 1, 2, \cdots, N$ denote the matrix indices in the defining representation)

$$A^{\mu}{}_{\nu} = \delta^{\mu}_{\nu} + \epsilon \beta^{\mu}{}_{\nu} + O(\epsilon^2)$$

$$= \begin{pmatrix} 1 + \epsilon\beta^1{}_1 & \epsilon\beta^1{}_2 & \cdots & \epsilon\beta^1{}_N \\ \epsilon\beta^2{}_1 & 1 + \epsilon\beta^2{}_2 & \cdots & \epsilon\beta^2{}_N \\ \vdots & \vdots & \vdots & \vdots \\ \epsilon\beta^N{}_1 & \epsilon\beta^N{}_2 & \cdots & 1 + \epsilon\beta^N{}_N \end{pmatrix} + O(\epsilon^2), \tag{4.2}$$

where β^μ_ν parameterize the group. Since ϵ is infinitesimal, it follows that

$$\det A = 1 + O(\epsilon) \neq 0, \tag{4.3}$$

so that A^{-1} exists as it should and has the form

$$\left(A^{-1}\right)^\mu{}_\nu = \delta^\mu_\nu - \epsilon\beta^\mu{}_\nu + O(\epsilon^2)$$

$$= \begin{pmatrix} 1 - \epsilon\beta^1{}_1 & -\epsilon\beta^1{}_2 & \cdots & -\epsilon\beta^1{}_N \\ -\epsilon\beta^2{}_1 & 1 - \epsilon\beta^2{}_2 & \cdots & -\epsilon\beta^2{}_N \\ \vdots & \vdots & \vdots & \vdots \\ -\epsilon\beta^N{}_1 & -\epsilon\beta^N{}_2 & \cdots & 1 - \epsilon\beta^N{}_N \end{pmatrix} + O(\epsilon^2). \tag{4.4}$$

If G is the real $GL(N)$ group, β would represent a real matrix (the matrix elements β^μ_ν would be real), while for the case of the group $U(N)$, the additional condition (1.49) (with the identification $A^\mu_\nu = \alpha_{\mu\nu}$, note also that $(A^T)^\mu{}_\nu = A^\nu{}_\mu$)

$$\sum_{\lambda=1}^N \left(A^\lambda{}_\mu\right)^* A^\lambda{}_\nu = \delta_{\mu\nu} + O(\epsilon^2), \tag{4.5}$$

requires that

$$\left(\beta^\nu{}_\mu\right)^* = -\beta^\mu{}_\nu, \tag{4.6}$$

for complex constants β^μ_ν. For the group $SU(N)$ we further need to impose the condition (1.56)

$$\det A = 1,$$

$$\text{or,} \quad \det\left(\mathbb{1} + \epsilon\beta + O(\epsilon^2)\right) = 1,$$

$$\text{or,} \quad \exp\text{Tr}\ln\left(\mathbb{1} + \epsilon\beta + O(\epsilon^2)\right) = 1,$$

$$\text{or,} \quad \exp\text{Tr}\left(\epsilon\beta + O(\epsilon^2)\right) = 1,$$

$$\text{or,} \quad \exp\left(\epsilon\,\text{Tr}\,\beta\right) = 1 + O(\epsilon^2),$$

$$\text{or,} \quad \text{Tr}\,\beta = \sum_{\lambda=1}^N \beta^\lambda{}_\lambda = 0. \tag{4.7}$$

Therefore, we can summarize that the matrix elements β^μ_ν satisfy

$$\beta^\mu_\nu = \text{real and arbitrary}, \qquad \text{if } G = \text{real } GL(N), \quad (4.8)$$

$$\left(\beta^\nu_\mu\right)^* = -\beta^\mu_\nu, \qquad \text{if } G = U(N), \quad (4.9)$$

$$\sum_{\lambda=1}^{N} \beta^\lambda_\lambda = 0 \text{ (in addition to (4.9))}, \quad \text{if } G = SU(N). \quad (4.10)$$

We note here that (4.10) holds even for $G = SL(N)$ without the condition (4.9). In this case, we call "a" (or the defining matrix A) to be an infinitesimal group element.

Let us next consider a general $n \times n$ dimensional matrix representation of the group G given by the matrix $U(a), a \in G$. If "a" corresponds to an infinitesimal group element, then correspondingly the $n \times n$ matrix representation $U(a)$ will have a form

$$U(a) = E_n + \epsilon \sum_{\mu,\nu=1}^{N} \beta^\mu_\nu X^\nu_\mu + O(\epsilon^2), \quad (4.11)$$

for some $n \times n$ matrices X^μ_ν which are known as the generators of the infinitesimal group and β^μ_ν parameterize the infinitesimal group. (Namely, there are, in general, N^2 such $n \times n$ matrices. Explicitly we can write them, for example, as $(X^\mu_\nu)_{ij}$ for $\mu, \nu = 1, 2, \cdots N$ and $i, j = 1, 2, \cdots, n$.)

For an arbitrary element $b \in G$ (not necessarily infinitesimal), we note that

$$U^{-1}(b)U(a)U(b) = U(b^{-1})U(a)U(b) = U(b^{-1}ab). \quad (4.12)$$

We can calculate explicitly

$$\left(b^{-1}ab\right)^\mu_\nu \to \left(B^{-1}AB\right)^\mu_\nu$$

$$= \sum_{\lambda,\rho=1}^{N} ((B^{-1})^\mu_\lambda A^\lambda_\rho B^\rho_\nu)$$

$$= \sum_{\lambda,\rho=1}^{N} (B^{-1})^\mu_\lambda \left(\delta^\lambda_\rho + \epsilon\beta^\lambda_\rho\right) a^\rho_\nu + O(\epsilon^2)$$

$$= \sum_{\lambda=1}^{N} (B^{-1})^\mu_\lambda B^\lambda_\nu + \epsilon \sum_{\lambda,\rho=1}^{N} (B^{-1})^\mu_\lambda \beta^\lambda_\rho B^\rho_\nu + O(\epsilon^2)$$

$$= \delta^\mu_\nu + \epsilon\tilde{\beta}^\mu_\nu + O(\epsilon^2), \quad (4.13)$$

where we have identified

$$\tilde{\beta}^\mu_{\ \nu} = \left(B^{-1}\beta B\right)^\mu_{\ \nu} = \sum_{\lambda,\rho=1}^N \left(B^{-1}\right)^\mu_{\ \lambda}\beta^\lambda_{\ \rho}B^\rho_{\ \nu}. \tag{4.14}$$

Equation (4.13) shows that $(b^{-1}ab)$ also defines an infinitesimal group element. Therefore, using (4.11) we conclude that

$$U(b^{-1}ab) = E_n + \epsilon \sum_{\mu,\nu=1}^N \tilde{\beta}^\mu_{\ \nu}X^\nu_{\ \mu} + O(\epsilon^2)$$

$$= E_n + \epsilon \sum_{\mu,\nu,\lambda,\rho=1}^N \left(B^{-1}\right)^\mu_{\ \lambda}\beta^\lambda_{\ \rho}B^\rho_{\ \nu}X^\nu_{\ \mu} + O(\epsilon^2)$$

$$= E_n + \epsilon \sum_{\mu,\nu=1}^N \beta^\mu_{\ \nu}\left(\sum_{\lambda,\rho=1}^N \left(B^{-1}\right)^\lambda_{\ \mu}B^\nu_{\ \rho}X^\rho_{\ \lambda}\right) + O(\epsilon^2), \tag{4.15}$$

where we have rearranged the summation indices $\mu \leftrightarrow \lambda, \nu \leftrightarrow \rho$ in the last step. Substituting this into (4.12) and using (4.11) we obtain

$$U^{-1}(b)\left(E_n + \epsilon \sum_{\mu,\nu=1}^N \beta^\mu_{\ \nu}X^\nu_{\ \mu} + O(\epsilon^2)\right)U(b)$$

$$= E_n + \epsilon \sum_{\mu,\nu=1}^N \beta^\mu_{\ \nu}\left(\sum_{\lambda,\rho=1}^N \left(B^{-1}\right)^\lambda_{\ \mu}B^\nu_{\ \rho}X^\rho_{\ \lambda}\right) + O(\epsilon^2),$$

or, $\quad \epsilon \sum_{\mu,\nu=1}^N \beta^\mu_{\ \nu}\left(U^{-1}(b)X^\nu_{\ \mu}U(b) - \sum_{\lambda,\rho=1}^N \left(B^{-1}\right)^\lambda_{\ \mu}B^\nu_{\ \rho}X^\rho_{\ \lambda}\right) = 0.$

$$\tag{4.16}$$

For the case of $G = GL(N)$, the parameters $\beta^\mu_{\ \nu}$ are arbitrary constants so that, in this case, we conclude from (4.16) that

$$U^{-1}(b)X^\mu_{\ \nu}U(b) = \sum_{\lambda,\rho=1}^N \left(B^{-1}\right)^\lambda_{\ \nu}B^\mu_{\ \rho}X^\rho_{\ \lambda}. \tag{4.17}$$

For $G = U(N)$, we have (see (4.6)) $\left(\beta^\nu_{\ \mu}\right)^* = -\beta^\mu_{\ \nu}$ so that

$$\mathrm{Re}\,\beta^\nu_{\ \mu} = -\mathrm{Re}\,\beta^\mu_{\ \nu}, \quad \mathrm{Im}\,\beta^\nu_{\ \mu} = \mathrm{Im}\,\beta^\mu_{\ \nu}. \tag{4.18}$$

Apart from these constraints, $\beta^\mu_{\ \nu}$ are arbitrary and (4.16) leads to (4.17) again. For the case of $G = SU(N)$, in addition to (4.6), we have $\sum\limits_{\lambda=1}^{N} \beta^\lambda{}_\lambda = 0$ (see (4.7)) so that we can require the generators $X^\mu{}_\nu$ to satisfy ("Tr" denotes summing over the indices μ, ν only and not the matrix trace)

$$\text{Tr}\, X = \sum_{\lambda=1}^{N} X^\lambda{}_\lambda = 0. \tag{4.19}$$

We note that we can always achieve this by redefining the generators as

$$X^\mu{}_\nu \to X^\mu{}_\nu - \frac{1}{N}\, \delta^\mu_\nu \sum_{\lambda=1}^{N} X^\lambda{}_\lambda, \tag{4.20}$$

so that we can consider the constant parameters $\beta^\mu{}_\nu$ to be arbitrary. (This discussion applies to $G = SL(N)$ as well without the constraint (4.6).) As a result, once again (4.16) leads to (4.17) and we conclude that (4.17) holds for $G = GL(N), U(N), SU(N)$ (and $SL(N)$).

Let us recall from (4.2), (4.4) and (4.11) that, for an infinitesimal group element b, we have (note that $b \to B$ in the defining $N \times N$ representation)

$$B^\mu{}_\nu = \delta^\mu_\nu + \epsilon \overline{\beta}^\mu{}_\nu + O(\epsilon^2),$$

$$\left(B^{-1}\right)^\mu{}_\nu = \delta^\mu_\nu - \epsilon \overline{\beta}^\mu{}_\nu + O(\epsilon^2),$$

$$U(b) = E_n + \epsilon \sum_{\mu,\nu=1}^{N} \overline{\beta}^\mu{}_\nu X^\nu{}_\mu + O(\epsilon^2), \tag{4.21}$$

so that we can write

$$U^{-1}(b) = U(b^{-1}) = E_n - \epsilon \sum_{\mu,\nu=1}^{N} \overline{\beta}^\mu{}_\nu X^\nu{}_\mu + O(\epsilon^2). \tag{4.22}$$

As a result, we can calculate

$$U^{-1}(b) X^\mu{}_\nu U(b) = \left(E_n - \epsilon \sum_{\lambda,\rho=1}^{N} \overline{\beta}^\lambda{}_\rho X^\rho{}_\lambda + O(\epsilon^2) \right) X^\mu{}_\nu$$

$$\times \left(E_n + \epsilon \sum_{\sigma,\tau=1}^{N} \overline{\beta}^\sigma{}_\tau X^\tau{}_\sigma + O(\epsilon^2) \right)$$

$$= X^\mu{}_\nu + \epsilon \sum_{\lambda,\rho=1}^{N} \overline{\beta}^\lambda{}_\rho \left(X^\mu{}_\nu X^\rho{}_\lambda - X^\rho{}_\lambda X^\mu{}_\nu \right) + O(\epsilon^2)$$

$$= X^\mu_{\ \nu} + \epsilon \sum_{\lambda,\rho=1}^{N} \overline{\beta}^\lambda_{\ \rho} [X^\mu_{\ \nu}, X^\rho_{\ \lambda}] + O(\epsilon^2). \qquad (4.23)$$

On the other hand, from (4.17) and (4.21) we also note that for an infinitesimal group element a

$$U^{-1}(b) X^\mu_{\ \nu} U(b) = \sum_{\lambda,\rho=1}^{N} (B^{-1})^\lambda_{\ \nu} B^\mu_{\ \rho} X^\rho_{\ \lambda} = \sum_{\lambda,\rho=1}^{N} B^\mu_{\ \rho} X^\rho_{\ \lambda} (B^{-1})^\lambda_{\ \nu}$$

$$= \sum_{\lambda,\rho=1}^{N} \left(\delta^\mu_\rho + \epsilon \overline{\beta}^\mu_{\ \rho} + O(\epsilon^2) \right) X^\rho_{\ \lambda} (\delta^\lambda_\nu - \epsilon \overline{\beta}^\lambda_{\ \nu} + O(\epsilon^2))$$

$$= X^\mu_{\ \nu} + \epsilon \sum_{\lambda,\rho=1}^{N} \overline{\beta}^\lambda_{\ \rho} \left(\delta^\mu_\lambda X^\rho_{\ \nu} - \delta^\rho_\nu X^\mu_{\ \lambda} \right) + O(\epsilon^2). \qquad (4.24)$$

Substituting (4.23) and (4.24) into (4.17) we obtain

$$\sum_{\lambda,\rho=1}^{N} \overline{\beta}^\lambda_{\ \rho} \left([X^\mu_{\ \nu}, X^\rho_{\ \lambda}] - \delta^\mu_\lambda X^\rho_{\ \nu} + \delta^\rho_\nu X^\mu_{\ \lambda} \right) = 0. \qquad (4.25)$$

As before we can argue that since the parameters $\overline{\beta}^\lambda_{\ \rho}$ are arbitrary (4.25) leads to

$$[X^\mu_{\ \nu}, X^\rho_{\ \lambda}] = \delta^\mu_\lambda X^\rho_{\ \nu} - \delta^\rho_\nu X^\mu_{\ \lambda}, \qquad (4.26)$$

for all cases of $G = GL(N), U(N), SU(N)$ (and $SL(N)$). Relation (4.26) describes how the Lie algebra for these groups can be obtained from the concept of an infinitesimal group, in particular defining the Lie algebra of $sl(N)$. This shows that the generators of the infinitesimal Lie group constitute the basis elements of the corresponding Lie algebra. We may wonder here as to what may be the difference between the Lie algebras, say for example, between $gl(N)$ and $u(N)$. To see this, let us next consider the converse problem of constructing the Lie group from the Lie algebra.

4.2 Lie groups from Lie algebras

To begin with let us consider the Lie group $G = U(N)$ so that any arbitrary group element $a \in G$ can be represented by a $N \times N$ unitary matrix A in the defining representation. It then follows that we can express

$$A = e^{\xi(a)}, \qquad (4.27)$$

for some $N \times N$ matrix $\xi_{(a)}$ which is anti-Hermitian, namely,

$$\xi_{(a)}^{\dagger} = -\xi_{(a)}, \quad \left(\xi_{(a)\ \mu}^{\nu}\right)^{*} = -\xi_{(a)\nu}^{\mu}, \tag{4.28}$$

where $\mu, \nu = 1, 2, \cdots, N$. Defining the $N \times N$ unitary matrix

$$C = e^{\frac{1}{M}\xi_{(a)}}, \tag{4.29}$$

for some integer M, we conclude that it must correspond to a group element $c \in G$. It follows from (4.27) that we can write

$$A = C^{M},$$

$$U(a) = U(c \cdot c \cdots c) = U(c)U(c) \cdots U(c) = (U(c))^{M}, \tag{4.30}$$

where $U(a)$ denotes a $n \times n$ matrix representation of $G = U(N)$.

If we now consider M to be a very large integer, then the $N \times N$ matrix C in (4.29) becomes infinitesimal and has the form

$$C^{\mu}_{\nu} = \delta^{\mu}_{\nu} + \frac{1}{M}\xi_{(a)\nu}^{\mu} + O\left(\frac{1}{M^{2}}\right), \tag{4.31}$$

so that we can write (see (4.11))

$$U(c) = E_{n} + \frac{1}{M}\sum_{\mu,\nu=1}^{N}\xi_{(a)\nu}^{\mu}X^{\nu}_{\mu} + O\left(\frac{1}{M^{2}}\right),$$

$$U(a) = (U(c))^{M} = \left(E_{n} + \frac{1}{M}\sum_{\mu,\nu=1}^{N}\xi_{(a)\nu}^{\mu}X^{\nu}_{\mu} + O\left(\frac{1}{M^{2}}\right)\right)^{M}, \tag{4.32}$$

where X^{μ}_{ν} represent the generators of the infinitesimal group for $G = U(N)$ and are $n \times n$ matrices (see the comment after (4.11)). As we have seen in (4.26) these constitute the basis elements of the Lie algebra $u(N)$ as $n \times n$ matrices. If we now take the limit $M \to \infty$, it follows from the second relation in (4.32) that we can write

$$U(a) = e^{\sum_{\mu,\nu=1}^{N}\xi_{(a)\nu}^{\mu}X^{\nu}_{\mu}}, \tag{4.33}$$

when (4.27) holds and where $\xi_{(a)\nu}^{\mu}$ can be thought of as numbers. This shows that a $n \times n$ matrix representation of the unitary group $G = U(N)$ can be obtained by exponentiating the basis elements of the Lie algebra $u(N)$ as $n \times n$ matrices.

For the case of $G = $ real $GL(N)$, we note that we may not be able to represent some of the elements $a \in G$ in the exponential form of (4.33) (for example, if $a \to A$ corresponds to a negative matrix). However, let us consider those elements $a \in G$ such that we can write it as

$$a \to A = e^{\xi_{(a)}}, \tag{4.34}$$

for some real matrix $\xi_{(a)}$. Then, the same reasoning as before leads to (4.33). However, the difference between the groups $U(N)$ and real $GL(N)$ lies in the fact that for the former, the parameters $\xi^{\mu}_{(a)\nu}$ are complex parameters satisfying $\left(\xi^{\nu}_{(a)\mu}\right)^* = -\xi^{\mu}_{(a)\nu}$ while for the latter $\xi^{\mu}_{(a)\nu}$ are arbitrary real parameters. As a result, if we are interested in a $n \times n$ unitary matrix representation satisfying

$$U^{\dagger}(a)U(a) = E_n, \tag{4.35}$$

the basis elements of the Lie algebras must satisfy (as $n \times n$ matrices)

$$(X^{\mu}_{\nu})^{\dagger} = X^{\mu}_{\nu}, \qquad \text{for } U(N),$$
$$(X^{\mu}_{\nu})^{\dagger} = -X^{\mu}_{\nu}, \qquad \text{for real } GL(N), \tag{4.36}$$

since (4.35) has to hold. Thus, we see that, for a unitary representation, the basis elements of the two Lie algebras would satisfy different Hermiticity properties even though they satisfy the same Lie algebra relation (4.26).

Let us next consider the real orthogonal group (rotation group) $SO(N) = R(N)$ discussed in section 1.2.5. In this case, the infinitesimal group element can be written as

$$a \to A_{\mu\nu} = \delta_{\mu\nu} + \epsilon\beta_{\mu\nu} + O(\epsilon^2), \tag{4.37}$$

where

$$\beta_{\mu\nu} = -\beta_{\nu\mu}, \tag{4.38}$$

are real parameters so that the group element satisfies (see (1.67))

$$A^T A = E_N. \tag{4.39}$$

Correspondingly, we can write the $n \times n$ dimensional matrices as (see (4.11))

$$U(a) = E_n + \frac{\epsilon}{2} \sum_{\mu,\nu=1}^{N} \beta_{\mu\nu}X_{\mu\nu} + O(\epsilon^2), \tag{4.40}$$

where the real $n \times n$ matrices $X_{\mu\nu}$ satisfy

$$X_{\mu\nu} = -X_{\nu\mu}. \tag{4.41}$$

This can always be achieved by replacing $X_{\mu\nu} \to \frac{1}{2}(X_{\mu\nu} - X_{\nu\mu})$ in view of (4.38) and the factor $\frac{1}{2}$ in (4.40) is there to avoid double counting.

We can now explicitly evaluate the relation (4.12) for the present case. Using (4.40) we obtain

$$U^{-1}(b)U(a)U(b) = E_n + \frac{\epsilon}{2} \sum_{\mu,\nu=1}^{N} \beta_{\mu\nu} U^{-1}(b) X_{\mu\nu} U(b) + O(\epsilon^2),$$
$$\tag{4.42}$$

for an arbitrary element $b \in SO(N)$. On the other hand, because of the group property we note again that

$$U^{-1}(b)U(a)U(b) = U(b^{-1})U(a)U(b) = U(b^{-1}ab) \tag{4.43}$$

and we can calculate (as in (4.13))

$$\left(b^{-1}ab\right)_{\mu\nu} \to \left(B^{-1}AB\right)_{\mu\nu}$$

$$= \left(B^T AB\right)_{\mu\nu} = \sum_{\lambda,\rho=1}^{N} (B^T)_{\mu\lambda} A_{\lambda\rho} B_{\rho\nu}$$

$$= \sum_{\lambda,\rho=1}^{N} (B^T)_{\mu\lambda} \left(\delta_{\lambda\rho} + \epsilon\beta_{\lambda\rho} + O(\epsilon^2)\right) B_{\rho\nu}$$

$$= \delta_{\mu\nu} + \epsilon \sum_{\lambda,\rho=1}^{N} \beta_{\lambda\rho} B_{\lambda\mu} B_{\rho\nu} + O(\epsilon^2)$$

$$= \delta_{\mu\nu} + \epsilon\tilde{\beta}_{\mu\nu} + O(\epsilon^2), \tag{4.44}$$

where we have idntified (see also (4.14))

$$\tilde{\beta}_{\mu\nu} = \sum_{\lambda,\rho=1}^{N} \beta_{\lambda\rho} B_{\lambda\mu} B_{\rho\nu}. \tag{4.45}$$

We conclude from (4.44) that $(b^{-1}ab)$ is an element of the infinitesimal group (even when b is arbitrary). Therefore, it follows from (4.40) that we can write

$$U^{-1}(b)U(a)U(b) = U(b^{-1}ab)$$

$$= E_n + \frac{\epsilon}{2} \sum_{\mu,\nu=1}^{N} \tilde{\beta}_{\mu\nu} X_{\mu\nu} + O(\epsilon^2)$$

$$= E_n + \frac{\epsilon}{2} \sum_{\mu,\nu=1}^{N} \beta_{\mu\nu} \left(\sum_{\lambda,\rho=1}^{N} B_{\mu\lambda} B_{\nu\rho} X_{\lambda\rho} \right) + O(\epsilon^2). \tag{4.46}$$

From eqs. (4.42) and (4.46) we obtain

$$U^{-1}(b) X_{\mu\nu} U(b) = \sum_{\lambda,\rho=1}^{N} B_{\mu\lambda} B_{\nu\rho} X_{\lambda\rho}, \tag{4.47}$$

which can be compared with (4.17).

If we now we choose the group element b to be infinitesimal, then we have (see (4.21))

$$B_{\mu\nu} = \delta_{\mu\nu} + \epsilon \overline{\beta}_{\mu\nu} + O(\epsilon^2), \quad \overline{\beta}_{\mu\nu} = -\overline{\beta}_{\nu\mu},$$

$$U(b) = E_n + \frac{\epsilon}{2} \sum_{\mu,\nu=1}^{N} \overline{\beta}_{\mu\nu} X_{\mu\nu} + O(\epsilon^2),$$

$$U^{-1}(b) = E_n - \frac{\epsilon}{2} \sum_{\mu,\nu=1}^{N} \overline{\beta}_{\mu\nu} X_{\mu\nu} + O(\epsilon^2). \tag{4.48}$$

Using these we can write

$$U^{-1}(b) X_{\mu\nu} U(b) = \left(E_n - \frac{\epsilon}{2} \sum_{\lambda,\rho=1}^{N} \overline{\beta}_{\lambda\rho} X_{\lambda\rho} + O(\epsilon^2) \right) X_{\mu\nu}$$

$$\times \left(E_n + \frac{\epsilon}{2} \sum_{\sigma,\tau=1}^{N} \overline{\beta}_{\sigma\tau} X_{\sigma\tau} + O(\epsilon^2) \right)$$

$$= X_{\mu\nu} + \frac{\epsilon}{2} \sum_{\lambda,\rho=1}^{N} \overline{\beta}_{\lambda\rho} [X_{\mu\nu}, X_{\lambda\rho}] + O(\epsilon^2), \tag{4.49}$$

On the other hand, we note that

$$\sum_{\lambda,\rho=1}^{N} B_{\mu\lambda} B_{\nu\rho} X_{\lambda\rho} = \sum_{\lambda,\rho=1}^{N} \left(\delta_{\mu\lambda} + \epsilon\overline{\beta}_{\mu\lambda} + O(\epsilon^2)\right) X_{\lambda\rho}$$

$$\times \left(\delta_{\nu\rho} + \epsilon\overline{\beta}_{\nu\rho} + O(\epsilon^2)\right)$$

$$= X_{\mu\nu} + \epsilon \sum_{\lambda,\rho=1}^{n} \overline{\beta}_{\lambda\rho} \left(\delta_{\nu\lambda} X_{\mu\rho} + \delta_{\mu\lambda} X_{\rho\nu}\right) + O(\epsilon^2)$$

$$= X_{\mu\nu} - \frac{\epsilon}{2} \sum_{\lambda,\rho=1}^{n} \overline{\beta}_{\lambda\rho} \left(\delta_{\mu\lambda} X_{\nu\rho} + \delta_{\nu\rho} X_{\mu\lambda} - \delta_{\mu\rho} X_{\nu\lambda} - \delta_{\nu\lambda} X_{\mu\rho}\right)$$

$$+ O(\epsilon^2), \tag{4.50}$$

where we have used the anti-symmetry properties $\overline{\beta}_{\mu\nu} = -\overline{\beta}_{\nu\mu}$ as well as $X_{\mu\nu} = -X_{\nu\mu}$. Substituting (4.49) and (4.50) into (4.47) we conclude that in the present case we have the Lie algebra of the form

$$[X_{\mu\nu}, X_{\lambda\rho}] = -\delta_{\mu\lambda} X_{\nu\rho} - \delta_{\nu\rho} X_{\mu\lambda} + \delta_{\mu\rho} X_{\nu\lambda} + \delta_{\nu\lambda} X_{\mu\rho}, \tag{4.51}$$

which can be compared with (3.24).

Note that if we write an arbitrary rotation (not necessarily infinitesimal) corresponding to the group element a as

$$A = e^{\xi(a)}, \quad A^T A = E_N, \tag{4.52}$$

for some real $N \times N$ matrix $\xi_{(a)}$, then we can determine as in (4.33) that the $n \times n$ matrix representation can be written as

$$U(a) = e^{\frac{1}{2} \sum_{\mu,\nu=1}^{n} \xi_{(a)\mu\nu} X_{\mu\nu}}, \tag{4.53}$$

where $X_{\mu\nu}$ denote the basis elements of the Lie algebra. The unitarity of the representation $U^\dagger(a)U(a) = E_n$ requires that

$$X_{\mu\nu}^\dagger = -X_{\mu\nu}, \tag{4.54}$$

since $\xi_{(a)\mu\nu}$ can be thought of as real constant parameters. Namely, the generators, in this case, will correspond to anti-Hermitian $n \times n$ matrices. This is the reason why it is often more convenient (and is commonly used in physics) to define

$$J_{\mu\nu} = i X_{\mu\nu}, \tag{4.55}$$

which are Hermitian, $J_{\mu\nu}^{\dagger} = J_{\mu\nu}$.

For the case of the symplectic group $Sp(2N)$, as we have seen in (1.103), we can write the matrices A in the defining $2N \times 2N$ representation (corresponding to the group element a) as

$$A^T \epsilon_{2N} A = \epsilon_{2N}, \tag{4.56}$$

where ϵ_{2N} denotes the $2N \times 2N$ dimensional anti-symmetric matrix defined in (1.101). The infinitesimal group element, in this case, can be written as

$$A_{\mu\nu} = \delta_{\mu\nu} + \epsilon \beta_{\mu\nu} + O(\epsilon^2), \quad \mu, \nu = 1, 2, \cdots, 2N, \tag{4.57}$$

for some $2N \times 2N$ matrix $\beta_{\mu\nu}$. The constraint (4.56) in this case, namely, $A^T \epsilon_{2N} A = \epsilon_{2N}$ leads to (as a matrix relation)

$$\epsilon_{2N} \beta + \beta^T \epsilon_{2N} = 0,$$

$$\text{or,} \quad (\epsilon_{2N} \beta)^T = (\epsilon_{2N} \beta), \tag{4.58}$$

where we have used the fact that $\epsilon_{2N}^T = -\epsilon_{2N}$ (namely, it is anti-symmetric, see (1.102)). Correspondingly, we can write for a $n \times n$ matrix representation

$$U(a) = E_{2n} + \frac{\epsilon}{2} \sum_{\mu,\nu=1}^{2N} (\epsilon_{2N} \beta)_{\mu\nu} X_{\mu\nu} + O(\epsilon^2), \tag{4.59}$$

where $X_{\mu\nu}$ denote $2n \times 2n$ matrices corresponding to the generators of the infinitesimal group which can be chosen to be symmetric,

$$X_{\mu\nu} = X_{\nu\mu}, \tag{4.60}$$

in view of relation (4.58). We can now conclude from an analysis similar to what we have discussed earlier that the generators satisfy the Lie algebra

$$[X_{\mu\nu}, X_{\lambda\rho}] = \epsilon_{2N,\mu\lambda} X_{\nu\rho} + \epsilon_{2N,\nu\lambda} X_{\mu\rho} - \epsilon_{2N,\lambda\mu} X_{\rho\nu} - \epsilon_{2N,\lambda\nu} X_{\rho\mu}, \tag{4.61}$$

for $\mu, \nu, \lambda, \rho = 1, 2, \cdots, 2N$.

Finally let us discuss the case of the Abelian group for which any two elements commute, namely, $ab = ba, a, b \in G$ where G is a N-dimensional Abelian group (for example, the N-dimensional transla-tion group $T(N)$). Since the group elements commute, their defining representation can be written in terms of N commuting parameters $\beta_\mu, \mu = 1, 2, \cdots, N$ and we can write an infinitesimal group element (in any other representation) as a $n \times n$ matrix of the form

$$U(a) = E_n + \epsilon \sum_{\mu=1}^{N} \beta_\mu X_\mu + O(\epsilon^2), \qquad (4.62)$$

for some real or complex parameters β_μ. Here X_μ denote the generators of the infinitesimal group and as before we can deduce the Lie algebra of the generators. In this case, the commuting nature of the group elements leads to

$$[X_\mu, X_\nu] = 0, \quad \mu, \nu = 1, 2, \cdots, N, \qquad (4.63)$$

namely, the Lie algebra is Abelian in this case. As an illustration, let us consider the 2×2 matrix representation of the one dimensional translation group T_1 (see (2.104))

$$U(a) = \begin{pmatrix} 1 & 0 \\ a & 1 \end{pmatrix}. \qquad (4.64)$$

In this case, we can write

$$U(a) = e^{aX}, \qquad (4.65)$$

where the generator of the infinitesimal group (or the Lie algebra element) is given by

$$X = \begin{pmatrix} 0 & 0 \\ 1 & 0 \end{pmatrix}, \quad [X, X] = 0. \qquad (4.66)$$

The Lie algebra is manifestly Abelian.

4.3 Baker-Campbell-Hausdorff formula

In the last section we obtained, in a circuitous way, the different kinds of Lie groups from their corresponding Lie algebras by exponentiating the basis elements of the Lie algebras. This can also be obtained more directly through the use of the Baker-Campbell-Hausdorf formula in the following way (see, for example, M. Hauser and J. T. Schwarz, *Lie groups and Lie algebras; notes in mathematics and its applications*).

Let A denote an associative algebra such as the matrix algebra $U(L)$ (namely, matrix representation of a Lie algebra L). For any

$x, y \in A$, we can express

$$e^x e^y = e^{\phi(x,y)},$$

or, $\quad \phi(x, y) = \ln\left(e^x e^y\right),$ \hfill (4.67)

where the solution of (4.67) for $\phi(x, y) \in A$ is known as the Baker-Campbell-Hausdorff formula. The low order terms in $\phi(x, y)$ can be determined explicitly as follows. Instead of (4.67), let us look at the expression

$$e^{tx} e^{ty} = e^{\phi(t,x,y)},$$ \hfill (4.68)

where t is a small parameter. We can now make a Taylor expansion as

$$\phi(t, x, y) = \sum_{n=0}^{\infty} t^n \phi_n(x, y).$$ \hfill (4.69)

On the other hand, we note that we can write

$$\phi(t, x, y) = \ln\left(e^{tx} e^{ty}\right) = \ln\left(1 + \left(e^{tx} e^{ty} - 1\right)\right)$$

$$= \sum_{n=1}^{\infty} \frac{(-1)^{n+1}}{n} \left(e^{tx} e^{ty} - 1\right)^n = \sum_{n=0}^{\infty} t^n \phi_n(x, y). \quad (4.70)$$

Therefore, matching the powers of t, the function $\phi_n(x, y)$ can be calculated order by order and has the form

$$\phi_n(x, y) = \sum_{m=1}^{\infty} \frac{(-1)^{m+1}}{m} \left(e^x e^y - 1\right)^m \Big|_n,$$ \hfill (4.71)

where the restriction implies that the sum of powers of x and y should add to n in the expression on the right hand side. We can now calculate explicitly the first few orders of the function $\phi_n(x, y)$ in a straightforward manner,

$$\phi_0(x, y) = 0,$$

$$\phi_1(x, y) = x + y,$$

$$\phi_2(x, y) = \frac{x^2}{2} + xy + \frac{y^2}{2} - \frac{1}{2}(x + y)^2 = \frac{1}{2}[x, y],$$

$$\phi_3(x, y) = \frac{1}{12}[x, [x, y]] - \frac{1}{12}[y, [x, y]],$$ \hfill (4.72)

and so on.

However, such a calculation of $\phi_n(x, y)$ becomes extremely complicated as we go to higher orders. On the other hand, the solution of (4.67) can be written compactly as (the derivation is technical)

$$\phi(x, y) = x + \int_0^1 ds\, \psi\left(e^{(ad\,x)} e^{s(ad\,y)}\right) y. \tag{4.73}$$

Here $(ad\,x)$ is the adjoint operation defined in (3.103), namely, $(ad\,x)y = [x, y]$ and

$$\psi(w) = \frac{1}{w-1}(w \ln w) = \left(1 + \frac{1}{w-1}\right) \ln\left(1 - (1-w)\right)$$

$$= -\left(1 + \frac{1}{w-1}\right) \sum_{n=1}^{\infty} \frac{1}{n}(1-w)^n$$

$$= -\sum_{n=1}^{\infty} \frac{1}{n}(1-w)^n + \sum_{n=0}^{\infty} \frac{1}{n+1}(1-w)^n$$

$$= 1 - \sum_{n=1}^{\infty} \frac{1}{n(n+1)}(1-w)^n. \tag{4.74}$$

Here $\psi(w)$ is known as the generator of Bernoulli numbers in the sense that

$$\psi(e^z) = \sum_{n=0}^{\infty} B_n \frac{z^n}{n!}, \tag{4.75}$$

where the coefficients B_n denote Bernoulli numbers, namely, $B_0 = 1, B_1 = \frac{1}{2}, B_2 = \frac{1}{6}, B_4 = -\frac{1}{30}, \cdots$.

Because of the definition of the adjoint operation, $\phi(x, y)$ is a sum of a series of commutators (and commutators of commutators etc.) involving x and y. The first few terms of the series are given by (see also (4.72))

$$\phi(x, y) = x + y + \frac{1}{2}[x, y] + \frac{1}{12}[x, [x, y]] - \frac{1}{12}[y, [x, y]] + \cdots. \tag{4.76}$$

As a special case of the Baker-Cambell-Hausdorff formula, let us consider the case when $[x, y] = c\mathbb{1}$, where c is a constant, or more generally

$$[x, [x, y]] = 0 = [y, [x, y]]. \tag{4.77}$$

In this case, all the higher commutators vanish and we have

$$\phi(x, y) = x + y + \frac{1}{2}[x, y], \tag{4.78}$$

and we obtain

$$e^x e^y = e^{\phi(x,y)} = e^{x+y+\frac{1}{2}[x,y]} = e^{\frac{1}{2}[x,y]} e^{x+y}. \tag{4.79}$$

For example, if we identify

$$x = i\alpha q, \quad y = i\beta p, \tag{4.80}$$

with α, β (in general, complex) constants and q, p denoting the coordinate and the conjugate momentum satisfying the canonical commutation relation $[q, p] = i\hbar \mathbb{1}$, we have

$$[x, y] = [i\alpha q, i\beta p] = -\alpha\beta[q, p] = -i\hbar\alpha\beta\mathbb{1}. \tag{4.81}$$

In this case, using (4.79) we have

$$e^{i\alpha q} e^{i\beta p} = e^{-\frac{i\hbar\alpha\beta}{2}} e^{i\alpha q + i\beta p}. \tag{4.82}$$

Similarly, for annihilation and creation operators $a, a^\dagger$ (of the bosonic harmonic oscillator) satisfying $[a, a^\dagger] = \mathbb{1}$, we obtain

$$e^{\alpha a} e^{\beta a^\dagger} = e^{\frac{1}{2}\alpha\beta} e^{\alpha a + \beta a^\dagger}, \quad e^{\beta a^\dagger} e^{\alpha a} = e^{-\frac{1}{2}\alpha\beta} e^{\alpha a + \beta a^\dagger}. \tag{4.83}$$

These relations can be equivalently written as

$$e^{\alpha a + \beta a^\dagger} = e^{\frac{1}{2}\alpha\beta} e^{\beta a^\dagger} e^{\alpha a} = e^{-\frac{1}{2}\alpha\beta} e^{\alpha a} e^{\beta a^\dagger}. \tag{4.84}$$

These relations are relevant in applications in quantum optics (in connection with coherent states).

If $U(L)$ denotes the matrix representation of a Lie algebra L, then we note, from (4.76) (as well as the properties of a Lie algebra), that for any $x, y \in U(L)$, we have $\phi(x, y) \in U(L)$. Therefore, if we define

$$G = e^{U(L)} = \{\text{the set consisting of } e^{U(x)}\}, \tag{4.85}$$

then this implies $GG \in G$ so that a product of the elements $e^x e^y$ is well defined and is contained in the set G. Clearly $e^0 = \mathbb{1}$ is the unit elelemnt with the inverse e^{-x} satisfying $e^{-x} e^x = \mathbb{1} = e^x e^{-x}$. Therefore, the set of G consisting of all e^x forms a group. This procedure of obtaining a group is known as exponentiation of the Lie algebra.

▶ **Example (Construction of unitary representation).** As we have mentioned above, a representation of the Lie group can be obtained from a given Lie algebra through exponentiation using the Baker-Campbell-Hausdorff formula. However, the explicit construction is, in general, quite difficult. Here we discuss such a construction for a simple example.

Let a, b, e denote three operators satisfying the commutation relations

$$[a, b] = -[b, a] = e.$$

$$[a, e] = -[e, a] = 0,$$

$$[b, e] = -[e, b] = 0. \tag{4.86}$$

It can be checked easily that the commutation relations satisfy Jacobi identity. For example,

$$[a, [a, b]] + [a, [b, a]] + [b, [a, a]] = [a, e] - [a, e] + 0 = 0. \tag{4.87}$$

Since the commutators are anti-symmetric and satisfy Jacobi identity, the elements a, b, e define a 3-dimensional Lie algebra. It follows now that

$$U(\alpha, \beta, \gamma) = e^{\alpha a + \beta b + \gamma e}, \tag{4.88}$$

generates a Lie group for real or complex constant parameters α, β, γ.

To see this, let us define

$$X_1 = \alpha_1 a + \beta_1 b + \gamma_1 e,$$

$$X_2 = \alpha_2 a + \beta_2 b + \gamma_2 e, \tag{4.89}$$

which leads to

$$[X_1, X_2] = \alpha_1 \beta_2 [a, b] + \alpha_2 \beta_1 [b, a] = (\alpha_1 \beta_2 - \alpha_2 \beta_1) e. \tag{4.90}$$

As a result, we have

$$[X_1, [X_1, X_2]] = [X_2, [X_1, X_2]] = 0, \tag{4.91}$$

as well as higher commutators in (4.76) and we can apply (4.78) to write

$$e^{X_1} e^{X_2} = e^{X_1 + X_2 + \frac{1}{2}[X_1, X_2]}$$

$$= e^{(\alpha_1 + \alpha_2)a + (\beta_1 + \beta_2)b + (\gamma_1 + \gamma_2 + \frac{1}{2}(\alpha_1 \beta_2 - \alpha_2 \beta_1))e} = e^{X_3}, \tag{4.92}$$

where

$$X_3 = \alpha_3 a + \beta_3 b + \gamma_3 e, \tag{4.93}$$

with

$$\alpha_3 = \alpha_1 + \alpha_2, \quad \beta_3 = \beta_1 + \beta_2, \quad \gamma_3 = \gamma_1 + \gamma_2 + \frac{1}{2}(\alpha_1 \beta_2 - \alpha_2 \beta_1). \tag{4.94}$$

Equivalently, we can write

$$U(\alpha_1, \beta_1, \gamma_1) U(\alpha_2, \beta_2, \gamma_2) = U(\alpha_3. \beta_3, \gamma_3), \tag{4.95}$$

We note from the definition (4.88) that

$$U(0, 0, 0) = \mathbb{1}, \tag{4.96}$$

and since by assumption the product of elements $U(\alpha, \beta, \gamma)$ is associative, the set of all elements forms a (Lie) group. However, in physics the unitary representation of a group is important, namely, we require

$$U^\dagger(\alpha, \beta, \gamma)U(\alpha, \beta, \gamma) = \mathbb{1}, \tag{4.97}$$

and we discuss two examples of how the unitarity condition can be implemented.

Case I: Let us assume that

$$b = a^\dagger, \quad e = \mathbb{1}, \tag{4.98}$$

so that (4.86) can be written in this case as

$$[a, a^\dagger] = \mathbb{1}, \tag{4.99}$$

with all others vanishing. As a result, we can think of $a, a^\dagger$ as the annihilation and creation operators of the bosonic harmonic oscillator and the unitarity condition (4.97) can be satisfied if we choose

$$\beta = -\alpha^*, \quad \gamma^* = -\gamma. \tag{4.100}$$

Therefore, we can write $\gamma = i\lambda$ for a real constant parameter λ and the group element (4.88) can be written in the form

$$U(\alpha, \beta, \gamma) = e^{\alpha a - \alpha^* a^\dagger + i\lambda \mathbb{1}}, \tag{4.101}$$

which is manifestly unitary. The group elements act on (harmonic oscillator) state vectors given by

$$|n\rangle = \frac{(a^\dagger)^n}{\sqrt{n!}} |0\rangle, \quad n = 0, 1, 2, \cdots, \tag{4.102}$$

where the vacuum state $|0\rangle$ satisfies $a|0\rangle = 0$.

Case II: As a second example, we can consider

$$a^\dagger = a, \quad b^\dagger = b, \quad e^\dagger = -e. \tag{4.103}$$

For example, we can identify $a = x, b = p, e = i\hbar\mathbb{1}$ so that the Lie algebra in (4.86) can be written as

$$[x, p] = i\hbar\mathbb{1}, \tag{4.104}$$

with all others vanishing and a group element (4.88) has the form

$$U(\alpha, \beta, \gamma) = e^{\alpha x + \beta p + i\hbar \gamma \mathbb{1}}. \tag{4.105}$$

From (4.103) we note that $x^\dagger = x, p^\dagger = p$ so that if we choose $\alpha^* = -\alpha, \beta^* = -\beta, \gamma^* = \gamma$, or equivalently

$$U(\alpha, \beta, \gamma) = e^{i(\alpha x + \beta p + \hbar \gamma \mathbb{1})}, \tag{4.106}$$

with α, β, γ real, the group element is formally manifestly unitary.

In quantum mechanics x denotes the coordinate operator while p represents the momentum operator, both of which are non-compact. Therefore, properties such as Hermiticity or unitarity depend on the space of functions on which the operators act. In the present case, the operators (x as well as $p = -i\hbar\frac{d}{dx}$) act on functions $f(x)$ with the coordinate taking values $-\infty < x < \infty$. However, as described in chapter 2 (see (2.112)-(2.123)), the function $f(x)$ has to be differentiable and has to vanish (or oscillate to zero) asymptotically.

◀

▶ **Example (Combination of rotations).** Although the Baker-Campbell-Hausdorff formula is too complicated for general application, in particular cases they can be used in a simple manner. We have already discussed one such simple case in the last example. Here we will describe how to combine rotations involving both the compact group $SO(3)$ as well as the non-compact group $SO(1,2)$.

Let us start with the compact group $SO(3)$ with the Lie algebra given by

$$[J_i, J_j] = i\epsilon_{ijk}J_k, \quad i, j, k = 1, 2, 3, \tag{4.107}$$

where ϵ_{ijk} denotes the three dimensional Levi-Civita tensor (and repeated indices are summed). As we have already seen, for $SO(3)$, the generators J_i are Hermitian so that $J_i^\dagger = J_i$. Following our discussions in this chapter, we note that a general group element of $SO(3)$ can be written as $e^{i\theta_i J_i}$ where repeated indices are summed and the constant parameters θ_i describe the components of rotation along the three different axes.

We consider next a product of three group elements of the form (see (4.82) or (4.83))

$$e^{\bar{\alpha}J_+} e^{\bar{\beta}J_-} e^{i\bar{\gamma}J_3}, \tag{4.108}$$

where $\bar{\alpha}, \bar{\beta}, \bar{\gamma}$ are, in general, real or complex constant parameters and, as before, $J_\pm = J_1 \pm iJ_2$. We note that since the product of group elements gives rise to a group element, we can, in general, write the product in (4.108) as

$$e^{\bar{\alpha}J_+} e^{\bar{\beta}J_-} e^{i\bar{\gamma}J_3} = e^{i\theta_i J_i}, \tag{4.109}$$

where $\theta_i = \theta_i(\bar{\alpha}, \bar{\beta}, \bar{\gamma})$.

To determine the parameters θ_i (and, therefore, the product in (4.109)) we note that the formula (4.109) holds for any $N \times N$ matrix representation of $SO(3)$ and, in particular, for the 2×2 representation where the generators are related to the three Pauli matrices as

$$J_i = \frac{1}{2}\sigma_i,$$

so that we have

$$J_+ = \begin{pmatrix} 0 & 1 \\ 0 & 0 \end{pmatrix}, \quad J_- = J_+^\dagger = \begin{pmatrix} 0 & 0 \\ 1 & 0 \end{pmatrix}, \quad J_3 = \frac{1}{2}\sigma_3 = \frac{1}{2}\begin{pmatrix} 1 & 0 \\ 0 & -1 \end{pmatrix}. \tag{4.110}$$

We note from (4.110) that in this 2×2 matrix representation

$$J_+^2 = 0 = J_-^2, \tag{4.111}$$

which leads to

$$e^{\bar{\alpha}J_+} = \mathbb{1} + \bar{\alpha}J_+ = \begin{pmatrix} 1 & \bar{\alpha} \\ 0 & 1 \end{pmatrix}, \quad e^{\bar{\beta}J_-} = \mathbb{1} + \bar{\beta}J_- = \begin{pmatrix} 1 & 0 \\ \bar{\beta} & 1 \end{pmatrix}. \tag{4.112}$$

Furthermore, since $J_3 = \frac{1}{2}\sigma_3$ is diagonal and $\sigma_3^2 = \mathbb{1}$, it follows that

$$e^{i\bar{\gamma}J_3} = \mathbb{1}\cos\frac{\bar{\gamma}}{2} + i\sigma_3\sin\frac{\bar{\gamma}}{2} = \begin{pmatrix} e^{\frac{i\bar{\gamma}}{2}} & 0 \\ 0 & e^{-\frac{i\bar{\gamma}}{2}} \end{pmatrix}. \tag{4.113}$$

With these simplifications we determine the left hand side of (4.109) to be

$$e^{\bar{\alpha}J_+} e^{\bar{\beta}J_-} e^{i\bar{\gamma}J_3} = \begin{pmatrix} 1 & \bar{\alpha} \\ 0 & 1 \end{pmatrix}\begin{pmatrix} 1 & 0 \\ \bar{\beta} & 1 \end{pmatrix}\begin{pmatrix} e^{\frac{i\bar{\gamma}}{2}} & 0 \\ 0 & e^{-\frac{i\bar{\gamma}}{2}} \end{pmatrix}$$

$$= \begin{pmatrix} (1+\bar{\alpha}\bar{\beta})e^{\frac{i\bar{\gamma}}{2}} & \bar{\alpha}e^{-\frac{i\bar{\gamma}}{2}} \\ \bar{\beta}e^{\frac{i\bar{\gamma}}{2}} & e^{-\frac{i\bar{\gamma}}{2}} \end{pmatrix}. \tag{4.114}$$

On the other hand, to evaluate the right hand side of (4.109), we note that we can write

$$
e^{i\theta_i J_i} = e^{\frac{i\theta}{2}\hat{\boldsymbol{\theta}}\cdot\boldsymbol{\sigma}}, \tag{4.115}
$$

where $\theta = \sqrt{\theta_1^2 + \theta_2^2 + \theta_3^2}$ denotes the magnitude of rotation and $\hat{\boldsymbol{\theta}} = \frac{1}{\theta}(\theta_1\hat{\mathbf{x}} + \theta_2\hat{\mathbf{y}} + \theta_3\hat{\mathbf{z}})$ represents the unit vector around which the rotation is performed. Using the properties of the Pauli matrices we can now determine

$$
e^{i\theta_i J_i} = \mathbb{1}\cos\frac{\theta}{2} + i\hat{\boldsymbol{\theta}}\cdot\boldsymbol{\sigma}\sin\frac{\theta}{2}
$$

$$
= \begin{pmatrix} \cos\frac{\theta}{2} + \frac{i\theta_3}{\theta}\sin\frac{\theta}{2} & \frac{i(\theta_1-i\theta_2)}{\theta}\sin\frac{\theta}{2} \\ \frac{i(\theta_1+i\theta_2)}{\theta}\sin\frac{\theta}{2} & \cos\frac{\theta}{2} - \frac{i\theta_3}{\theta}\sin\frac{\theta}{2} \end{pmatrix}. \tag{4.116}
$$

Comparing with (4.114) we can determine

$$
\cos\frac{\theta}{2} = \frac{1}{2}\left((1+\bar{\alpha}\bar{\beta})e^{\frac{i\bar{\gamma}}{2}} + e^{-\frac{i\bar{\gamma}}{2}}\right),
$$

$$
\frac{i(\theta_1 - i\theta_2)}{\theta}\sin\frac{\theta}{2} = \bar{\alpha}e^{-\frac{i\bar{\gamma}}{2}},
$$

$$
\frac{i(\theta_1 + i\theta_2)}{\theta}\sin\frac{\theta}{2} = \bar{\beta}e^{\frac{i\bar{\gamma}}{2}},
$$

$$
\frac{\theta_3}{\theta}\sin\frac{\theta}{2} = -\frac{i}{2}\left((1+\bar{\alpha}\bar{\beta})e^{\frac{i\bar{\gamma}}{2}} - e^{-\frac{i\bar{\gamma}}{2}}\right). \tag{4.117}
$$

Here we have combined a product of group elements into a single element and consequently have determined $\theta_i = \theta_i(\bar{\alpha}, \bar{\beta}, \bar{\gamma})$. However, the converse procedure of separating a group elements into a product of group elements (in a normal ordered form), as we have described in (4.83) and (4.84), is often more useful. Here we will discuss this procedure in connection with the non-compact group $SO(1,2)$. As we have already discussed (see (3.81)-(3.91)), the Lie algebra of $so(1,2)$ is isomorphic to the angular momentum algebra (assuming that the algebra is defined over complex fields which we are assuming in our discussions) where not all the generators are Hermitian. The algebra can, in fact, be described in terms of the creation and annihilation operators of the one dimensional oscillator in the following manner. Let us choose

$$
J_3 = \frac{1}{2}\left(a^\dagger a + \frac{1}{2}\right), \quad J_+ = \frac{i}{2}(a^\dagger)^2, \quad J_- = \frac{i}{2}a^2, \tag{4.118}
$$

with $[a, a^\dagger] = 1$. (In the language of (3.57), we can identify $A^1{}_1 = \frac{1}{2}(a^\dagger a + 1/2) = -A^2{}_2, A^1{}_2 = \frac{i}{2}(a^\dagger)^2, A^2{}_1 = \frac{i}{2}a^2$.) Then, it can be checked that the generators $J_3, J_\pm$ satisfy the angular momentum algebra

$$
[J_3, J_\pm] = \pm J_\pm, \quad [J_+, J_-] = 2J_3. \tag{4.119}
$$

Although the algebra is the same, as we have mentioned, the difference from the case of $so(3)$ arises from the fact that here not all generators (basis elements) are Hermitian. In particular, we have

$$
(J_\pm)^\dagger = -J_\mp, \tag{4.120}
$$

so that the generators J_1, J_2 are not Hermitian

$$J_1 = \frac{1}{2}(J_+ + J_-) = \frac{i}{4}\left(a^2 + (a^\dagger)^2\right) = -J_1^\dagger,$$

$$J_2 = \frac{1}{2i}(J_+ - J_-) = -\frac{1}{4}\left(a^2 - (a^\dagger)^2\right) = -J_2^\dagger. \tag{4.121}$$

In fact, we note that in this case, the quadratic Casimir operator is given by

$$I_2 = \mathbf{J}^2 = J_1^2 + J_2^2 + J_3^2$$

$$= -\frac{1}{16}\left(a^2 + (a^\dagger)^2\right)^2 + \frac{1}{16}\left(a^2 - (a^\dagger)^2\right)^2 + \frac{1}{4}\left(a^\dagger a + \frac{1}{2}\right)^2$$

$$= -\frac{3}{16}\,\mathbb{1}, \tag{4.122}$$

where we have used the commutation relations between the annihilation and creation operators. Recalling that the eigenvalues of the Casimir operator (for angular momentum) can be written as $j(j+1)$ (see (3.122)), in this case, we have the unusual value of $j = -\frac{1}{4}$. In fact, this case corresponds to an infinite dimensional unitary representation of the non-compact group $SO(1,2)$ for which $x_3^2 - x_1^2 - x_2^2$ defines the invariant length (see also discussion following (3.90)).

For applications in some branches in physics including quantum optics, let us see how a group element of the form

$$e^{i(\theta_1 J_1 + \theta_2 J_2)}, \tag{4.123}$$

can be decomposed into a normal ordered product (with respect to the creation and annihilation operators, namely, in a form where the creation operators are on the left of the annihilation operators) of group elements as described in the earlier example. To do this, let us rewrite

$$e^{i(\theta_1 J_1 + \theta_2 J_2)} = e^{\frac{i}{2}\left((\theta_1 - i\theta_2)J_+ + (\theta_1 + i\theta_2)J_-\right)}$$

$$= e^{-\frac{1}{2}(\theta_1 - i\theta_2)(a^\dagger)^2 - \frac{1}{2}(\theta_1 + i\theta_2)a^2}. \tag{4.124}$$

Comparing with the right hand side, we note that this case corresponds to $\theta_3 = 0$ and $\theta = \sqrt{\theta_1^2 + \theta_2^2}$. Therefore, we can write (see (4.109))

$$e^{i(\theta_1 J_1 + \theta_2 J_2)} = e^{\frac{i}{2}\left((\theta_1 - i\theta_2)J_+ + (\theta_1 + i\theta_2)J_-\right)} = e^{\bar{\alpha}J_+}\, e^{\bar{\beta}J_-}\, e^{i\bar{\gamma}J_3},$$

or, $\quad e^{-\frac{1}{2}(\theta_1 - i\theta_2)(a^\dagger)^2 - \frac{1}{2}(\theta_1 + i\theta_2)a^2} = e^{\frac{i\bar{\alpha}}{2}(a^\dagger)^2}\, e^{\frac{i\bar{\beta}}{2}a^2}\, e^{\frac{i\bar{\gamma}}{4}(a^\dagger a + \frac{1}{2})},$

or, $\quad e^{\frac{\beta}{2}(a^\dagger)^2 + \frac{\alpha}{2}a^2} = e^{\frac{i\bar{\alpha}}{2}(a^\dagger)^2}\, e^{\frac{i\bar{\beta}}{2}a^2}\, e^{\frac{i\bar{\gamma}}{4}(a^\dagger a + \frac{1}{2})}, \tag{4.125}$

where we have defined

$$\alpha = \frac{i(\theta_1 + i\theta_2)}{2}, \quad \beta = \frac{i(\theta_1 - i\theta_2)}{2}, \quad \alpha\beta = -\frac{\theta^2}{4}. \tag{4.126}$$

Comparing with (4.117) we obtain

$$\bar{\alpha} = \frac{i(\theta_1 - i\theta_2)}{\theta}\tan\frac{\theta}{2} = \sqrt{\frac{\beta}{-\alpha}}\tan\sqrt{-\alpha\beta},$$

$$\bar{\beta} = \frac{i(\theta_1 + i\theta_2)}{2\theta}\sin\theta = \frac{1}{2}\sqrt{\frac{\alpha}{-\beta}}\sin 2\sqrt{-\alpha\beta},$$

$$\bar{\gamma} = 2i\log\left(\cos\sqrt{-\alpha\beta}\right). \tag{4.127}$$

Note that acting on an energy eigenstate $|n\rangle$ of the harmonic oscillator, the last factor on the right hand side of (4.125) is a constant so that the operators are indeed normal ordered. Relation (4.125) can be thought of as the analog of (4.84) for the group $SO(1,2)$.

◄

4.4 Ray representation

In quantum mechanics, the Schrödinger wave function $\psi(x)$ has a phase ambiguity in the sense that both $\psi(x)$ and $\psi'(x) = e^{i\theta}\psi(x)$, for a real constant θ, represent the same physical state (since the probability density $\rho(x) = |\psi(x)|^2$ is independent of the phase). Therefore, for applications in physics, we can also allow a phase ambiguity in the representation $U(g)$ of a group G in the composition relation (see (2.2))

$$U(g_1)U(g_2) = e^{i\omega(g_1,g_2)} U(g_1 g_2), \quad \forall g_1, g_2 \in G, \tag{4.128}$$

for some real phase function $\omega(g_1, g_2)$. If $\omega(g_1, g_2) = 0$ for all group elements, then it reduces to the standard representation of the group. However, when $\omega(g_1, g_2) \neq 0$ for all the elements of the group, (4.128) leads to a ray representation of the group G. We note that $U(g)$ has to satisfy the associative law (see (2.6))

$$(U(g_1)U(g_2))\, U(g_3) = U(g_1)\, (U(g_2)U(g_3)), \tag{4.129}$$

and since (4.128) leads to

$$
\begin{aligned}
(U(g_1)U(g_2))\, U(g_3) &= e^{i\omega(g_1,g_2)} U(g_1 g_2) U(g_3) \\
&= e^{i\omega(g_1,g_2)+i\omega(g_1 g_2,g_3)} U(g_1 g_2 g_3), \\
U(g_1)\, (U(g_2)U(g_3)) &= e^{i\omega(g_2,g_3)} U(g_1) U(g_2 g_3) \\
&= e^{i\omega(g_2,g_3)+i\omega(g_1,g_2 g_3)} U(g_1 g_2 g_3), \tag{4.130}
\end{aligned}
$$

for a ray representation we must have

$$\omega(g_1 g_2, g_3) - \omega(g_1, g_2 g_3) = \omega(g_2, g_3) - \omega(g_1, g_2). \tag{4.131}$$

$\omega(g_1, g_2)$ is called the cocycle of the group G and (4.131) is known as the cocycle condition. (In deriving (4.130) we have assumed the associativity of the group composition $(g_1g_2)g_3 = g_1(g_2g_3) = g_1g_2g_3$ as in (1.2).)

Let us next define a new function

$$V(g) = e^{i\theta(g)}U(g),\tag{4.132}$$

where θ denotes a real constant parameter. (We can view (4.132) as a group transformation without specifying its physical significance.) The new functions would satisfy the composition relation

$$
\begin{aligned}
V(g_1)V(g_2) &= e^{i(\theta(g_1)+\theta(g_2))}U(g_1)U(g_2) \\
&= e^{i(\theta(g_1)+\theta(g_2)+\omega(g_1,g_2))}U(g_1g_2) \\
&= e^{i\widetilde{\omega}(g_1,g_2)}V(g_1g_2),
\end{aligned}\tag{4.133}
$$

where we have used (4.128) and have identified

$$\widetilde{\omega}(g_1, g_2) = \omega(g_1, g_2) + \theta(g_1) + \theta(g_2) - \theta(g_1g_2).\tag{4.134}$$

It follows now that if the cocycle $\omega(g_1, g_2)$ satisfies

$$\omega(g_1, g_2) = \theta(g_1g_2) - \theta(g_1) - \theta(g_2), \quad \forall g_1, g_2 \in G,\tag{4.135}$$

for some $\theta(g)$, then, we will have (see (4.134))

$$\widetilde{\omega}(g_1, g_2) = 0,\tag{4.136}$$

so that $V(g)$ would lead to the standard group representation of G. In this case, the cocycle is called trivial. For some non-compact groups, the study of unitary ray representations is essential in applications in physics. A familiar example arises in the study of representations of the Galilean group (see Bargmann). However, it is much easier, in general, to study the ray representation of the Lie algebra rather than that of the Lie group. On the other hand, this is achieved at the expense of modifying (extending) the standard Lie algebra in a manner which we illustrate with the following example.

Let us consider the case of 2-dimensional translation group T_2 ($T(2)$). As in the case of T_1, a unitary representation of T_2 can be given by (see (2.72) and (2.76))

$$U(\alpha_1, \alpha_2) = \exp\left(\alpha_1 \frac{\partial}{\partial x_1} + \alpha_2 \frac{\partial}{\partial x_2}\right),$$ (4.137)

of two real variables x_1 and x_2 with two real parameters α_1 and α_2 operating on functions $f(x_1, x_2)$ with norm given by

$$||f||^2 = \int\limits_{-\infty}^{\infty}\!\!\int dx_1 dx_2 \, |f(x_1, x_2)|^2 < \infty.$$ (4.138)

The corresponding Lie algebra consists of the basis elements

$$D_1 \equiv \frac{\partial}{\partial x_1}, \quad D_2 \equiv \frac{\partial}{\partial x_2},$$ (4.139)

which satisfy the Abelian Lie algebra

$$[D_1, D_2] = 0.$$ (4.140)

To obtain a unitary ray representation of T_2, the simplest approach is to consider again the annihilation and creation operators a and $a^\dagger$ respectively (of the one dimensional harmonic oscillator) satisfying (see (2.130) in connection with the one dimensional translation where we have identified $a \to \alpha$ and so on)

$$[a, a^\dagger] = \mathbb{1}, \quad [a, a] = 0 = [a^\dagger, a^\dagger].$$ (4.141)

Let α denote a constant complex parameter and let us consider the transformations

$$a \to a' = a + \alpha\mathbb{1}, \quad a^\dagger \to a'^\dagger = a^\dagger + \alpha^*\mathbb{1},$$ (4.142)

where α^* denotes the complex conjugate of α. It follows from (4.141) and (4.142) that

$$[a', a'^\dagger] = \mathbb{1}, \quad [a', a'] = 0 = [a'^\dagger, a'^\dagger],$$ (4.143)

so that we obtain a representation of the corresponding three dimensional group through the relation

$$U(\alpha) = \exp\left(\alpha^* a - \alpha a^\dagger\right), \tag{4.144}$$

which formally corresponds to a unitary representation. (This is also invariant under the transformation (4.142).)

We note that for a different constant complex parameter β, we can write

$$U(\beta) = \exp\left(\beta^* a - \beta a^\dagger\right). \tag{4.145}$$

If we make the identifications

$$A = \alpha^* a - \alpha a^\dagger, \quad B = \beta^* a - \beta a^\dagger, \tag{4.146}$$

using (4.141), we obtain

$$[A, B] = [\alpha^* a - \alpha a^\dagger, \beta^* a - \beta a^\dagger] = -\left(\alpha^* \beta - \alpha \beta^*\right) \mathbb{1}. \tag{4.147}$$

Furthermore, using the Baker-Campbell-Hausdorff formula, namely,

$$e^A e^B = e^{A+B} e^{\frac{1}{2}[A,B]} = e^{\frac{1}{2}[A,B]} e^{A+B}, \tag{4.148}$$

when $[A, B]$ is a constant and is proportional to the identity operator, we now obtain

$$U(\alpha)U(\beta) = e^A e^B = e^{\frac{1}{2}[A,B]} e^{A+B} = e^{i\omega(\alpha,\beta)} U(\alpha + \beta), \tag{4.149}$$

where we have identified

$$\omega(\alpha, \beta) = \frac{i}{2}\left(\alpha^* \beta - \alpha \beta^*\right). \tag{4.150}$$

Thus, (4.144) truly furnishes a unitary ray representation of the three dimensional group. Moreover, the connection of this representation with that of T_2 can be made if we make the identifications (see (4.137) and (4.139))

$$\alpha = \alpha_1 - i\alpha_2, \quad a \equiv D = \frac{1}{2}\left(D_1 - iD_2\right). \tag{4.151}$$

where $\alpha_1 = \text{Re}\,\alpha, \alpha_2 = \text{Im}\,\alpha$. Namely, we can rewrite (4.144) as

$$U(\alpha) = \exp\left(\alpha_1 D_1 + \alpha_2 D_2\right). \tag{4.152}$$

However, D_1 and D_2 cannot commute any more (see (4.140)) since

$$[a, a^\dagger] = [D, D^\dagger] = \mathbb{1}, \tag{4.153}$$

which translates into the commutation relation

$$[D_1, D_2] = 2i\mathbb{1}. \tag{4.154}$$

As a result, the Lie algebra, in the present case consists of three basis elements D_1, D_2 and $\mathbb{1}$. The addition of an extra element (in this case, $\mathbb{1}$) to the standard Lie algebra of a Lie group, with some modification in the commutation relations, to obtain a ray representation for the group is quite general. When the extra element commutes with all the other elements of the algebra, it is known as the *Abelian extension* or the *central extension*.

4.5 References

V. Bargmann, The Annals of Mathematics 59 (1954) 1.

M. Hauser and J. T. Schwarz, *Lie groups and Lie algebras; notes in mathematics and its applications*, Gordon and Breach (1968).

CHAPTER 5

Irreducible tensor representations and Young tableau

In chapters 2 and 3, we have constructed simple representations of various Lie groups and Lie algebras. In this chapter, we will discuss how more complicated representations can be constructed and how the Young tableau or the Young diagrams (named after the British mathematician Alfred Young) prove to be of great help in this direction. For simplicity we will restrict ourselves to the $U(N)$ and $SU(N)$ groups.

5.1 Irreducible tensor representations of the $U(N)$ group

Let us change notations (compared with, say, chapter 2) and write a nonsingular $N \times N$ matrix in this section as (namely, we are representing the matrix A in the defining (fundamental) representation also by the same symbol as the group element a to avoid proliferation of symbols)

$$
a = \begin{pmatrix} a_1{}^1 & a_1{}^2 & a_1{}^3 & \cdots & a_1{}^N \\ \vdots & \vdots & \vdots & \vdots & \vdots \\ a_N{}^1 & a_N{}^2 & a_N{}^3 & \cdots & a_N{}^N \end{pmatrix}
\tag{5.1}
$$

with $a_{ji} = a_i{}^j$ (see, for example, (2.18) and (2.21) where a_{ij} was denoted as α_{ij} to avoid confusion with the annihilation and creation operators) for reasons to become clear soon. We note that a general linear transformation of an N-dimensional covariant vector can now be written as

$$
x_\mu \to x'_\mu = a_\mu{}^\nu x_\nu,
\tag{5.2}
$$

where $\mu, \nu = 1, 2, \cdots, N$ and summation over repeated indices (ν) is understood. We have already noted (see the discussion at the end of section 1.2) that the general linear group $GL(N)$ consists of the set of $N \times N$ nonsingular matrices a (namely, $\det a \neq 0$ so that a^{-1} exists). Therefore, we can denote an N dimensional representation of $GL(N)$ by the $N \times N$ dimensional matrices (the meaning of the subscript "1" will become clear within the context shortly)

$$U_1(a) = a. \tag{5.3}$$

In this case, $U_1(a)$ may be regarded as describing a linear transformation of the covariant vector x_μ in this N-dimensional space by (repeated indices are summed)

$$U_1(a): \quad x \to U_1(a)x,$$

$$\text{or,} \quad x_\mu \to (U_1(a)x)_\mu = x'_\mu = a_\mu{}^\nu x_\nu. \tag{5.4}$$

It can be easily checked, as in chapter 2, that these matrices satisfy all the properties of a group and, therefore, provide a representation. For example,

$$(U_1(e)x)_\mu = (\mathbb{1}_N)_\mu{}^\nu x_\nu = \delta_\mu{}^\nu x_\nu = x_\mu, \tag{5.5}$$

so that we can identify $U_1(e) = E_N = \mathbb{1}_N$. Similarly, we have

$$(U_1(a)U_1(b)x)_\mu = (U_1(a)\,(U_1(b)x))_\mu = a_\mu{}^\nu\,(U_1(b)x)_\nu$$

$$= a_\mu{}^\nu b_\nu{}^\lambda x_\lambda = (ab)_\mu{}^\lambda x_\lambda = (U_1(ab)x)_\mu, \tag{5.6}$$

where we have used the usual definition of the matrix product $(ab)_\mu{}^\lambda = a_\mu{}^\nu b_\nu{}^\lambda$ (with the convention that repeated indices are being summed). This then implies that $U_1(a)U_1(b) = U_1(ab)$ and that the matrices define an $N \times N$ dimensional matrix representation (or simply an N dimensional representation) of $GL(N)$. This is conventionally known as the defining representation or the fundamental representation for any group and is obtained by simply considering the linear transformation of the covariant vector x_μ. (Namely, the covariant vector x_μ provides the simplest representation space for $U(N)$ (or $SU(N)$))

However, we are interested mostly in the groups $U(N)$ and $SU(N)$ and we recall that the $U(N)$ group is defined by the set of all $N \times N$ nonsingular matrices a satisfying the unitarity condition (see (1.48))

$$a^\dagger a = E_N = aa^\dagger. \tag{5.7}$$

Furthermore, if we impose the additional condition

$$\det a = 1, \tag{5.8}$$

on the matrices they lead to a representation of the group $SU(N)$.

Just as (5.4) defines the transformation of a covariant vector, we may similarly consider an N-dimensional contravariant vector x^μ which transforms inversely as

$$x^\mu \to (U_{-1}(a)x)^\mu = x'^\mu = x^\nu \left(a^{-1}\right)_\nu^{\;\mu} = \left(a^{-1}\right)_\nu^{\;\mu} x^\nu. \tag{5.9}$$

As in (5.5)-(5.6) we can check that

$$U_{-1}(a) = a^{-1}, \tag{5.10}$$

also defines a representation (since a is assumed to be nonsingular a^{-1} exists). On the other hand, (5.7) for $U(N)$ and $SU(N)$ implies that

$$\left(a^{-1}\right)_\nu^{\;\mu} = \left(a^\dagger\right)_\nu^{\;\mu} = \left(a_\mu^{\;\nu}\right)^*, \tag{5.11}$$

so that we can write (5.9) also as

$$x^\mu \to (U_{-1}(a)x)^\mu = x'^\mu = \left(a_\mu^{\;\nu}\right)^* x^\nu. \tag{5.12}$$

Comparing with (5.4) we see that this allows us to identify the contravariant vector as the complex conjugate of the covariant vector, namely,

$$x^\mu = (x_\mu)^*, \tag{5.13}$$

so that we can also identify

$$U_{-1}(a) = (U_1(a))^*, \tag{5.14}$$

to be the complex conjugate representation of the defining representation $U_1(a)$. This is also consistent with (5.10) together with the unitarity condition (5.7).

We can now construct more complicated representations as follows. Let x_μ and y_μ be two N-dimensional covariant vectors. Then the second rank covariant tensor $T_{\mu\nu} = x_\mu y_\nu$, $\mu, \nu = 1, 2, \cdots, N$ transforms under $U(N)$ or $SU(N)$ as $(\alpha, \beta = 1, 2, \cdots, N)$

$$T_{\mu\nu} \to a_\mu^{\;\alpha} a_\nu^{\;\beta} T_{\alpha\beta}. \tag{5.15}$$

Since $T_{\mu\nu}$ has N^2 independent components, the transformation should generate a N^2- dimensional or a $N^2 \times N^2$ matrix representation of $U(N)$ or $SU(N)$. This is easily seen by recognizing that we can arrange the N^2 components of the tensor $T_{\mu\nu}$ into a column basis vector of the form $T_A = T_{(\mu,\nu)}, A = 1, 2, \cdots, N^2$, where, say for example, $T_1 = T_{11}, T_2 = T_{12}, \cdots, T_N = T_{1N}, T_{N+1} = T_{21}, \cdots, T_{N^2} = T_{NN}$ so that the transformation matrix (summation over $B = 1, 2, \cdots, N^2$ is understood)

$$T_A \to U_{AB}(a)T_B, \quad U_{AB}(a) = U_{(\mu,\nu)(\alpha,\beta)}(a) = a_\mu{}^\alpha a_\nu{}^\beta, \tag{5.16}$$

is a $N^2 \times N^2$ matrix. We can readily verify the validity $U_{AB}(a)U_{BC}(b) = U_{AC}(ab)$ (repeated indices are summed) so that this forms a representation of the group.

However, it is easy to show that such a representation is not irreducible in the following way. Let us define, from $T_{\mu\nu}$, the symmetric and the anti-symmetric tensors $S_{\mu\nu}$ and $A_{\mu\nu}$ respectively as

$$S_{\mu\nu} = \frac{1}{2}\left(T_{\mu\nu} + T_{\nu\mu}\right) = S_{\nu\mu},$$

$$A_{\mu\nu} = \frac{1}{2}\left(T_{\mu\nu} - T_{\nu\mu}\right) = -A_{\nu\mu}, \tag{5.17}$$

so that we can also write

$$T_{\mu\nu} = S_{\mu\nu} + A_{\mu\nu}. \tag{5.18}$$

These tensors also transform as in (5.15) which can be seen from

$$S_{\mu\nu} = \frac{1}{2}\left(T_{\mu\nu} + T_{\nu\mu}\right) \to \frac{1}{2}a_\mu{}^\alpha a_\nu{}^\beta T_{\alpha\beta} + \frac{1}{2}a_\nu{}^\beta a_\mu{}^\alpha T_{\beta\alpha}$$

$$= a_\mu{}^\alpha a_\nu{}^\beta \frac{1}{2}\left(T_{\alpha\beta} + T_{\beta\alpha}\right)$$

$$= a_\mu{}^\alpha a_\nu{}^\beta S_{\alpha\beta}. \tag{5.19}$$

Similarly, we can show that

$$A_{\mu\nu} \to a_\mu{}^\alpha a_\nu{}^\beta A_{\alpha\beta}. \tag{5.20}$$

Therefore, both $S_{\mu\nu}$ and $A_{\mu\nu}$ also provide representations for the group of dimensions $\frac{1}{2}N(N+1)$ and $\frac{1}{2}N(N-1)$ respectively. Namely, $S_{\mu\nu}$ acts on vectors in a $\frac{1}{2}N(N+1)$ dimensional vector space V_1 while $A_{\mu\nu}$ acts on vectors in a $\frac{1}{2}N(N-1)$ dimensional vector space V_2. Furthermore, V_1 and V_2 define invariant spaces, namely, under the transformation, the vectors in these two spaces do not mix (see, for example, section 2.2, particularly, the discussion towards the end of that section). If we write the basis vectors representing $T_{\mu\nu}$ in terms of the corresponding basis vectors of $S_{\mu\nu}$ and $A_{\mu\nu}$, then these basis vectors would define invariant subspaces of $T_{\mu\nu}$ so that $T_{\mu\nu}$ is fully reducible as also can be seen from (5.18). Counting the dimensions of the representations we see as well that

$$N^2 = \frac{1}{2} N(N+1) + \frac{1}{2} N(N-1). \tag{5.21}$$

Furthermore, the representations given by $S_{\mu\nu}$ and $A_{\mu\nu}$ can be shown to be irreducible, namely, there exist no other nontrivial sub-tensors (of lower dimension) of V_1 and V_2 which are invariant under $U(N)$ (or $SU(N)$). Therefore, this shows that $U(N)$ (or $SU(N)$) has two irreducible (second rank covariant tensor) representations of dimensions $\frac{1}{2} N(N+1)$ and $\frac{1}{2} N(N-1)$.

For the mixed tensor $T_\mu{}^\nu = x_\mu y^\nu$, the transformation property under $U(N)$ (or $SU(N)$) is obtained to be (see (5.4) and (5.9), repeated indices are summed)

$$T_\mu{}^\nu \to (T')_\mu{}^\nu = a_\mu{}^\alpha \left(a^{-1}\right)_\beta{}^\nu T_\alpha{}^\beta, \tag{5.22}$$

and yields a $N^2 \times N^2$ matrix representation. However, as before this representation is not irreducible. To obtain the irreducible representations, we note that we do not have a symmetric and anti-symmetric decomposition for a mixed tensor. Rather, let us define (summation over repeated indices is assumed)

$$T = T_\mu{}^\mu, \tag{5.23}$$

which is invariant under $U(N)$ $(SU(N))$ since

$$\begin{aligned} T \to T' = (T')_\mu{}^\mu &= a_\mu{}^\alpha \left(a^{-1}\right)_\beta{}^\mu T_\alpha{}^\beta \\ &= a_\mu{}^\alpha \left(a^{-1}\right)_\beta{}^\mu T_\alpha{}^\beta = \delta_\beta^\alpha T_\alpha{}^\beta \\ &= T_\alpha{}^\alpha = T. \end{aligned} \tag{5.24}$$

Namely, T provides a trivial one-dimensional representation of $U(N)$ $(SU(N))$.

As a result, we see that

$$\widetilde{T}_\mu{}^\nu = T_\mu{}^\nu - \frac{1}{N} \delta_\mu^\nu T, \tag{5.25}$$

transforms exactly as $T_\mu{}^\nu$ under $U(N)$ $(SU(N))$. However, it satisfies the traceless condition (recall that $\mu = 1, 2, \cdots, N$), namely,

$$\widetilde{T}_\mu{}^\mu = \left(T_\mu{}^\mu - \frac{1}{N} \delta_\mu^\mu T\right) = T - T = 0. \tag{5.26}$$

Consequently it defines a $(N^2 - 1)$ dimensional matrix representation of $U(N)$ $(SU(N))$. Furthermore, from the definition in (5.25) we note that

$$T_\mu{}^\nu = \widetilde{T}_\mu{}^\nu + \frac{1}{N} \delta_\mu^\nu T, \tag{5.27}$$

leads to the correct number of dimensions for the representation of $T_\mu{}^\nu$, namely, $N^2 = (N^2 - 1) + 1$ as expected. We can verify that the traceless tensor $\widetilde{T}_\mu{}^\nu$ is irreducible.

The situation becomes more complicated for tensors of larger rank. As an example, let us consider a tensor of the form $T_{\mu\nu\lambda} = x_\mu y_\nu z_\lambda$ which gives a N^3-dimensional representation space. This space is again reducible and the decomposition of this space into invariant tensor spaces is more involved as can be seen in the following way. Let σ denote a generic permutation of the indices μ, ν and λ such as $\mu \to \nu \to \lambda \to \mu$. Let us define the two third rank tensors by summing over all possible permutations as

$$S_{\mu\nu\lambda} = \frac{1}{3!} \sum_\sigma T_{\sigma(\mu)\sigma(\nu)\sigma(\lambda)}$$

$$\equiv \frac{1}{3!} \left(T_{\mu\nu\lambda} + T_{\nu\lambda\mu} + T_{\lambda\mu\nu} + T_{\nu\mu\lambda} + T_{\mu\lambda\nu} + T_{\lambda\nu\mu} \right),$$

$$A_{\mu\nu\lambda} = \frac{1}{3!} \sum_\sigma (-1)^\sigma \, T_{\sigma(\mu)\sigma(\nu)\sigma(\lambda)}$$

$$\equiv \frac{1}{3!} \left(T_{\mu\nu\lambda} + T_{\nu\lambda\mu} + T_{\lambda\mu\nu} - T_{\nu\mu\lambda} - T_{\mu\lambda\nu} - T_{\lambda\nu\mu} \right), \tag{5.28}$$

where the weight factor $(-1)^\sigma$ introduces a negative sign for every pair of indices that are interchanged. We note that $S_{\mu\nu\lambda}$ is manifestly totally symmetric (under the exchange of any pair of indices) while $A_{\mu\nu\lambda}$ is manifestly totally anti-symmetric. The total number of independent components of these tensors are given respectively by

$$D(S) = \frac{1}{3!} N(N+1)(N+2),$$

$$D(A) = \frac{1}{3!} N(N-1)(N-2), \tag{5.29}$$

and we note that $D(S) + D(A) = \frac{1}{3} N(N^2 + 2) \leq N^3$ (equality holds only for the trivial case $N = 1$), so that these two classes of tensors alone cannot reproduce the whole set of $T_{\mu\nu\lambda}$. We need to consider more complicated tensors other than the familiar classes of totally symmetric and totally anti-symmetric tensors. For this we proceed as follows.

Let us consider a third rank (mixed symmetry) tensor $M_{\mu\nu\lambda}$ which is obtained from $T_{\mu\nu\lambda}$ by first symmetrizing in the first two indices (μ, ν) and then anti-symmetrizing in the first and the last indices (μ, λ) as

$$M_{\mu\nu\lambda} = \frac{1}{3!} \left((T_{\mu\nu\lambda} + T_{\nu\mu\lambda}) - (T_{\lambda\nu\mu} + T_{\nu\lambda\mu}) \right). \tag{5.30}$$

This tensor satisfies the symmetry relations

$$M_{\mu\nu\lambda} = -M_{\lambda\nu\mu},$$
$$M_{\mu\nu\lambda} + M_{\nu\lambda\mu} + M_{\lambda\mu\nu} = 0. \tag{5.31}$$

Note that the first relation in (5.31) says $M_{\mu\nu\lambda}$ does not contain the totally symmetric tensor $S_{\mu\nu\lambda}$ while the second relation (together with the first relation) implies the absence of the totally anti-symmetric tensor $A_{\mu\nu\lambda}$ in $M_{\mu\nu\lambda}$.

Similarly, we can construct another third rank (mixed symmetry) tensor $M'_{\mu\nu\lambda}$ which is obtained from $T_{\mu\nu\lambda}$ by first symmetrizing in the $(\mu\lambda)$ indices and then anti-symmetrizing in the $(\mu\nu)$ indices. More explicitly, this tensor has the form

$$M'_{\mu\nu\lambda} = \frac{1}{3!} \left((T_{\mu\nu\lambda} + T_{\lambda\nu\mu}) - (T_{\nu\mu\lambda} + T_{\lambda\mu\nu}) \right). \tag{5.32}$$

Correspondingly, this tensor satisfies the relations

$$M'_{\mu\nu\lambda} = -M'_{\nu\mu\lambda},$$
$$M'_{\mu\nu\lambda} + M'_{\nu\lambda\mu} + M'_{\lambda\mu\nu} = 0. \tag{5.33}$$

Both $M_{\mu\nu\lambda}$ and $M'_{\mu\nu\lambda}$ have the same number of independent components given by

$$D(M) = D(M') = \frac{1}{3} N(N^2 - 1). \tag{5.34}$$

It is easier now to check that the tensor $T_{\mu\nu\lambda}$ can now be expressed uniquely in terms of these four third rank tensors as

$$T_{\mu\nu\lambda} = S_{\mu\nu\lambda} + A_{\mu\nu\lambda} + (M_{\mu\nu\lambda} + M'_{\mu\nu\lambda}) - (M_{\nu\lambda\mu} + M'_{\nu\lambda\mu}), \tag{5.35}$$

leading to the correct number of total dimension, namely,

$$N^3 = D(S) + D(A) + D(M) + D(M'). \tag{5.36}$$

Let us note here that although we constructed the (mixed symmetry) tensor $M_{\mu\nu\lambda}$ by first symmetrizing in the $(\mu\nu)$ indices and then anti-symmetrizing in $(\mu\lambda)$, the operations are commutative. Namely, we could have started by anti-symmetrizing in the $(\mu\lambda)$ indices and then symmetrizing in $(\mu\nu)$ and the final result would not be the same, but would lead to an equivalent result in the following way. For example, constructed this way, the (mixed symmetry) tensor can be written as

$$\widetilde{M}_{\mu\nu\lambda} = \frac{1}{3!}\left((T_{\mu\nu\lambda} - T_{\lambda\nu\mu}) + (T_{\nu\mu\lambda} - T_{\lambda\mu\nu})\right), \tag{5.37}$$

which satisfies the relations

$$\widetilde{M}_{\mu\nu\lambda} = \widetilde{M}_{\nu\mu\lambda},$$

$$\widetilde{M}_{\mu\nu\lambda} + \widetilde{M}_{\nu\lambda\mu} + \widetilde{M}_{\lambda\mu\nu} = 0. \tag{5.38}$$

The two relations in (5.38) imply the absence of the totally symmetric tensor $S_{\mu\nu\lambda}$ in $\widetilde{M}_{\mu\nu\lambda}$. The number of independent components of $\widetilde{M}_{\mu\nu\lambda}$ is exactly the same as that of $M_{\mu\nu\lambda}$ and can be expressed as the linear superposition

$$\widetilde{M}_{\mu\nu\lambda} = \frac{1}{2}\left((M_{\mu\nu\lambda} + M_{\nu\mu\lambda}) + (M'_{\mu\nu\lambda} + M'_{\nu\mu\lambda})\right), \tag{5.39}$$

simply corresponding to a change in basis. All these tensors are known to be irreducible. It is clear now that constructing higher rank irreducible tensor representations algebraically is a formidable task. However, such irreducible tensors can be naturally written in a diagrammatic form through Young tableaux which we discuss next.

5.2 Young tableau

A Young tableau or a Young diagram is a combinatoric diagram which appears in various areas of mathematics and physics. It is a collection of rows of boxes (with entries in them) left justified and drawn such that a lower row never contains more boxes than an upper row. So, for example, if a Young tableau (diagram) has ℓ rows, then it is algebraically designated by $(f_1, f_2, \cdots, f_\ell)$ where f_i denotes the number of boxes in the i th row and they satisfy the relation

$$f_1 \geq f_2 \geq f_3 \geq \cdots \geq f_\ell \geq 0. \tag{5.40}$$

The irreducible representations of covariant tensors for the groups $U(N)$ and $SU(N)$ can be naturally written in terms of Young diagrams in the following manner. In this case, we identify the number of rows to coincide with $\ell = N$ and identify the entry in a box with a tensor index. The tensor structure represented by a Young diagram is constructed by first symmetrizing in the indices along a given row and then anti-symmetrizing in the indices along a given column. In this way, a Young diagram represents a unique tensor representation.

For example, the simplest Young diagram for $U(N)$ (more generally for $GL(N)$ or $SL(N)$) corresponding to the fundamental representation has only one row with only one box with a vector index

$$x_\mu : \boxed{\mu} \ = (1, \overbrace{0, 0, \cdots, 0}^{N-1}) = [1], \tag{5.41}$$

where the square bracket representation simply (neglects the zero elements) assumes that the rest of the $(N-1)$ entries vanish corresponding to the fact that the diagram has only one row with one box. (Since there is only one index, the question of symmetry or anti-symmetry does not arise.) Similarly, turning to irreducible tensor representations of second rank, we have already seen that there are two of them (see (5.17)) and we can represent them respectively as

$$S_{\mu\nu} : \boxed{\mu \ \nu} \ = (2, \overbrace{0, 0, \cdots, 0}^{N-1}) = [2],$$

$$A_{\mu\nu} : \begin{array}{|c|} \hline \mu \\ \hline \nu \\ \hline \end{array} \quad = (1, 1, \overbrace{0, 0, \cdots, 0}^{N-2}) = [1, 1]. \tag{5.42}$$

The symmetry properties of these tensors are manifest. Since $S_{\mu\nu}$ is represented by a single row with two boxes with indices μ, ν, it is symmetric in those indices while $A_{\mu\nu}$ is denoted by two rows with one box in each column (containing indices μ, ν), it is anti-symmetric in the indices.

Similarly, we can write the Young diagram for the third rank tensor $M_{\mu\nu\lambda}$ as

$$M_{\mu\nu\lambda} : \begin{array}{|c|c|} \hline \mu & \nu \\ \hline \lambda \\ \cline{1-1} \end{array} \ = (2, 1, \overbrace{0, 0, \cdots, 0}^{N-2}) = [2, 1], \tag{5.43}$$

while that for $M'_{\mu\nu\lambda}$ has the diagrammatic representation

$$M'_{\mu\nu\lambda} : \begin{array}{|c|c|} \hline \mu & \lambda \\ \hline \nu \\ \cline{1-1} \end{array} \ = (2, 1, \overbrace{0, 0, \cdots, 0}^{N-2}) = [2, 1], \tag{5.44}$$

As another example, let us consider a tensor $T_{\mu\nu\lambda\alpha\beta\gamma\tau}$ with the Young tableau

$$
\begin{array}{|c|c|c|}
\hline
\mu & \nu & \lambda \\
\hline
\alpha & \beta \\
\cline{1-2}
\gamma & \tau \\
\cline{1-2}
\end{array}
\quad = (3, 2, 2, \overbrace{0, 0, \cdots, 0}^{N-3}) = [3, 2, 2], \tag{5.45}
$$

which can be obtained by first symmetrizing all rows (namely, symmetrizing in the indices μ, ν, λ and α, β as well as in γ, τ) and then anti-symmetrizing in all columns (namely, in μ, α, γ and ν, β, τ). Equivalently, we can first anti-symmetrize in all columns and then symmetrize in all the rows. This can be generalized for any covariant tensor. The general Young tableau for a covariant irreducible tensor is given by N symbols $[f_1, f_2, \cdots, f_N]$ satisfying

$$
f_1 \geq f_2 \geq f_3 \geq \cdots \geq f_N \geq 0. \tag{5.46}
$$

Note that we do not need more than N symbols which can be seen as follows. Consider a totally anti-symmetric tensor $A_{\mu_1\mu_2\cdots\mu_{N+1}}$ with $N + 1$ indices. The corresponding Young tableau may be described as $[\overbrace{1, 1, \cdots, 1}^{N+1}]$ with $f_1 = f_2 = \cdots = f_{N+1} = 1$. However, in the N-dimensional case that we are discussing, the tensor indices $\mu_1, \mu_2, \cdots, \mu_{N+1}$ can only take N distinct values $1, 2, \cdots, N$. Therefore, at least two of the tensor indices must have the same value. On the other hand, since the tensor is anti-symmetric in all its indices, in this case it must identically vanish, namely, $A_{\mu_1\mu_2\cdots\mu_{N+1}} = 0$. The same argument also holds for any irreducible tensor with the Young tableau $[f_1, f_2, \cdots, f_M]$ with $M > N$ and $f_M > 0$.

The dimension of an irreducible tensor representation can also be calculated using the Young diagram (for the representation) in the following way. First, we introduce the concept of a hook length of a given box which is a unique integer associated with a box in a given diagram. It simply corresponds to the sum of the number of boxes to its right (along the given row), the number of boxes below it (along the given column) plus one (corresponding to the box under consideration). Then the hook length associated with any given diagram corresponds simply to the product of all the hook lengths of boxes in that diagram. Next, for $SU(N)$, let us label the top most left box as N and increase this integer by unity for every box to its right and decrease it by unity for every box below it. This gives a unique labelling of the diagram by integers and the product of all these integers divided by the hook length associated with the diagram gives the dimensionality associated with the tensor representation associated with that diagram. For example, following these rules, we can immediately calculate the dimension of the respresentation associated with the mixed symmetry tensors in (5.43) and (5.44) to be

$$D(M) = D(M') = \frac{N(N+1)(N-1)}{3 \times 1 \times 1} = \frac{1}{3} N(N^2 - 1), \quad (5.47)$$

which coincides with (5.34). Similarly, for the general tensor representation in (5.53), the dimension of the representation is obtained to be

$$\frac{N(N+1)(N+2)(N-1)N(N-2)(N-1)}{5 \times 4 \times 1 \times 3 \times 2 \times 2 \times 1}$$

$$= \frac{1}{240} N^2(N-1)(N^2-1)(N^2-4). \quad (5.48)$$

As we will point out towards the end of this section and in the next section, only covariant tensors are sufficient to describe irreducible tensors in the case of $SU(N)$ group, while both covariant and contravariant tensors are necessary for the $U(N)$ group. Because of this we need to generalize the Young symbol $[f_1, f_2, \cdots, f_N]$ to admit negative integer values for some of the f_is. As a simple example, consider the conjugate representation denoted by a contravariant vector x^μ (see discussion in (5.9)-(5.14)). This has the simple Young diagram representation

$$x^\mu : \quad \boxed{\mu}^{\,*} \; = \; (\overbrace{0, 0, \cdots, 0}^{N-1}, -1) = [-1] = [1]^*. \quad (5.49)$$

The (-1) here refers to one contravariant index which transforms inversely from a covariant index. Similarly, we can verify that the traceless tensor $\widetilde{T}_\mu{}^\nu$ (see (5.25)) is irreducible and we can represent this as

$$\widetilde{T}_\mu{}^\nu = (1, \overbrace{0, 0, \cdots, 0}^{N-2}, -1) = ([1], [-1]) = ([1], [1]^*). \quad (5.50)$$

As another example, let us consider the mixed tensor $T_{\mu\nu}^{\alpha\beta}$ satisfying

$$T_{\mu\nu}^{\alpha\beta} = -T_{\mu\nu}^{\beta\alpha} = T_{\nu\mu}^{\alpha\beta}, \quad (5.51)$$

as well as the traceless condition

$$\sum_{\lambda=1}^{N} T_{\lambda\nu}^{\lambda\beta} = 0. \quad (5.52)$$

We may represent such a tensor more conveniently as

$$\left(\boxed{\mu\,|\,\nu}\,,\ \boxed{\begin{matrix}\alpha\\\beta\end{matrix}}^{\!*}\right) = (2,\overbrace{0,0,\cdots,0}^{N-3}-1,-1\,) = ([2],[1,1]^{*}),$$

(5.53)

allowing negative integer values for f_is corresponding to contravariant indices. More generally, let the mixed tensor $T^{\nu_1\nu_2\cdots\nu_q}_{\mu_1\mu_2\cdots\mu_p}$ correspond to the Young symbol $(f_1, f_2, \cdots, f_n)$ for the covariant indices $(\mu_1, \mu_2, \cdots, \mu_p)$, while the contravariant indices $(\nu_1, \nu_2, \cdots, \nu_q)$ correspond to the Young symbol $(g_1, g_2, \cdots, g_m)$so that we can represent it as

$$((f_1, f_2, \cdots, f_n), (g_1, g_2, \cdots, g_m)^{*}).$$

(5.54)

Written this way, f_i, g_j satisfy the conditions

$$f_1 \geq f_2 \geq \cdots \geq f_n > 0,$$
$$g_1 \geq g_2 \geq \cdots \geq g_m > 0.$$

(5.55)

Moreover, we require the mixed tensor to satisfy the traceless condition

$$\sum_{\lambda=1}^{N} T^{\lambda\nu_2\cdots\nu_q}_{\lambda\mu_2\cdots\mu_p} = 0,$$

(5.56)

for any pair of indices (μ_j, ν_k), for $j = 1, 2, \cdots, p$ and $k = 1, 2, \cdots, q$. As we will argue, it is sufficient to consider only the cases satisfying $n+m \leq N$. Introducing the negative numbers $f_{N-m+1}, f_{N-m+2}, \cdots, f_N$ by

$$f_{N-m+1} = -g_m, \quad f_{N-m+2} = -g_{m-1}, \quad \cdots, \quad f_N = -g_1, \quad (5.57)$$

it follows that (see (5.55))

$$0 > f_{N-m} \geq f_{N-m+1} \geq f_{N-m+2} \geq \cdots \geq f_N.$$

(5.58)

We designate the tensor to have the generalized Young symbol

$$(f_1, f_2, \cdots, f_n, \overbrace{0, 0, \cdots, 0}^{N-n-m}, f_{N-m+1}, \cdots, f_N),$$

(5.59)

satisfying

$$f_1 \geq f_2 \geq \cdots \geq f_n > \overbrace{0, 0, \cdots, 0}^{N-n-m} > f_{N-m+1} \geq \cdots \geq f_N. \quad (5.60)$$

The reason why we can restrict ourselves to the cases $n + m \leq N$ is that we can reduce the problem by lowering the contravariant indices $\nu_1, \nu_2, \cdots, \nu_q$ by products of Levi-Civita tensors $\epsilon_{\nu_j \alpha_2 \cdots \alpha_N}$ to make it a covariant tensor. To give an example, let us consider the mixed tensor $\overline{T}^{\alpha\beta}_{\mu\nu}$ satisfying (it is anti-symmetric in the covariant as well as the contravariant indices)

$$\overline{T}^{\alpha\beta}_{\mu\nu} = -\overline{T}^{\beta\alpha}_{\mu\nu} = -\overline{T}^{\alpha\beta}_{\nu\mu},$$

$$\overline{T}^{\mu\beta}_{\mu\nu} = 0, \tag{5.61}$$

where the repeated index μ is being summed in the second relation. This tensor would correspond to the generalized Young symbol (in the earlier notation $m = n = 2$ in this case)

$$(1, 1, \overbrace{0, 0, \cdots, 0}^{N-4}, -1, -1). \tag{5.62}$$

It is clear that, for $N = 3$, this symbol is not of the form (f_1, f_2, f_3) and, therefore, it must vanish identically in this case. To see this, let us introduce a new lower rank tensor $\overline{T}^{\lambda}_{\gamma}$ by (repeated indices are summed)

$$\overline{T}^{\lambda}_{\gamma} = \frac{1}{4} \epsilon_{\alpha\beta\gamma} \epsilon^{\mu\nu\lambda} \overline{T}^{\alpha\beta}_{\mu\nu}, \tag{5.63}$$

where ϵ_{ijk} denote three dimensional Levi-Civita tensors. Using the identity satisfied by the Levi-Civita tensor

$$\epsilon^{\mu\nu\lambda} \epsilon_{\alpha\beta\gamma} = \delta^{\mu}_{\alpha}(\delta^{\nu}_{\beta}\delta^{\lambda}_{\gamma} - \delta^{\nu}_{\gamma}\delta^{\lambda}_{\beta}) - \delta^{\mu}_{\beta}(\delta^{\nu}_{\gamma}\delta^{\lambda}_{\alpha} - \delta^{\nu}_{\alpha}\delta^{\lambda}_{\gamma})$$
$$+ \delta^{\mu}_{\gamma}(\delta^{\nu}_{\alpha}\delta^{\lambda}_{\beta} - \delta^{\nu}_{\beta}\delta^{\lambda}_{\alpha}), \tag{5.64}$$

we obtain from (5.63) (repeated indices are summed)

$$\overline{T}^{\lambda}_{\gamma} = \frac{1}{4} \left(\delta^{\lambda}_{\gamma} \overline{T}^{\mu\nu}_{\mu\nu} - \overline{T}^{\mu\lambda}_{\mu\gamma} - \overline{T}^{\lambda\mu}_{\mu\gamma} + \delta^{\lambda}_{\gamma} \overline{T}^{\nu\mu}_{\mu\nu} + \overline{T}^{\nu\lambda}_{\gamma\nu} - \overline{T}^{\lambda\nu}_{\gamma\nu} \right)$$

$$= 0, \tag{5.65}$$

because of the traceless condition in (5.61). Relation (5.63) is invertible and it follows from (5.65) that

$$\overline{T}^{\alpha\beta}_{\mu\nu} = \epsilon_{\mu\nu\lambda} \epsilon^{\alpha\beta\gamma} \overline{T}^{\lambda}_{\gamma} = 0, \tag{5.66}$$

so that this tensor indeed vanishes identically for $N = 3$.

The importance of the generalized (mixed) tensors $T^{\nu_1 \nu_2 \cdots \nu_q}_{\mu_1 \mu_2 \cdots \mu_p}$ together with the generalization of the Young tableau lies in the fact that all irreducible representations of the $U(N)$ group can be obtained in this way, as has been shown by H. Weyl. He also calculates the formula for the character dimension of these irreducible representations. Let a represent a $N \times N$ unitary matrix corresponding to the group element $a \in U(N)$. Then, the matrix a can be diagonalized by another $N \times N$ unitrary matrix S so that (we are using a to denote the group element as well as its defining matrix representation and we note that $S \in U(N)$)

$$S^{-1}aS = a_D = \begin{pmatrix} \xi_1 & 0 & 0 & 0 & \cdots \\ 0 & \xi_2 & 0 & 0 & \cdots \\ 0 & 0 & \xi_3 & 0 & \cdots \\ 0 & 0 & 0 & \ddots & \cdots \end{pmatrix}, \qquad S^\dagger S = E_N. \quad (5.67)$$

Let $V(a)$ denote a $n \times n$ matrix representation of the $U(N)$ group corresponding to the group element $a \in U(N)$. It follows, therefore, that

$$V(a_D) = V(S^{-1}aS) = V(S^{-1})V(a)V(S). \quad (5.68)$$

Taking the trace of both sides yields

$$\operatorname{Tr} V(a_D) = \operatorname{Tr}\left(V(S^{-1})V(a)V(S)\right) = \operatorname{Tr}\left(V(S)V(S^{-1})V(a)\right)$$
$$= \operatorname{Tr}\left(E_n V(a)\right) = \operatorname{Tr} V(a), \quad (5.69)$$

where we have used the cyclicity of the trace. The character of a representation $V(a)$ of the group $U(N)$ can now be defined as

$$\chi_V(a) = \operatorname{Tr} V(a) = \operatorname{Tr} V(a_D) = \chi_V(a_D), \quad (5.70)$$

which is a function of the diagonal elements $\xi_1, \xi_2, \cdots, \xi_N$ of a_D in (5.67) satisfying $|\xi_j| = 1, j = 1, 2, \cdots, N$ (recall that the matrix a is unitary). Namely, we can write

$$\chi_V(a) = \chi_V(\xi_1, \xi_2, \cdots, \xi_N). \quad (5.71)$$

We note here parenthetically that for the defining representation, the character is simply given by $\chi(a) = \xi_1 + \xi_2 + \cdots + \xi_N$.

Suppose the representation $V(a)$ is irreducible with the Young tableau corresponding to $(f_1, f_2, \cdots, f_N)$ satisfying $f_1 \geq f_2 \geq \cdots \geq f_N$. Let us define

$$\ell_j = f_j + (N - j), \quad (5.72)$$

so that we have

$$\ell_1 > \ell_2 > \cdots > \ell_N. \tag{5.73}$$

Let us define

$$K(\ell_1, \ell_2, \cdots, \ell_N) = \det \begin{vmatrix} (\xi_1)^{\ell_1} & (\xi_1)^{\ell_2} & \cdots & (\xi_1)^{\ell_N} \\ (\xi_2)^{\ell_1} & (\xi_2)^{\ell_2} & \cdots & (\xi_2)^{\ell_N} \\ \vdots & \vdots & \cdots & \vdots \\ (\xi_N)^{\ell_1} & (\xi_N)^{\ell_2} & \cdots & (\xi_N)^{\ell_N} \end{vmatrix}. \tag{5.74}$$

Then, Weyl's character formula is defined in terms of this determinant as

$$\chi_V(\ell_1, \ell_2, \cdots, \ell_N) = \frac{K(\ell_1, \ell_2, \cdots, \ell_N)}{K((N-1), (N-2), \cdots 1, 0)}. \tag{5.75}$$

From this formula, Weyl also determines the dimension of the irreducible representation to be given by

$$D_V(f_1, f_2, \cdots, f_N) = \frac{\prod\limits_{\mu < \nu}^{N} (\ell_\mu - \ell_\nu)}{1! 2! \cdots (N-1)!}. \tag{5.76}$$

by noting that the dimension of a representation can be thought of as

$$D_V(f_1, f_2, \cdots, f_N) = \operatorname{Tr} V(e) = \chi_V(e) = \chi_V(1, 1, \cdots, 1). \tag{5.77}$$

5.3 Irreducible tensor representations of the $SU(N)$ group

For the $SU(N)$ group we note that the $N \times N$ unitary matrix $a \in U(N)$ satisfies the additional condition $\det a = 1$. In this case, for irreducible representations we need not consider the contravariant (or mixed) tensors (since the tensor indices can be raised using the Levi-Civita tensor leading to an equivalent representation). Moreover, as we will show next we can always set $f_N = 0$, if we wish. Therefore, irreducible representations of the $SU(N)$ group can be realized by covariant tensors $T_{\mu_1 \mu_2 \cdots \mu_n}$ with the Young tableau $(f_1, f_2, \cdots, f_{N-1}, 0)$ with $f_1 \geq f_2 \geq \cdots \geq f_{N-1} \geq 0$.

    These facts can be understood as follows. Let $A_{\mu_1 \mu_2 \cdots \mu_N}$ denote a totally anti-symmetric covariant tensor with N indices $\mu_1, \mu_2, \cdots, \mu_N$. Using the totally anti-symmetric Levi-Civita tensor in N dimensions (with $\epsilon_{123\cdots N} = 1$), we can also write this tensor as (in N dimensions any completely anti-symmetric Nth rank tensor is proportional to the Levi-Civita tensor)

$$A_{\mu_1\mu_2\cdots\mu_N} = \epsilon_{\mu_1\mu_2\cdots\mu_N}\,\phi, \tag{5.78}$$

for some scalar ϕ. Under the action of $a \in U(N)$, the tensor transforms as (repeated indices are summed, see (5.15))

$$\begin{aligned}
A_{\mu_1\mu_2\cdots\mu_N} \to A'_{\mu_1\mu_2\cdots\mu_N} &= a^{\alpha_1}_{\mu_1}a^{\alpha_2}_{\mu_2}\cdots a^{\alpha_N}_{\mu_N}A_{\alpha_1\alpha_2\cdots\alpha_N} \\
&= a^{\alpha_1}_{\mu_1}a^{\alpha_2}_{\mu_2}\cdots a^{\alpha_N}_{\mu_N}\,\epsilon_{\alpha_1\alpha_2\cdots\alpha_N}\,\phi \\
&= (\det a)\,\epsilon_{\mu_1\mu_2\cdots\mu_N}\phi = (\det a)\,A_{\mu_1\mu_2\cdots\mu_N}, \tag{5.79}
\end{aligned}$$

where we have used the definition of the determinant of a matrix in terms of the Levi-Civita tensor, namely,

$$a^{\alpha_1}_{\mu_1}a^{\alpha_2}_{\mu_2}\cdots a^{\alpha_N}_{\mu_N}\,\epsilon_{\alpha_1\alpha_2\cdots\alpha_N} = (\det a)\,\epsilon_{\mu_1\mu_2\cdots\mu_N}.$$

Note that this tensor has the Young tableau notation of $[1, 1, 1, \cdots, 1]$ (N unit elements). If $a \in SU(N)$ so that $\det a = 1$, then it follows from (5.79) that

$$A'_{\mu_1\mu_2\cdots\mu_N} = A_{\mu_1\mu_2\cdots\mu_N}, \tag{5.80}$$

which describes the trivial representation of $SU(N)$ (namely, ϕ is indeed a scalar). As a result, this shows that for $SU(N)$ the representation with the Young diagram $[1, 1, 1, \cdots, 1]$ (N unit elements) is equivalent to that with the Young diagram $[0, 0, 0, \cdots, 0]$. More generally we can show that any tensor T_A with Young tableau $(f_1 + \Delta, f_2 + \Delta, \cdots, f_N + \Delta)$ for any integer Δ has the transform given by $(\det a)^\Delta$ times the transform of the tensor with the Young tableau $(f_1, f_2, \cdots, f_N)$. In particular, for the $SU(N)$ group, the two tensors transform exactly in the same way since $\det a = 1$, namely, they are equivalent representations. In other words, for the $SU(N)$ group

$$(f_1 + \Delta, f_2 + \Delta, \cdots, f_N + \Delta) \simeq (f_1, f_2, \cdots, f_N), \tag{5.81}$$

for any integer Δ. In particular, if we choose $\Delta = -f_N$, this implies that we can effectively set $f_N = 0$ as we have mentioned earlier.

To illustrate this and to indicate how contravariant (or mixed) tensors are equivalent to covariant ones, let us consider the group $SU(3)$ with $N = 3$. Let us consider the traceless mixed second rank tensor $T^\mu_\nu, \mu, \nu = 1, 2, 3$ so that it has the Young tableau given by

$(1, 0, -1)$. If we choose $\Delta = 1$, then we see from (5.81) that we have an equivalent representation given by the Young tableau $(2, 1, 0)$. The correspondence is indeed established by identifying (repeated indices are summed)

$$M'_{\mu\nu\lambda} = \epsilon_{\mu\nu\alpha} T^\alpha_\lambda, \tag{5.82}$$

satisfying (see (5.33))

$$M'_{\mu\nu\lambda} = -M'_{\nu\mu\lambda}, \quad M'_{\mu\nu\lambda} + M'_{\nu\lambda\mu} + M'_{\lambda\mu\nu} = 0. \tag{5.83}$$

The second relation in (5.83) holds because of the (three dimensional) identity

$$\epsilon_{\mu\nu\alpha} T^\alpha_\lambda + \epsilon_{\nu\lambda\alpha} T^\alpha_\mu + \epsilon_{\lambda\mu\alpha} T^\alpha_\nu = \epsilon_{\mu\nu\lambda} T^\alpha_\alpha = 0, \tag{5.84}$$

since, by assumption, the tensor T^μ_ν is traceless. The tensor $M'_{\mu\nu\lambda}$, as we have seen in (5.44), has the associated Young tableau

$$M'_{\mu\nu\lambda} : \quad \begin{array}{|c|c|} \hline \mu & \lambda \\ \hline \nu \\ \cline{1-1} \end{array} \quad . \tag{5.85}$$

Conversely, the second rank mixed tensor T^μ_ν can be expressed in terms of the covariant third rank tensor $M'_{\mu\nu\lambda}$ (by inverting the relation (5.82)) as

$$T^\mu_\nu = \frac{1}{2} \epsilon^{\mu\alpha\beta} M'_{\alpha\beta\nu}, \tag{5.86}$$

where $\epsilon^{\mu\nu\lambda} = \epsilon_{\mu\nu\lambda}$ numerically. In this way, one can argue that it is sufficient to study the representation in terms of covariant tensor in the case of $SU(N)$. (Remember that $\det a \neq 1$ in the case of $U(N)$ and, therefore, this argument does not generalize to $U(N)$.)

For the group $SU(2)$ similar arguments lead to the conclusion that all irreducible representations can be obtained from the totally symmetric tensors $S_{\mu_1\mu_2\cdots\mu_n}$ where $\mu_1, \mu_2, \cdots, \mu_n = 1, 2$. The dimension of the representation, in this case, follows to correspond to (see discussion in the paragraph containing eqs. (5.47)-(5.48))

$$D(n) = \frac{(n+1)!}{n!} = n + 1, \tag{5.87}$$

which can be identified with $2j + 1$ for $j = \frac{n}{2}$, $n = 0, 1, 2, \cdots$ representing the familiar multiples of half integer values for the angular momentum j.

5.4 Product representation and branching rule

Let $U_1(a)$ and $U_2(a)$ respectively denote $d_1 \times d_1$ and $d_2 \times d_2$ irreducible matrix representations of the $U(N)$ group. Then, the tensor product

$$W(a) = U_1(a) \otimes U_2(a), \tag{5.88}$$

also defines a representation of $U(N)$ with dimension $d_1 d_2$ (namely, $(d_1 d_2) \times (d_1 d_2$ matrices). That the tensor product defines a representation is easily seen from (5.88)

$$
\begin{aligned}
W(a)W(b) &= (U_1(a) \otimes U_2(a)) \, (U_1(b) \otimes U_2(b)) \\
&= U_1(a)U_1(b) \otimes U_2(a)U_2(b) = U_1(ab) \otimes U_2(ab) \\
&= W(ab).
\end{aligned} \tag{5.89}
$$

The dimensionality follows from

$$W(e) = U_1(e) \otimes U_2(e) = E_{d_1 d_2} = \mathbb{1}_{d_1 d_2}, \tag{5.90}$$

where $E_{d_1 d_2} = \mathbb{1}_{d_1 d_2}$ represents the $d_1 d_2$ dimensional unit matrix. However, the product representation $W(a)$ is not, in general, irreducible.

Let us recall that for two independent (covariant) vectors x_μ and y_ν, their product defines a tensor $T_{\mu\nu} = x_\mu y_\nu$ which can be decomposed into the irreducible forms (see discussion following (5.15))

$$T_{\mu\nu} = S_{\mu\nu} + A_{\mu\nu}, \tag{5.91}$$

where $S_{\mu\nu}$ and $A_{\mu\nu}$ denote the symmetric and the anti-symmetric combinations respectively (see (5.17)). Regarding the product as $x_\mu \otimes y_\nu$, for more generalized settings, the discussion carries over without any modification. We can assign as before

$$x_\mu = \boxed{\mu}, \qquad S_{\mu\nu} = \boxed{\mu\ \ \nu}, \qquad A_{\mu\nu} = \boxed{\begin{matrix} \mu \\ \nu \end{matrix}}. \tag{5.92}$$

Dropping the indices μ, ν, we can write the decomposition in (5.91) symbolically as

$$\square \otimes \square = \boxed{\ \ \ } \oplus \boxed{\begin{matrix} \ \\ \ \end{matrix}}. \tag{5.93}$$

Similarly, if x_μ, y_ν, z_λ denote three independent vectors, then we have already noted in (5.35) that we can express

$$T_{\mu\nu\lambda} = S_{\mu\nu\lambda} + A_{\mu\nu\lambda} + (M_{\mu\nu\lambda} + M'_{\mu\nu\lambda}) - (M_{\nu\lambda\mu} + M'_{\nu\lambda\mu}). \quad (5.94)$$

If we consider the product of vectors as a tensor product $x_\mu \otimes y_\nu \otimes z_\lambda$. we can again write the decomposition in (5.94) as

$$\square \otimes \square \otimes \square = \square\square\square \oplus \begin{array}{c}\square\square\\\square\end{array} \oplus \begin{array}{c}\square\square\\\square\end{array} \oplus \begin{array}{c}\square\\\square\\\square\end{array}, \quad (5.95)$$

and so on.

In general the product of two irreducible representations can be decomposed into a direct sum of irreducible representations which can be shown by considering their character. For the irreducible representations $U_1(a)$ and $U_2(a)$ of the unitary group $U(N)$, their product $W(a) = U_1(a) \otimes U_2(a)$ can be written in a suitable basis in the fully reducible form

$$S^{-1}W(a)S = \begin{pmatrix} \boxed{W_1(a)} & 0 & 0 & \cdots & 0 \\ 0 & \boxed{W_2(a)} & 0 & \cdots & 0 \\ 0 & 0 & \boxed{W_3(a)} & \cdots & 0 \\ 0 & 0 & 0 & \ddots & \boxed{W_n(a)} \end{pmatrix},$$

$$(5.96)$$

for irreducible matrices $W_1(a), W_2(a), \cdots, W_n(a)$. Taking the trace of this relation we obtain

$$\text{Tr}\, W(a) = \sum_{j=1}^{n} \text{Tr}\, W_j(a),$$

$$\text{or,} \quad \chi(a) = \sum_{j=1}^{n} \chi_j(a). \quad (5.97)$$

Namely, the character of the product representation $\chi(a)$ is simply the sum of the characters $\chi_j(a)$ of the irreducible representations contained in the representation. On the other hand, we also know that we can write the character of the product representation as

$$\chi(a) = \text{Tr}\, W(a) = \text{Tr}\,(U_1(a) \otimes U_2(a))$$

$$= \chi(U_1(a))\chi(U_2(a)), \quad (5.98)$$

namely, as a product of the characters of the two irreducible representations. It follows from (5.97) and (5.98) that

$$\chi(a) = \chi(U_1(a))\chi(U_2(a)) = \sum_{j=1}^{n} \chi_j(a). \tag{5.99}$$

This also implies that

$$\text{Dim}\,(U_1(a))\,\text{Dim}\,(U_2(a)) = \sum_{j=1}^{n} \text{Dim}_j\,(a), \tag{5.100}$$

where Dim_j denotes the dimension of the jth irreducible block. Namely, there must be a matching of dimensions for the decomposition of a product representation into sums of irreducible representations.

To illustrate this, let us return to the example discussed in (5.93). In the present case, it is far easier to compute the characters of the individual representations directly rather than using Weyl's formula (5.75). For the purpose of the calculation of characters we can choose the basis in which a is diagonal (see (5.67))

$$a = \begin{pmatrix} \xi_1 & 0 & 0 & \cdots & 0 \\ 0 & \xi_2 & 0 & \cdots & 0 \\ 0 & 0 & \xi_3 & \cdots & 0 \\ 0 & 0 & 0 & \ddots & \xi_N \end{pmatrix}, \quad |\xi_\mu| = 1, \quad \mu = 1, 2, \cdots, N,$$

$$\tag{5.101}$$

so that the covariant vectors transform as (repeated indices are not summed)

$$x_\mu \rightarrow \xi_\mu x_\mu, \quad \mu = 1, 2, \cdots, N. \tag{5.102}$$

Therefore, the symmetric and anti-symmetric products transform as (for fixed μ, ν)

$$(x_\mu y_\nu \pm x_\nu y_\mu) \rightarrow \xi_\mu \xi_\nu (x_\mu y_\nu \pm x_\nu y_\mu), \tag{5.103}$$

and this leads to

$$\chi\left(\Box\right) = \sum_{\mu=1}^{N} \xi_\mu,$$

$$\chi\left(\Box\Box\right) = \sum_{\mu \geq \nu} \xi_\mu \xi_\nu = \frac{1}{2}\left(\left(\sum_{\mu=1}^{N} \xi_\mu\right)^2 + \sum_{\mu=1}^{N} (\xi_\mu)^2\right),$$

$$\chi\left(\square\!\square\atop\square\right) = \sum_{\mu>\nu}\xi_\mu\xi_\nu = \frac{1}{2}\left(\left(\sum_{\mu=1}^{N}\xi_\mu\right)^2 - \sum_{\mu=1}^{N}(\xi_\mu)^2\right). \tag{5.104}$$

This shows indeed that

$$\chi\left(\square\right)\chi\left(\square\right) = \chi\left(\square\square\right) + \chi\left(\square\atop\square\right). \tag{5.105}$$

Moreover, if we set $\xi_1 = \xi_2 = \cdots = \xi_N = 1$ (see (5.77)), then their dimensions can be computed from (5.104) correctly as (they also coincide with the earlier method discussed for calculating dimensions)

$$\mathrm{Dim}\,\square = \sum_{\mu=1}^{N} 1 = N, \qquad \mathrm{Dim}\,\square\square = \frac{1}{2}(N^2 + N),$$

$$\mathrm{Dim}\,{\square\atop\square} = \frac{1}{2}(N^2 - N). \tag{5.106}$$

For the general case, there exists a simple mechanical way of decomposing the product of two irreducible representations. For simplicity, we consider the case of

$$\square\!\square\atop\square \;\otimes\; \square\;. \tag{5.107}$$

For this, we label $\square = \boxed{a}$ and construct the product as

$$\square\!\square\atop\square \;\otimes\; \boxed{a} \;=\; \square\!\square\,\boxed{a}\atop\square \;\oplus\; {\square\!\square\atop\square\,\boxed{a}} \;\oplus\; {\square\!\square\atop\square\atop\boxed{a}}\;, \tag{5.108}$$

namely, we add $\boxed{a}$ to the Young tableau $\square\!\square\atop\square$ without violating the condition $f_1 \geq f_2 \geq \cdots \geq f_N \geq 0$ (see (5.46)). Such a rule can be obtained by computing the product of their characters and decomposing it into a sum of characters of irreducible representations (or by matching the dimensions as in (5.100)). For example,

$$\square\square \otimes \boxed{a} = \square\square\,\boxed{a} \oplus {\square\square\atop\boxed{a}}\;,$$

$$\square\atop\square \;\otimes\; \boxed{a} = {\square\,\boxed{a}\atop\square} \oplus {\square\atop\square\atop\boxed{a}}\;, \tag{5.109}$$

correponding to the decompositions of (see (5.29) and (5.34) as well as (5.106), for example, for the dimensions)

$$\chi\left(\square\square\right)\chi\left(\square\right)=\chi\left(\square\square\square\right)+\chi\left(\begin{array}{c}\square\square\\\square\end{array}\right),$$

and

$$\chi\left(\begin{array}{c}\square\\\square\end{array}\right)\chi\left(\square\right)=\chi\left(\begin{array}{c}\square\square\\\square\end{array}\right)+\chi\left(\begin{array}{c}\square\\\square\\\square\end{array}\right). \qquad (5.110)$$

We can, similarly, calculate

$$\boxed{}\otimes\boxed{}\otimes\boxed{a}=\left(\boxed{}\oplus\begin{array}{c}\square\\\square\end{array}\right)\otimes\boxed{a}$$

$$=\left(\boxed{}\otimes\boxed{a}\right)\oplus\left(\begin{array}{c}\square\\\square\end{array}\otimes\boxed{a}\right)$$

$$=\left(\boxed{\,a}\oplus\begin{array}{c}\square\square\\\boxed{a}\end{array}\right)\oplus\left(\begin{array}{cc}\square\square&\boxed{a}\\\end{array}\oplus\begin{array}{c}\square\\\square\\\boxed{a}\end{array}\right), $$

$$(5.111)$$

to obtain the decomposition given already as the decomposition of tensor $T_{\mu\nu\lambda}$ into its irreducible arguments (see (5.35)).

This procedure can be generalized to more complicated products and we give only a few more examples in the following

$$\begin{array}{c}\square\\\square\end{array}\otimes\boxed{a\ a}=\begin{array}{c}\square\ \boxed{a\ a}\\\square\end{array}\oplus\begin{array}{c}\square\ \boxed{a}\\\square\\\boxed{a}\end{array}. \qquad (5.112)$$

Note that in the decomposition of the product on the right hand side, we do not allow the diagram

$$\begin{array}{c}\square\\\square\\\boxed{a}\\\boxed{a}\end{array}\,, \qquad (5.113)$$

since, two $\boxed{a}$'s are initially in a totally symmetric configuration of $\boxed{a\ a}$. Similarly for products involving an anti-symmetric diagram, we proceed as follows: we write ⊟ as $\boxed{\begin{smallmatrix} a \\ b \end{smallmatrix}}$, and compute

$$\square \otimes \boxed{\begin{smallmatrix} a \\ b \end{smallmatrix}} = \boxed{\begin{smallmatrix} \square & a \\ & b \end{smallmatrix}} \oplus \boxed{\begin{smallmatrix} \square & a \\ b \end{smallmatrix}} \oplus \boxed{\begin{smallmatrix} \square \\ a \\ b \end{smallmatrix}} \,. \tag{5.114}$$

Once again, we do not allow the diagram

$$\boxed{\begin{smallmatrix} \square & a & b \\ \square \end{smallmatrix}}, \tag{5.115}$$

since $\boxed{a}$ and $\boxed{b}$ must be in the anti-symmetric configuration. Furthermore, we note that if $N \le 3$, then we can omit the last diagram

$$\boxed{\begin{smallmatrix} \square \\ \square \\ a \\ b \end{smallmatrix}} = (1,1,1,1), \tag{5.116}$$

as it violates the allowed Young tableau symbol (f_1, f_2, f_3) for $U(3)$ (namely, there is no nontrivial fourth rank anti-symmetric tensor in three dimensions).

Another subject of importance in physics is the branching rule, namely, how does a representation of a given group decompose into representations of its subgroups. Let us consider the covariant vector x_μ, $\mu = 1, 2, = N$ which corresponds to the basis of the irreducible N-dimensional matrix representation of the $U(N)$ group with the Young tableau $(1, \underbrace{0, 0, \cdots, 0}_{N-1})$ with "0" occurring $(N-1)$ times. But if we restrict ourselves to its subgroup $U(N-1)$, then it is no longer irreducible. Indeed

$$x_j, \qquad j = 1, 2, \cdots, N-1, \tag{5.117}$$

provides an irreducible $(N-1)$ dimensional representation of the $U(N-1)$ group, while $\phi = x_N$ remains invariant under $U(N-1)$. We can express this as

$$U(N) \to U(N-1),$$
$$(\square)_N = (\square)_{N-1} \oplus \{1\}_{N-1},$$
or, $\quad N = (N-1) \oplus 1, \tag{5.118}$

where $\{1\}$ represents the one-dimensional trivial representation of $U(N-1)$. Similarly, the symmetric tensor of $U(N)$

$$S_{\mu\nu} = S_{\nu\mu}, \qquad \mu,\nu = 1, 2, \cdots, N, \tag{5.119}$$

decomposes into the symmetric tensor

$$S_{jk} = S_{kj} = \left(\boxed{}\right)_{N-1}, \qquad j,k = 1, 2 \cdots, N-1, \tag{5.120}$$

and the vector

$$\phi_j = S_{jN} = \left(\square\right)_{N-1}, \qquad j = 1, 2, \cdots, N-1, \tag{5.121}$$

as well as the scalar $\phi_0 = S_{NN} = \{1\}_{N-1}$. In this case, we can write

$$U(N) \to U(N-1),$$

$$\left(\boxed{}\right)_N = \left(\boxed{}\right)_{N-1} \oplus \left(\square\right)_{N-1} \oplus \{1\}_{N-1}. \tag{5.122}$$

By reducing the characters of $U(N)$ as a sum of characters of $U(N-1)$ group with $\xi_N = 1$, Weyl has shown that the general case is obtained as follows

$$U_N((f_1, f_2 \cdots, f_N)) = \sum_{f_1^{(1)}, f_2^{(1)} \cdots, f_{N-1}^{(1)}} \oplus U_{N-1}((f_1^{(1)}, f_2^{(1)}, \cdots, f_{N-1}^{(1)})), \tag{5.123}$$

where the summation is over all $f_1^{(1)}, f_2^{(1)}, \cdots, f_{N-1}^{(1)}$ satisfying

$$f_1 \geq f_1^{(1)} \geq f_2 \geq f_2^{(1)} \geq f_3 \geq \cdots \geq f_{N-1} \geq f_{N-1}^{(1)} \geq f_N. \tag{5.124}$$

Here $(f_1^{(1)}, f_2^{(1)}, \cdots, f_{N-1}^{(1)})$ represents the Young diagram of $U(N-1)$ or the first level of reduction. Replacing N by $N-1$, we can further decompose $U(N-1)$ into $U(N-2)$ etc to obtain that an irreducible representation $(f_1, f_2 \cdots, f_N)$ of $U(N)$ can be decomposed to

$$U(N) \to U(N-1) \to U(N-2) \to \cdots \to U(1), \tag{5.125}$$

by considering the inverted pyramid

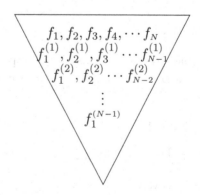

where $f_1^{(j)} \geq f_1^{(j+1)} \geq f_2^{(j)} \geq f_2^{(j+2)} \geq \cdots \geq f_{N-j}^{(j)}$ for $j = 0, 1, 2, \cdots, N-1$ and we have identified $f_i^{(0)} = f_i$, $i = 1, 2, \cdots, N$.

As an example, consider the case of $SU(2)$ group. As we have already noted (at the end of section 5.3), in this case, any irreducible representation can be obtained from the totally symmetric tensor $S_{\mu_1, \mu_2 \ldots \mu_f}$ corresponding to Young tableau of

$$\boxed{\mu_1}\ \boxed{\mu_2}\ \boxed{\cdots}\ \boxed{\mu_f} \qquad\qquad (5.126)$$
$$\leftarrow f\ \text{boxes} \rightarrow$$

with dimension $D(f) = f + 1$ (see (5.87) with $f = n$). In order to apply the branching rule, we embed this representation of $SU(2)$ into the $U(2)$ group to correspond to $(f, 0)$ with $f_1 = f, f_2 = 0$. Then, using the reduction $U(2) \rightarrow U(1)$, we see from (5.123) that we can write (recall that for $SU(N)$ we can always set $f_N = 0$ so that the Young tableau would have only $(N - 1)$ nontrivial entries)

$$U_2((f, 0)) = \sum_{f \geq f^{(1)} \geq 0} \oplus U_1((f^{(1)})). \qquad (5.127)$$

Setting $f = 2j$, $(j = 0, \frac{1}{2}, \frac{1}{3}, \frac{3}{2}, \cdots)$ with $D(f) = f + 1 = 2j + 1$ (see (5.87) and the discussion afterwards), and $f^{(1)} = j + m$, $(m = j, j - 1, \cdots, -j)$, this reproduces the familiar angular momentum basis vectors $|j, m\rangle$ as

$$|j, m\rangle = \begin{array}{c}\boxed{f, 0}\\ \boxed{f^{(1)}}\end{array}\!\!\!\!\!\nabla, \quad f = 2j,\ f' = j + m,\ m = j, j - 1, \cdots - j. \qquad (5.128)$$

In fact, the basis state $|j, m\rangle$ corresponds to a tensor component

$$S_{\underbrace{1 \cdots 1}_{j+m}\ \underbrace{2 \cdots 2}_{j-m}}, \qquad\qquad (5.129)$$

for the totally symmetric tensor $S_{\mu_1 \mu \ldots \mu_f}$, $\mu_1, \mu_2, \ldots, \mu_f = 1\ \text{or}\ 2$.

5.5 Representations of $SO(N)$ groups

Let us recall the fact that the orthogonal group $O(N)$ is the group of all linear transformations (see section 1.2.5 as well as section 2.1)

$$x_\mu \rightarrow \sum_{\nu=1}^{N} a_{\mu\nu} x_\nu \left(\equiv \sum_{\nu=1}^{N} a_\mu^{\ \nu} x_\nu \right), \qquad (5.130)$$

which leave the quadratic form

$$x_1^2 + x_2^2 + \cdots + x_N^2, \tag{5.131}$$

invariant. The $N \times N$ matrix $(a)_{\mu\nu}$ satisfies, as in (1.67),

$$a^T a = E_N, \tag{5.132}$$

where a^T denotes the transform matrix of "a". If we impose the additional condition of

$$\text{Det } a = 1, \tag{5.133}$$

on the matrix "a", then the resulting restricted subgroup is called the special orthogonal group $SO(N)$. If all the $a_{\mu\nu}$'s are real, $i.e.$

$$a_{\mu\nu}^* = a_{\mu\nu}, \qquad \mu, \nu = 1, 2, \cdots, N, \tag{5.134}$$

then the group is written as $O(N, \mathbb{R})$ or $SO(N, \mathbb{R})$ accordingly. However, if $a_{\mu\nu}$'s are complex numbers, then the resulting group is denoted as $O(N, \mathbb{C})$ or $SO(N, \mathbb{C})$. We note that $O(N, \mathbb{R})$ and $SO(N, \mathbb{R})$ are compact groups, while $O(N, \mathbb{C})$ and $SO(N, \mathbb{C})$ are non-compact. Another important group is the non-compact group $O(p, q)$ which is the group of all real linear transformations which leave the quadratic form

$$(x_1^2 + x_2^3 + \cdots + x_p^2) - (x_{p+1}^2 + \cdots + x_{p+q}^2), \tag{5.135}$$

invariant for real coordinates $x_1, x_2, \cdots, x_{p+q}$.

We will first discuss tensor representations of $O(N)$ and $SO(N)$ groups, which contain both $SO(N, \mathbb{R})$ and $SO(N, \mathbb{C})$ as special cases. Since $a^{-1} = a^T$ (see (5.132)), we see that the contravariant vector x^μ transforms as (see (5.9))

$$x^\mu \rightarrow (x')^\mu = \sum_{\nu=1}^N (a^{-1})_\nu{}^\mu \, x^\nu = \sum_{\nu=1}^N (a^T)_\nu{}^\mu \, x^\nu$$

$$= \sum_{\nu=1}^N a_\mu{}^\nu \, x^\nu = \sum_{\nu=1}^N a_{\mu\nu} x^\nu. \tag{5.136}$$

Therefore, it transforms exactly in the same way as the covariant the covariant vector x_μ. As a result, we may identify $x^\mu \equiv x_\mu$ so that there is no distinction between the covariant and contravariant vectors. Consequently, we need consider only covariant tensors

$T_{\mu_1\mu_2\cdots\mu_n}$. However, we note for example that the irreducible mixed tensor $T_{\mu\nu}^{\lambda} = T_{\lambda\mu\nu}$ must satisfy the traceless condition

$$\sum_{\lambda=1}^{n} T_{\lambda\nu}^{\lambda} = 0 = \sum_{\lambda=1}^{n} T_{\mu\lambda}^{\lambda} = 0,$$

$$\text{or,} \quad \sum_{\lambda=1}^{n} T_{\lambda\lambda\nu} = 0 = \sum_{\lambda=1}^{n} T_{\mu\lambda\lambda}, \tag{5.137}$$

and so on. In general, if $T_{\mu_1\mu_2\cdots\mu_n}$ denotes an irreducible tensor of the $O(N)$ group, we must impose the traceless conditions such as

$$\sum_{\lambda=1}^{n} T_{\lambda\lambda\mu_3\cdots\mu_n} = \sum_{\lambda=1}^{n} T_{\lambda\mu_2\lambda\mu_4\cdots\mu_n} = \sum_{\lambda=1}^{n} T_{\mu_1\mu_2\lambda\lambda\mu_5\cdots\mu_n} = 0, \tag{5.138}$$

etc for any pair of indices μ_j and $\mu_k, j \neq k$ of the tensor.

Let us now restrict ourselves to the $SO(N)$ subgroup. As in the discussion of the $SU(N)$ group, the completely anti-symmetric Levi-Civita tensor $\epsilon_{\mu_1\mu_2\cdots\mu_n} = \epsilon^{\mu_1\mu_2\cdots\mu_n}$ is invariant under $SO(N)$ (recall that $\det a = 1$ in this case). Together with the fact that there is no distinction between the covariant and the contravariant tensors, we can further restrict the irreducible tensors of $SO(N)$ as follows. We have now two different cases to discuss, depending upon N being an even or an odd integer. Let us consider first the case of $N = 2M + 1$ to be an odd integer.

For simplicity, consider the case of $M = 2$ and $N = 5$. (The case $M = 1$ and $N = 3$ has been discussed in detail earlier following (5.81).) Note that the totally anti-symmetric tensor $A_{\mu\nu\lambda}$, $\mu, \nu, \lambda = 1, 2, \cdots, 5$, has the Young tableau representation

$$\begin{array}{|c|}
\hline
\mu \\
\hline
\nu \\
\hline
\lambda \\
\hline
\end{array} \quad . \tag{5.139}$$

However, $A_{\mu\nu\lambda}$ is equivalent to another anti-symmetric tensor $\phi_{\alpha\beta} = -\phi_{\beta\alpha}$ given by

$$\phi_{\alpha\beta} = \frac{1}{3!} \sum_{\mu\nu\lambda=1}^{5} \epsilon_{\alpha\beta\mu\nu\lambda} A_{\mu\nu\lambda}. \tag{5.140}$$

Conversely, we can express

$$A_{\mu\nu\lambda} = \frac{1}{2!} \sum_{\alpha,\beta=1}^{5} \epsilon_{\mu\nu\lambda\alpha\beta} \phi_{\alpha\beta}. \tag{5.141}$$

As a result, we see that we can make the Young tableaux identification

$$
\begin{array}{|c|}
\hline
\mu \\
\hline
\nu \\
\hline
\lambda \\
\hline
\end{array}
\simeq
\begin{array}{|c|}
\hline
\alpha \\
\hline
\beta \\
\hline
\end{array}
. \tag{5.142}
$$

Correspondingly, we see that we need consider only irreducible tensors with smaller Young tableau $(f_1, f_2, \cdots, f_M)$ of at most M entries instead of the larger Young tableaux $(f_1, f_2, \cdots, f_N)$ with $N = 2M+1$ entries.

Next, let us discuss the case of $N = 2M$ being even. For this case, using the Levi-Civita tensor, we can also show that we need consider only tensors with smaller Young tableaux $(f_1, f_2, \cdots f_M)$ of at most M entries. However, we encounter a little complication of the following nature. Again for simplicity, we discuss only the simplest case of $M = 2$ and $N = 4$. (The case $M = 1$ and $N = 2$ has been discussed at length earlier in this chapter.) If $G_{\mu\nu} = -G_{\nu\mu}$, $\mu, \nu = 1, 2, 3, 4$, is an anti-symmetric tensor with the Young tableaux

$$
\begin{array}{|c|}
\hline
\mu \\
\hline
\nu \\
\hline
\end{array}
, \tag{5.143}
$$

then we can define another antisymmetric tensor

$$
{}^*G_{\mu\nu} = \frac{1}{2i} \sum_{\alpha,\beta=1}^{4} \epsilon_{\mu\nu\alpha\beta} G_{\alpha\beta}, \tag{5.144}
$$

with the same Young tableaux. However, a difficulty arises in that $G_{\mu\nu}$ and ${}^*G_{\mu\nu}$ are not irreducible under $SO(4)$. For example, we note that if we define

$$
F_{\mu\nu}^{(\pm)} = \frac{1}{2}(G_{\mu\nu} \pm {}^*G_{\mu\nu}), \tag{5.145}
$$

then it satisfies

$$
{}^*F_{\mu\nu}^{(\pm)} = \frac{1}{2!} \sum_{\alpha,\beta=1}^{4} \epsilon_{\mu\nu\alpha\beta} F_{\alpha\beta}^{(\pm)} = \pm F_{\mu\nu}^{(\pm)}. \tag{5.146}
$$

The tensors $F_{\mu\nu}^{(\pm)}$ are called selfdual or anti-selfdual tensors of $SO(4)$ and are irreducible tensors of $SO(4)$. In comparison to $G_{\mu\nu}$, which has 6-independent components, $F_{\mu\nu}^{(\pm)}$ have only 3-independent components each. This is slightly more familiar from electromagnetism where the relevant group is $SO(3,1)$ (the Lorentz group) instead of the $SO(4)$ group being discussed here. In conclusion, we note that for the case of $N = 2M$, if the tensor $T_{\mu_1 \cdots \mu_n}$ has the Young tableau of form $(f_1, f_2, \cdots f_M)$ with $f_M > 0$, then we must impose similar duality restrictions on the tensor in order to determine its irreducibility.

So far, we have discussed the groups $O(N)$ are $SO(N)$ and without making any distinction between $SO(N, \mathbb{R})$ are $SO(N, \mathbb{C})$. For the case of $SO(N, \mathbb{R})$, if $T_{\mu_1 \mu_2 \cdots \mu_n}$ is an irreducible tensor and transform as

$$T_{\mu_1 \cdots \mu_n} \to T'_{\mu_1 \cdots \mu_n} = \sum_{\nu_1 \nu_2 \cdots \nu_n = 1}^{N} a_{\mu_1 \nu_1} a_{\mu_2 \nu_2} \cdots a_{\mu_n \nu_n} T_{\nu_1 \cdots \nu_n}, \quad (5.147)$$

then its complex conjugate $T^*_{\mu_1 \mu_2 \cdots \mu_n}$ transforms exactly in the same way as $T_{\mu_1 \cdots \mu_n}$ (since $a_{\mu\nu}$ are real in this case). In other words, they give the same equivalent irreducible representation of $SO(N, \mathbb{R})$. However, in the case of $SO(N, \mathbb{C})$, this is not case and the two give inequivalent representations. Namely, if $U(a)$ is a $n \times n$ irreducible matrix representation of $SO(N, \mathbb{C})$, then its complex conjugate

$$W(a) = U^*(a) \quad \text{(equivalently } U(a^*)), \quad (5.148)$$

gives an inequivalent irreducible representation of the same dimension, called the complex conjugate representation. We will not go into any further discussion on this topic.

It may be instructive to discuss briefly the case of $SO(3, \mathbb{R})$, where we need consider only the completely symmetric tensor $S_{\mu_1 \mu_2 \cdots \mu_n}$ with $\mu_1, \mu_2 \cdots, \mu_n = 1, 2, 3$ satisfying the traceless condition

$$\sum_{\lambda=1}^{3} S_{\lambda \lambda \mu_3 \cdots \mu_n} = 0, \quad (5.149)$$

for $n \geq 2$. The dimension of such a representation (the number of independent components of such a tensor) is exactly computed to be

$$D = 2n + 1. \quad (5.150)$$

For example, for $n = 2$, the tensor $S_{\mu\nu} = S_{\nu\mu}$ satisfying $\sum_{\mu=1}^{3} S_{\mu\mu} = 0$ has ($\frac{1}{2} \times 3 \times 4 - 1 = 5 = 2 \times 2 + 1$) 5 independent components. Comparing (5.150) with the dimensions of representations of the standard angular momentum algebra of quantum mechanics ($D = 2j + 1$), we see that $n = j = 0, 1, 2, 3, \cdots$, i.e. these tensors realize all integer angular momentum states, but not the cases of half-integer states with $j = \frac{1}{2}, \frac{3}{2}, \cdots$, in contrast to the case of $SU(2)$ discussed already. Properly speaking, the half-integer representations are not representations of $SO(3)$, rather they are the representations of the covering group spin(3), since $SO(3)$ is not simply connected. Nevertheless, in view of their importance in physics, we will discuss briefly the so-called double-valued spinor representations of $SO(N)$ group (for $N = 3$) in the following section.

5.6 Double valued representation of $SO(3)$

Let us recall that a general rotation of the three dimensional (Euclidean) vector can be written as

$$\begin{pmatrix} x \\ y \\ z \end{pmatrix} \rightarrow \begin{pmatrix} x' \\ y' \\ z' \end{pmatrix} = R(\theta, \phi, \psi) \begin{pmatrix} x \\ y \\ z \end{pmatrix}, \qquad (5.151)$$

where the (3×3) rotation matrix $R(\theta, \phi, \psi)$ can be parameterized in terms of the (three) Euler's angles ψ, ϕ and ψ as

$$R(\theta, \phi, \psi) =$$

$$\begin{pmatrix} cs\phi cs\psi - cs\theta sn\phi sn\psi & -sn\phi cs\psi - cs\theta cs\phi sn\psi & sn\theta sn\psi \\ -cs\theta sn\psi + cs\theta sn\phi cs\psi & -sn\phi sn\psi + cs\theta cs\phi cs\psi & -sn\theta cs\psi \\ sn\theta sn\phi & sn\theta cs\phi & cs\theta \end{pmatrix}.$$

$$(5.152)$$

Here we have abbreviated "cosine" by "cs" and "sine" by "sn" for simplicity. Furthrmore, we note that the ranges of the Euler angles are given by $0 \le \theta \le \pi, 0 \le \phi \le 2\pi, 0 \le \psi \le 2\pi$. The rotation matrix in (5.152) can also be written as a product simpler matrices

$$R(\theta, \phi, \psi) = R_3(\psi) R_1(\theta) R_3(\phi), \qquad (5.153)$$

where R_1, R_3 denote respectively rotations around the x, z axes, namely,

$$R_1(\theta) = \begin{pmatrix} 1 & 0 & 0 \\ 0 & \cos\theta & -\sin\theta \\ 0 & \sin\theta & \cos\theta \end{pmatrix},$$

$$R_3(\phi) = \begin{pmatrix} \cos\phi & -\sin\phi & 0 \\ \sin\phi & \cos\phi & 0 \\ 0 & 0 & 1 \end{pmatrix}. \qquad (5.154)$$

Considering infinitesimal rotations, we have the generators of the three rotations around the x, y, z axes (or the basis elements of the Lie algebra) given by

$$J_1 = \begin{pmatrix} 0 & 0 & 0 \\ 0 & 0 & i \\ 0 & -i & 0 \end{pmatrix}, \quad J_2 = \begin{pmatrix} 0 & 0 & i \\ 0 & 0 & 0 \\ -i & 0 & 0 \end{pmatrix},$$

$$J_3 = \begin{pmatrix} 0 & i & 0 \\ -i & 0 & 0 \\ 0 & 0 & 0 \end{pmatrix}, \tag{5.155}$$

and they can be easily checked to satisfy the Lie algebra of $so(3)$, namely,

$$[J_i, J_j] = i\epsilon_{ijk}J_k, \quad i, j, k = 1, 2, 3. \tag{5.156}$$

It is now straightforward to verify that we can write

$$R_1(\theta) = e^{i\theta J_1}, \quad R_3(\phi) = e^{i\phi J_3}, \tag{5.157}$$

so that we can write (see (5.153) and (5.157))

$$R(\theta, \phi, \psi) = e^{i\psi J_3} e^{i\theta J_1} e^{i\phi J_3}. \tag{5.158}$$

It is worth noting here that this discussion has been within the context of the defining (3×3) matrices. For a general representation of $SO(3)$, we can replace the generators J_1, J_2, J_3 by the corresponding $(2j + 1) \times (2j + 1)$ dimensional matrices.

Let us next note that for rotations around the z-axis

$$R_3(\phi) = e^{i\phi J_3}, \tag{5.159}$$

since ϕ and $\phi + 2\pi$ realize the same rotation around the z-axis, we must have

$$R_3(\phi + 2\pi) = R_3(\phi), \quad \text{or}, \quad e^{2\pi i J_3} = \mathbb{1}, \tag{5.160}$$

which can be satisfied only for integer values of the angular momentum eigenvalues j. We can verify that (5.160) holds for the 3×3 matrix representation given in (5.155). However, for the two dimensional (spinor) representation of the Lie algebra, we have

$$J_i = \frac{1}{2}\sigma_i, \quad i = 1, 2, 3, \tag{5.161}$$

where σ_i denote the three Pauli matrices. In this case we have

$$R_3(\phi) = e^{\frac{i}{2}\phi\sigma_3} = \left(\cos\frac{\phi}{2}\right)\mathbb{1} + i\left(\sin\frac{\phi}{2}\right)\sigma_3$$

$$= \begin{pmatrix} e^{\frac{i}{2}\phi} & 0 \\ 0 & e^{-\frac{i}{2}\phi} \end{pmatrix}. \tag{5.162}$$

In this case, it is clear that

$$R_3(\phi + 2\pi) = -R_3(\phi), \quad R_3(\phi + 4\pi) = R_3(\phi), \tag{5.163}$$

namely, it represents a double valued representation.

5.7 References

H. Weyl, *The Classical Groups*, Princeton University Press, Princeton, NJ (1939).

L. C. Biedenharn and J. D. Louck, *Angular Momentum in Quantum Physics*, Cambridge University Press (1981).

CHAPTER 6

Clifford algebra

6.1 Clifford algebra

We start with the defining relation for a Clifford algebra (named after the British mathematician William Kingdon Clifford) which is generated by matrices $\gamma_1, \gamma_2 \cdots, \gamma_N$ satisfying the algebraic relation

$$\gamma_\mu \gamma_\nu + \gamma_\nu \gamma_\mu = 2\delta_{\mu\nu} \mathbb{1}, \tag{6.1}$$

with $\mu, \nu = 1, 2, \cdots, N$. Setting $\mu = \nu$ in (6.1), we note that

$$\gamma_\mu^2 = \mathbb{1}, \tag{6.2}$$

for any fixed $\mu = 1, 2, \cdots, N$. Also, if $\mu \neq \nu$, it follows from (6.1) that

$$\gamma_\mu \gamma_\nu = -\gamma_\nu \gamma_\mu, \tag{6.3}$$

so that γ_μ and γ_ν with $\mu \neq \nu$ anti-commute with each other. The matrices γ_μ are known as the generators of the Clifford algebra (6.1). The study of Clifford algebra is important in physics in connection with Dirac equation which we will discuss in the next chapter.

6.1.1 Dimension of the representation. The matrix dimensions of the generators γ_μ (and, therefore, a representation of the Clifford algebra) can be determined from the relations (6.1)-(6.3) as follows. In this section we will restrict ourselves to irreducible representations and reducible representations will be discussed in the next section. For the irreducible representations, there are at least two ways of determining the dimension and here we will discuss one of the methods. The second and more direct determination will be carried out in the next section.

Let us assume that the matrices γ_μ provide an irreducible representation of the Clifford algebra (6.1). From the matrices $\mathbb{1}$ and γ_μ, we can construct the set of matrices

$$\mathbb{1}, \; \gamma_\mu, \; \gamma_\mu\gamma_\nu \; (\mu<\nu), \; \gamma_\mu\gamma_\nu\gamma_\lambda \; (\mu<\nu<\lambda), \; \cdots, \; \gamma_1\gamma_2\gamma_3\cdots\gamma_N. \qquad (6.4)$$

Any higher order product of the γ_μ matrices clearly reduces to one of these structures because of (6.2) and (6.3). The number of such matrices in each category is easily calculated to be respectively (the binomial coefficients)

$$1, \; N, \; \frac{1}{2!}N(N-1), \; \frac{1}{3!}N(N-1)(N-2), \; \cdots, \; 1, \qquad (6.5)$$

so that the total number of such matrices in the set (6.4) is given by

$$1 + N + \frac{1}{2!}N(N-1) + \frac{1}{3!}N(N-1)(N-2) + \cdots = 2^N. \qquad (6.6)$$

We can easily verify that the matrices in (6.4) are linearly independent in the following manner. Let us denote the set of matrices collectively as Γ_A with $A = 0, 1, 2, \cdots, 2^N - 1$ and note that

$$(\Gamma_A)^2 = \epsilon_A \mathbb{1}, \qquad (6.7)$$

with $\epsilon_A = \pm 1$ for any $A = 0, 1, 2, \cdots, 2^N - 1$. For example, we see that for $\mu \neq \nu$

$$(\gamma_\mu\gamma_\nu)^2 = \gamma_\mu\gamma_\nu\gamma_\mu\gamma_\nu = \gamma_\mu(-\gamma_\mu\gamma_\nu)\gamma_\nu = -\mathbb{1}, \qquad (6.8)$$

Equation (6.7) implies that Γ_A^{-1} exists and can be identified with (recall that $\epsilon_A^2 = 1$)

$$\Gamma_A^{-1} = \epsilon_A \Gamma_A. \qquad (6.9)$$

Also, for any two Γ_A and Γ_B, there exists Γ_C in the set of matrices satisfying

$$\Gamma_A\Gamma_B = \epsilon_{AB}\Gamma_C, \qquad (6.10)$$

with $\epsilon_{AB} = \pm 1$. Moreover, $\Gamma_C = \mathbb{1}$ is possible only for $\Gamma_A = \Gamma_B$. For example, we note that if $\Gamma_A = \gamma_\lambda$ and $\Gamma_B = \gamma_\mu\gamma_\nu$ with $\mu < \nu$, then

$$\Gamma_A\Gamma_B = \gamma_\lambda\gamma_\mu\gamma_\nu = \begin{cases} \gamma_\nu, & \text{if } \lambda = \mu \\ -\gamma_\mu, & \text{if } \lambda = \nu, \\ \gamma_\lambda\gamma_\mu\gamma_\nu & \text{if } \lambda < \mu < \nu, \\ -\gamma_\mu\gamma_\lambda\gamma_\nu & \text{if } \mu < \lambda < \nu, \\ \gamma_\mu\gamma_\nu\gamma_\lambda, & \text{if } \mu < \nu < \lambda. \end{cases} \qquad (6.11)$$

Further, for a fixed Γ_B, a set consisting of all $\Gamma_A\Gamma_B$ covers the original set of matrices in (6.4) $\Gamma_A = \mathbb{1}, \gamma_\mu, \gamma_\mu\gamma_\nu, \cdots, \gamma_1\gamma_2\gamma_3\cdots\gamma_N$, up to signs.

To proceed we have to now make a distinction between the cases of $N = 2M$ (even) and $N = 2M + 1$ (odd) for the following reason. Let us denote

$$\Lambda = \gamma_1\gamma_2\gamma_3\cdots\gamma_N. \tag{6.12}$$

It now follows from (6.3) that

$$\gamma_\mu\Lambda = \begin{cases} \Lambda\gamma_\mu, & \text{if } N = 2M + 1 \text{(odd)}, \\ -\Lambda\gamma_\mu, & \text{if } N = 2M \text{(even)} \end{cases}. \tag{6.13}$$

In the case of $N = 2M + 1$, then, this implies that

$$[\Lambda, \Gamma_A] = 0, \tag{6.14}$$

for any Γ_A in the set (6.4). Moreover, we have that

1. $\mathrm{Tr}\,\Gamma_A = 0$, unless $\Gamma_A = \mathbb{1}$ for the case if $N = 2M$ (even),

2. $\mathrm{Tr}\,\Gamma_A = 0$, unless $\Gamma_A = \mathbb{1}$ or $\Gamma_A = \Lambda$ for the case of $N = 2M + 1$ (odd).

In order to show this, we note that for any fixed Γ_A such that $\Gamma_A \neq \mathbb{1}$ for N even, and $\Gamma_A \neq \mathbb{1}, \Lambda$ for N odd, we can always find some Γ_B satisfying

$$\Gamma_A = -\Gamma_B\Gamma_A\Gamma_B^{-1}, \quad \text{or,} \quad \Gamma_A\Gamma_B = -\Gamma_B\Gamma_A. \tag{6.15}$$

This can be checked directly, for example, if $\Gamma_A = \gamma_\mu$, we can choose $\Gamma_B = \gamma_\nu$ where $\nu \neq \mu$ and they will anti-commute. (Similarly, if $\Gamma_A = \gamma_\mu\gamma_\nu, \mu \neq \nu$, we can choose $\Gamma_B = \gamma_\mu$ or γ_ν and the two will anticommute and so on.) Taking the trace of both sides in (6.15), we obtain

$$\mathrm{Tr}\,\Gamma_A = -\mathrm{Tr}\,(\Gamma_B\Gamma_A\Gamma_B^{-1}) = -\mathrm{Tr}\,(\Gamma_A\Gamma_B^{-1}\Gamma_B)$$
$$= -\mathrm{Tr}\,(\Gamma_A\mathbb{1}) = -\mathrm{Tr}\,\Gamma_A = 0, \tag{6.16}$$

where we have used the cyclicity under trace. This leads to the fact that Γ_A's are linearly independent for the case of N even. For example, suppose that there exist constants C_A satisfying

$$\sum_{A=0}^{2^N-1} C_A\Gamma_A = 0, \tag{6.17}$$

then, multiplying with an arbitrary Γ_B and taking the trace of both sides, we obtain

$$\sum_{A=0}^{2^N-1} C_A \, \text{Tr} \, (\Gamma_A \Gamma_B) = 0. \tag{6.18}$$

On the other hand, when $N = 2M$ (even), we note that $\text{Tr} \, (\Gamma_A \Gamma_B) = \text{Tr} \, (\epsilon_{AB} \Gamma_C) = 0$ (see from (6.10) and (6.16)) unless $\Gamma_C = \mathbb{1}$ which is possible only if $\Gamma_A = \Gamma_B$. Therefore, in this case we can write

$$\sum_{A=0}^{2^N-1} C_A \, \text{Tr} \, (\Gamma_A \Gamma_B) = \sum_{A=0}^{2^N-1} C_A \delta_{AB} = C_B = 0, \tag{6.19}$$

for any B. As a result, the constant coefficients in (6.17) cannot be nontrivial which proves the linear independence of the set of matrices Γ_A for $N = 2M$ (even). Since we have 2^N linearly independent matrices Γ_A, the Burnside theorem implies that Γ_A's must be $2^M \times 2^M$ matrices if $N = 2M$ (even) and if Γ_A's are irreducible. However, we will prove the same result more directly in the next section.

For the case of $N = 2M + 1$ (odd), we have already noted in (6.14) that

$$[\Lambda, \Gamma_A] = 0, \tag{6.20}$$

for all Γ_A in the set (6.4). If the set of matrices Γ_A is irreducible, then by Schur's lemma we can identify

$$\Lambda = c_{2M+1} \mathbb{1}. \tag{6.21}$$

Using the anti-commutativity of the matrices γ_μ (see (6.3)), it follows that

$$\Lambda^2 = (\gamma_1 \gamma_2 \cdots \gamma_{2M+1})^2 = (-1)^{M(2M+1)} \mathbb{1}, \tag{6.22}$$

so that we determine (from (6.21))

$$c_{2M+1}^2 = (-1)^{M(2M+1)}, \quad \text{or,} \quad c_{2M+1} = (-1)^{\frac{M(2M+1)}{2}}. \tag{6.23}$$

As a result, $c_{2M+1} = \pm 1$ or $c_{2M+1} = \pm\sqrt{-1}$ depending on the value of M. Using (6.21) as well as (6.2), it follows now that

$$\Lambda = \gamma_1 \gamma_2 \cdots \gamma_{2M+1} = c_{2M+1} \mathbb{1},$$

$$\text{or,} \quad \gamma_{2M+1} = c_N^{-1} \gamma_1 \gamma_2 \cdots \gamma_{2M}. \tag{6.24}$$

Namely, when $N = 2M + 1$ (odd), the algebra reduces to that of an even dimensional ($N = 2M$) Clifford algebra generated by $\gamma_1, \gamma_2, \cdots, \gamma_{2M}$. As a result, the matrices γ_μ (Γ_A) are now realized as $2^M \times 2^M$ matrices for an irreducible realization.

6.1.2 Reducible representation. Let us next show that a reducible representation of the Clifford algebra is fully reducible. Let us assume that γ_μ denotes a reducible representation of the Clifford algebra (6.1) of the lower triangular form (see section 2.2)

$$\gamma_\mu = \begin{pmatrix} \gamma_\mu^{(1)} & 0 \\ N_\mu' & \gamma_\mu^{(2)} \end{pmatrix}, \tag{6.25}$$

where $\gamma_\mu^{(1)}$ and $\gamma_\mu^{(2)}$ denote irreducible representations of (6.1). Namely, they satisfy

$$\gamma_\mu^{(1)}\gamma_\nu^{(1)} + \gamma_\nu^{(1)}\gamma_\mu^{(1)} = 2\delta_{\mu\nu}\mathbb{1}^{(1)},$$
$$\gamma_\mu^{(2)}\gamma_\nu^{(2)} + \gamma_\nu^{(2)}\gamma_\mu^{(2)} = 2\delta_{\mu\nu}\mathbb{1}^{(2)}. \tag{6.26}$$

(The off-diagonal elements also have to satisfy a relation such as (2.94), with N' identified with $\widetilde{N}$, in order that γ_μ denotes a representation of (6.1). However, this is not important for our discussion.) We can now construct the set of matrices Γ_A using $\mathbb{1}$ and γ_μ which will all be lower triangular. Let us next define a block diagonal (fully reducible) representation of (6.1) of the form

$$\widetilde{\gamma}_\mu = \begin{pmatrix} \gamma_\mu^{(1)} & 0 \\ 0 & \gamma_\mu^{(2)} \end{pmatrix}, \tag{6.27}$$

so that we can construct a second set of matrices $\widetilde{\Gamma}_A$ from $\mathbb{1}$ and $\widetilde{\gamma}_\mu$ which will all be block diagonal.

Let us now construct the matrix

$$S = \sum_{A=0}^{2^N-1} \left(\widetilde{\Gamma}_A\right)^{-1} \Gamma_A. \tag{6.28}$$

It follows now that, for any fixed B,

$$\left(\widetilde{\Gamma}_B\right)^{-1} S\Gamma_B = \sum_{A=0}^{2^N-1} \left(\widetilde{\Gamma}_A\widetilde{\Gamma}_B\right)^{-1} (\Gamma_A\Gamma_B). \tag{6.29}$$

On the other hand, we know (see (6.10)) that

$$\Gamma_A\Gamma_B = \epsilon_{AB}\Gamma_C, \quad \widetilde{\Gamma}_A\widetilde{\Gamma}_B = \epsilon_{AB}\widetilde{\Gamma}_C, \tag{6.30}$$

for the same ϵ_{AB} and the same C (depending on A for a fixed B). As a result, it follows from (6.29) and (6.30) that

$$\left(\widetilde{\Gamma}_B\right)^{-1} S\Gamma_B = \sum_{C=0}^{2^N-1} \left(\widetilde{\Gamma}_C\right)^{-1} \Gamma_C = S. \tag{6.31}$$

Since the matrices Γ_A are lower triangular and the matrices $(\tilde{\Gamma}_A)^{-1} = \epsilon_A \tilde{\Gamma}_A$ are block diagonal, by a direct straightforward calculation we can determine from (6.28) that

$$S = \begin{pmatrix} 2^N \mathbb{1}^{(1)} & 0 \\ \overline{N} & 2^N \mathbb{1}^{(2)} \end{pmatrix}, \tag{6.32}$$

for some rectangular matrix $\overline{N}$ whose exact form is not important. It follows now that

$$\det S \neq 0, \tag{6.33}$$

so that the inverse S^{-1} exists and we obtain from (6.31) that

$$\tilde{\Gamma}_B = S\Gamma_B S^{-1}, \tag{6.34}$$

for all B. This proves the full reducibility of the representation since we can always find a similarity matrix S which will bring γ_μ in (6.25) into the block diagonal form (see section 2.2). A similar argument can be carried through if we choose the matrix γ_μ in (6.25) to be upper triangular instead. This proves that any reducible representation of the Clifford algebra is fully reducible.

We will now describe an alternate and direct way of calculating the dimension of an irreducible representation. As is clear from the above discussion, this is sufficient to determine the dimension of any reducible representation (since they are fully reducible). Let us assume that the generators γ_μ are $d \times d$ irreducible matrices. For an arbitrary $d \times d$ matrix Y, let us define

$$S = \sum_{A=0}^{2^N-1} (\Gamma_A)^{-1} Y \Gamma_A. \tag{6.35}$$

It now follows that for a fixed B

$$(\Gamma_B)^{-1} S \Gamma_B = \sum_{A=0}^{2^N-1} (\Gamma_A \Gamma_B)^{-1} Y (\Gamma_A \Gamma_B)$$

$$= \sum_{C=0}^{2^N-1} (\Gamma_C)^{-1} Y \Gamma_C = S,$$

$$\text{or,} \quad S\Gamma_B = \Gamma_B S, \tag{6.36}$$

where we have used $\Gamma_A \Gamma_B = \epsilon_{AB} \Gamma_C$ in the intermediate step. Since S commutes with all the matrices in the set, by Schur's lemma we conclude that S must be proportional to the identity matrix so that we can write

$$S = \sum_{A=0}^{2^N-1} (\Gamma_A)^{-1} Y \Gamma_A = \lambda \mathbb{1},$$

$$\text{or,} \quad \text{Tr}\, S = \sum_{A=0}^{2^N-1} \text{Tr}\, Y = \lambda d,$$

$$\text{or,} \quad \lambda = \frac{2^N}{d}\, \text{Tr}\, Y. \tag{6.37}$$

As a result, the matrix S has the form

$$S = \sum_{A=0}^{2^N-1} (\Gamma_A)^{-1} Y \Gamma_A = \left(\frac{2^N}{d}\, \text{Tr}\, Y\right) \mathbb{1}. \tag{6.38}$$

Taking the (j,m) matrix element of both sides in (6.38) we obtain (summation over the repeated indices k,l is understood)

$$\sum_{A=0}^{2^N-1} (\Gamma_A)^{-1}_{jk}\, Y_{kl}\, (\Gamma_A)_{lm} = \frac{2^N}{d}\, \delta_{jm} Y_{kk} = \frac{2^N}{d}\, \delta_{jm} \delta_{kl} Y_{kl},$$

$$\text{or,} \quad \sum_{A=0}^{2^N-1} (\Gamma_A)^{-1}_{jk}\, (\Gamma_A)_{lm} = \frac{2^N}{d}\, \delta_{jm} \delta_{kl}, \tag{6.39}$$

where $j, k, l, m = 1, 2, \cdots, d$ and we have used the fact that Y is an arbitrary $d \times d$ matrix. If we set $j = k, l = m$ and sum over these two indices, we obtain

$$\sum_{A=0}^{2^N-1} \text{Tr}\, (\Gamma_A)^{-1}\, \text{Tr}\, \Gamma_A = 2^N. \tag{6.40}$$

There are two cases to consider, namely, N even and N odd. For $N = 2M$ (even), we know that $\text{Tr}\, \Gamma_A = 0$ except for $\Gamma_0 = \mathbb{1}$ for which $\text{Tr}\, \Gamma_0 = \text{Tr}\, (\Gamma_0)^{-1} = d$. Therefore, in this case (6.40) leads to

$$d^2 = 2^N = 2^{2M}, \quad \text{or,} \quad d = 2^M, \tag{6.41}$$

which coincides with our earlier result (see discussion after (6.18)). For the case of $N = 2M + 1$, we have $\text{Tr}\, \Gamma_A = 0$ unless $\Gamma_A = \mathbb{1}$ or $\Gamma_A = \Lambda$ and in each of the two cases $\text{Tr}\, \Gamma_A \propto d$ (see (6.24)). Consequently, (6.40) leads to

$$2d^2 = 2^N = 2^{2M+1}, \quad \text{or,} \quad d = 2^M, \tag{6.42}$$

as we have already seen (see discussions after (6.24)).

6.1.3 Irreducible representation and its uniqueness.

We will next show that an irreducible representation of the Clifford algebra is unique for the case of $N = 2M$ (even), while there exist two inequivalent irreducible representations of the same dimensionality for the case of $N = 2M+1$ (odd). Let γ_μ and $\widetilde{\gamma}_\mu$ be two inequivalent representations of the Clifford algebra of dimensionlity d. We can construct two sets of matrices Γ_A and $\widetilde{\Gamma}_A$ from $(\mathbb{1}, \gamma_\mu)$ and $(\mathbb{1}, \widetilde{\gamma}_\mu)$ respectively. Then, for any arbitrary $d \times d$ matrix Y, let us define

$$S = \sum_{A=0}^{2^N-1} (\Gamma_A)^{-1} Y \widetilde{\Gamma}_A. \tag{6.43}$$

As before (see (6.36)), we can show that, for an arbitrary fixed B, the matrix S satisfies

$$(\Gamma_B)^{-1} S \widetilde{\Gamma}_B = S, \quad \text{or,} \quad S\widetilde{\Gamma}_B = \Gamma_B S. \tag{6.44}$$

If Γ_B and $\widetilde{\Gamma}_B$ denote inequivalent representations, then we must have $S = 0$ identically by the second Schur's lemma so that (repeated indices k, l are summed)

$$\sum_{A=0}^{2^N-1} (\Gamma_A)^{-1} Y \widetilde{\Gamma}_A = 0,$$

$$\text{or,} \quad \sum_{A=0}^{2^N-1} (\Gamma_A)_{jk}^{-1} \left(\widetilde{\Gamma}_A\right)_{lm} Y_{kl} = 0,$$

$$\text{or,} \quad \sum_{A=0}^{2^N-1} (\Gamma_A)_{jk}^{-1} \left(\widetilde{\Gamma}_A\right)_{lm} = 0, \tag{6.45}$$

for arbitrary $j, k, l, m = 1, 2, \cdots, d$ since the elements Y_{kl} are arbitrary. In particular, choosing $j = k$ and $l = m$ in (6.45) and summing over these indices we obtain

$$\sum_{A=0}^{2^N-1} \text{Tr} (\Gamma_A)^{-1} \text{Tr} \widetilde{\Gamma}_A = 0. \tag{6.46}$$

For the case $N = 2M$, only $\Gamma_0 = \widetilde{\Gamma}_0 = \mathbb{1}$ give a nonzero contribution $\text{Tr} \Gamma_A = \text{Tr} \widetilde{\Gamma}_A = d\delta_{A0}$, so that (6.46) leads to

$$d^2 = 0, \tag{6.47}$$

which is not compatible for a nontrivial d. Therefore, this proves that the Clifford algebra with an even N admits only one (unique) irreducible representation. However, in the case of $N = 2M + 1$, both $\Gamma_0 = \mathbb{1}$ and $\Gamma_{2N-1} = \Lambda$ lead to nontrivial traces. Using (6.21), we note that we can write $\Lambda = c\mathbb{1}, \widetilde{\Lambda} = \widetilde{c}\mathbb{1}$ so that (6.46), in this case, leads to

$$d^2 + c^{-1}\widetilde{c}d^2 = 0,$$

or, $\quad c^{-1}\widetilde{c} = -1,$

or, $\quad \widetilde{c} = -c,$ \hfill (6.48)

which also implies that

$$\widetilde{\Lambda} = -\Lambda. \hfill (6.49)$$

This shows that, for $N = 2M + 1$ (odd), inequivalent representations of the Clifford algebra with the same dimensionality are possible.

In fact, we note that given an irreducible representation γ_μ, we can find another irreducible representation $\widetilde{\gamma}_\mu$ simply given by

$$\widetilde{\gamma}_\mu = -\gamma_\mu, \hfill (6.50)$$

which, in odd dimensions ($N = 2M + 1$) indeed leads to

$$\widetilde{\Lambda} = \widetilde{\gamma}_1\widetilde{\gamma}_2 \cdots \widetilde{\gamma}_{2M+1} = (-1)^{2M+1}\gamma_1\gamma_2 \cdots \gamma_{2M+1} = -\Lambda. \hfill (6.51)$$

That the two representations are irreducible can be seen simply from the following argument. Suppose that there exists a nonsingular $d \times d$ matrix (similarity transformation) S satisfying (relating the two representations)

$$\widetilde{\gamma}_\mu = S^{-1}\gamma_\mu S, \hfill (6.52)$$

then, this would imply that

$$\widetilde{\Lambda} = \widetilde{\gamma}_1\widetilde{\gamma}_2 \cdots \widetilde{\gamma}_{2M+1} = S^{-1}\gamma_1\gamma_2 \cdots \gamma_{2M+1}S,$$

or, $\quad \widetilde{\Lambda} = S^{-1}\Lambda S = cS^{-1}\mathbb{1}S = c\mathbb{1} = \Lambda, \hfill (6.53)$

which would contradict (6.51). This shows that there cannot be a similarity transformation connecting the two representations and, therefore, they are indeed inequivalent. However, there cannot be a third inequivalent representation beyond these two, which can be seen as follows. Suppose there exists another irreducible representation $\overline{\gamma}_\mu$

so that $(\gamma_\mu, \tilde{\gamma}_\mu, \overline{\gamma}_\mu)$ define three inequivalent representations of the Clifford algebra. In this case, applying the analysis of (6.48)-(6.49) pairwise, we obtain the relations

$$\tilde{\Lambda} = -\Lambda, \quad \overline{\Lambda} = -\tilde{\Lambda}, \quad \Lambda = -\overline{\Lambda}, \tag{6.54}$$

which are clearly inconsistent.

In conclusion, we see that $N = 2M$ (even) admits only a unique irreducible representation of dimensionality 2^M while there exist two inequivalent representations with the same dimensionality 2^M for $N = 2M+1$ (odd). In the latter case, the two representations may be simply related as $\tilde{\gamma}_\mu = -\gamma_\mu$. For example, let us consider the case of $M = 1$. For $N = 2M = 2$, we note that the three 2×2 ($d = 2^1 = 2$) Pauli matrices

$$\sigma_1 = \begin{pmatrix} 0 & 1 \\ 1 & 0 \end{pmatrix}, \quad \sigma_2 = \begin{pmatrix} 0 & -i \\ i & 0 \end{pmatrix}, \quad \sigma_3 = \begin{pmatrix} 1 & 0 \\ 0 & -1 \end{pmatrix}, \tag{6.55}$$

satisfy the Clifford algebra ($\gamma_\mu = \sigma_\mu, \mu = 1, 2, 3$)

$$\sigma_\mu \sigma_\nu + \sigma_\nu \sigma_\mu = 2\delta_{\mu\nu} \mathbb{1}, \tag{6.56}$$

and provide a unique (2-dimensional irreducible) representation. However, for $N = 2M + 1 = 3$, in addition to the Pauli matrices (which, of course, provide a 2-dimensional irreducible representation), a second inequivalent representation of the Clifford algebra is given by $\tilde{\sigma}_\mu = -\sigma_\mu$ and we note that

$$\Lambda = \sigma_1 \sigma_2 \sigma_3 = i \mathbb{1}, \quad \tilde{\Lambda} = \tilde{\sigma}_1 \tilde{\sigma}_2 \tilde{\sigma}_3 = -i \mathbb{1} = -\Lambda. \tag{6.57}$$

6.2 Charge conjugation

Let us next discuss the charge conjugation matrix C associated with a Clifford algebra. This is a rather important concept in relativistic quantum field theories involving fermions and is defined by the relation

$$C^{-1} \gamma_\mu C = -\gamma_\mu^T, \quad \text{or,} \quad C\gamma_\mu^T C^{-1} = -\gamma_\mu, \tag{6.58}$$

where γ_μ^T denotes the transpose of the matrix γ_μ (generators of the Clifford algebra). Since γ_μ satisfies the Clifford algebra (6.1), it follows that $\tilde{\gamma}_\mu = -\gamma_\mu^T = C^{-1} \gamma_\mu C$ also satisfies the Clifford algebra

$$\tilde{\gamma}_\mu \tilde{\gamma}_\nu + \tilde{\gamma}_\nu \tilde{\gamma}_\mu = 2\delta_{\mu\nu} \mathbb{1}. \tag{6.59}$$

Therefore, the charge conjugation matrix C can be thought of as a similarity matrix between the two representations. To study the properties of this matrix, we note that by taking the transpose of the definition in (6.58) we obtain

$$\gamma_\mu = -\left(C^{-1}\gamma_\mu C\right)^T = -C^T\gamma_\mu^T\left(C^{-1}\right)^T = C^T C^{-1}\gamma_\mu C(C^{-1})^T,$$

$$\text{or,} \quad \gamma_\mu = C^T C^{-1}\gamma_\mu \left(C^T C^{-1}\right)^{-1},$$

$$\text{or,} \quad [C^T C^{-1}, \gamma_\mu] = 0. \tag{6.60}$$

It follows from this that $C^T C^{-1}$ commutes with the entire set of matrices Γ_A and, therefore, must be proportional to the identity matrix. Therefore, writing

$$C^T C^{-1} = \eta \mathbb{1}, \tag{6.61}$$

where η is a constant scalar parameter, we obtain

$$C^T = \eta C,$$

$$\text{or,} \quad C = \eta C^T = \eta^2 C,$$

$$\text{or,} \quad \eta^2 = 1, \tag{6.62}$$

so that $\eta = \pm 1$.

We can determine the value of η as follows. Let us consider (6.39) and write it in the form

$$\sum_{A=0}^{2^N-1} (\Gamma_A)_{jk}^{-1} \left(\Gamma_A^T\right)_{ml} = \frac{2^N}{d} \delta_{jm}\delta_{kl}. \tag{6.63}$$

Multiplying this with $C_{km}C_{ln}^{-1}$ and summing over the repeated indices, (6.63) leads to

$$\sum_{A=0}^{2^N-1} \left(\Gamma_A^{-1}C\Gamma_A^T C^{-1}\right)_{jn} = \frac{2^N}{d}\left(C^T C^{-1}\right)_{jn} = \frac{2^N}{d}\eta\delta_{jn}, \tag{6.64}$$

where we have used (6.62). We note that a Γ_A involves a product of the matrices γ_μ of the form $\gamma_{\mu_1}\gamma_{\mu_2}\cdots\gamma_{\mu_\ell}$ with $\mu_1 < \mu_2 < \cdots < \mu_\ell$. For a fixed ℓ, there will be $\frac{N!}{(N-\ell)!\ell!}$ such matrices in Γ_A. For every such Γ_A with a fixed ℓ, we have (see (6.1) as well as the second relation in

(6.58))

$$C\Gamma_A^T C^{-1} = C\gamma_{\mu_\ell}^T C^{-1} C\gamma_{\mu_{\ell-1}}^T C^{-1} \cdots C\gamma_{\mu_2}^T C^{-1} C\gamma_{\mu_1}^T C^{-1}$$

$$= (-1)^\ell \gamma_{\mu_\ell}\gamma_{\mu_{\ell-1}} \cdots \gamma_{\mu_2}\gamma_{\mu_1}$$

$$= (-1)^\ell (-1)^{\frac{\ell(\ell-1)}{2}} \gamma_{\mu_1}\gamma_{\mu_2} \cdots \gamma_{\mu_{\ell-1}}\gamma_{\mu_\ell}$$

$$= (-1)^{\frac{\ell(\ell+1)}{2}} \Gamma_A. \tag{6.65}$$

Using this in (6.64) and replacing the sum over A by a sum over ℓ, we obtain

$$\sum_{A=0}^{2^N-1} \left(\Gamma_A^{-1} C\Gamma_A^T C^{-1}\right)_{jn} = \frac{2^N}{d}\eta\delta_{jn},$$

or, $\quad \displaystyle\sum_{\ell=0}^{N} \frac{N!}{(N-\ell)!\ell!}(-1)^{\frac{\ell(\ell+1)}{2}}\delta_{jn} = \frac{2^N}{d}\eta\delta_{jn},$

or, $\quad \eta = \dfrac{d}{2^N}\displaystyle\sum_{\ell=0}^{N}\frac{N!}{(N-\ell)!\ell!}(-1)^{\frac{\ell(\ell+1)}{2}}. \tag{6.66}$

To proceed further let us consider the case of even N. We recall the identities

$$(1+a)^N + (1-a)^N = \sum_{\ell=0}^{N} \frac{N!}{(N-\ell)!\ell!}\left(a^\ell + (-a)^\ell\right)$$

$$= 2\sum_{m=0}^{\frac{N}{2}} \frac{N!}{(N-2m)!(2m)!}a^{2m},$$

$$(1+a)^N - (1-a)^N = \sum_{\ell=0}^{N} \frac{N!}{(N-\ell)!\ell!}\left(a^\ell - (-a)^\ell\right)$$

$$= 2\sum_{m=0}^{\frac{N}{2}-1} \frac{N!}{(N-2m-1)!(2m+1)!}a^{2m+1}. \tag{6.67}$$

Decomposing the sum in (6.66) into even and odd values of ℓ and using (6.41) as well as (6.67) we obtain (recall that $d = 2^{\frac{N}{2}}$ for even

N, see (6.41))

$$\eta = \frac{d}{2^N} \sum_{m=0}^{\frac{N}{2}} \frac{N!}{(N-2m)!(2m)!} (-1)^{m(2m+1)}$$

$$+ \frac{d}{2^N} \sum_{m=0}^{\frac{N}{2}-1} \frac{N!}{(N-2m-1)!(2m+1)!} (-1)^{(2m+1)(m+1)}$$

$$= \frac{1}{2^{\frac{N}{2}+1}} \left[\left((1+\sqrt{-1})^N + (1-\sqrt{-1})^N \right) \right.$$

$$\left. + \sqrt{-1} \left((1+\sqrt{-1})^N - (1-\sqrt{-1})^N \right) \right]$$

$$= \frac{1}{2^{\frac{N}{2}+1}} \left[(1+\sqrt{-1})^{N+1} + (1-\sqrt{-1})^{N+1} \right]$$

$$= \frac{1}{2^{\frac{N}{2}+1}} \left[\left(\sqrt{2} e^{\frac{i\pi}{4}} \right)^{N+1} + \left(\sqrt{2} e^{-\frac{i\pi}{4}} \right)^{N+1} \right]$$

$$= \sqrt{2} \cos \frac{\pi}{4}(N+1). \tag{6.68}$$

Here we have used $\sqrt{-1} = i$ in the intermediate step, but the expression is symmetric in $i \leftrightarrow -i$. For even N, therefore, we obtain

$$\eta = \begin{cases} (-1)^m & \text{for} \quad N = 4m, \\ (-1)^{m+1} & \text{for} \quad N = 4m+2. \end{cases} \tag{6.69}$$

The case of odd N, however, is a bit more complicated. Since there now exist two inequivalent irreducible representations γ_μ and $\widetilde{\gamma}_\mu$, there is no guarantee that $-\gamma_\mu^T$ would belong to the original irreducible space of γ_μ. If it does, then there will still be a charge conjugation matrix C satisfying (6.58), namely,

$$C^{-1}\gamma_\mu C = -\gamma_\mu^T. \tag{6.70}$$

On the other hand, if $-\gamma_\mu^T$ belongs to the second irreducible representation, the charge conjugation matrix has to satisfy instead (see (6.50))

$$C^{-1}\widetilde{\gamma}_\mu C = -\gamma_\mu^T,$$

$$\text{or,} \quad C^{-1}\gamma_\mu C = \gamma_\mu^T. \tag{6.71}$$

Repeating the earlier analysis leads to

$$\begin{aligned} N = 4m+1: \quad & C^{-1}\gamma_\mu C = \gamma_\mu^T, \quad && C^T = (-1)^m C, \\ N = 4m+3: \quad & C^{-1}\gamma_\mu C = -\gamma_\mu^T, \quad && C^T = (-1)^{m+1} C. \end{aligned} \tag{6.72}$$

As an example, let us consider the case of $m = 0$ and $N = 4m+3 = 3$. In this case, as we have already seen (see (6.55)) we can identify γ_μ with the three 2×2 Pauli matrices $\gamma_\mu = \sigma_\mu$. In this case, we obtain

$$C = \sigma_2 = -C^T, \tag{6.73}$$

which satisfies (6.72) (the second relation).

Let us conclude this section with the following observation. Let us consider $\gamma_\mu^\dagger$ which is the Hermitian conjugate of γ_μ. Taking the Hermitian conjugate of (6.1), we see that $\gamma_\mu^\dagger$ also satisfies the Clifford algebra

$$\gamma_\mu^\dagger \gamma_\nu^\dagger + \gamma_\nu^\dagger \gamma_\mu^\dagger = 2\delta_{\mu\nu} \mathbb{1}. \tag{6.74}$$

Therefore, for $N = 2M$ (even), there must exist a similarity transformation T relating the two

$$\gamma_\mu^\dagger = T^{-1} \gamma_\mu T. \tag{6.75}$$

We know that we can always choose γ_μs to be Hermitian with a suitable choice of basis, namely,

$$\gamma_\mu^\dagger = \gamma_\mu, \tag{6.76}$$

for $\mu = 1, 2, \cdots, N$ (this holds even for odd N which we will not go into). For example, for $N = 2M$ we can write

$$\gamma_\mu = \begin{pmatrix} 0 & \sigma_\mu \\ \sigma_\mu^\dagger & 0 \end{pmatrix}, \tag{6.77}$$

with $2^{M-1} \times 2^{M-1}$ matrices σ_μ (not the Pauli matrices) satisfying

$$\sigma_\mu \sigma_\nu^\dagger + \sigma_\nu \sigma_\mu^\dagger = \sigma_\mu^\dagger \sigma_\nu + \sigma_\nu^\dagger \sigma_\mu = 2\delta_{\mu\nu} \mathbb{1}, \tag{6.78}$$

so that (6.1) holds. For example, for the case $M = 2, N = 4$, we can identify the 2×2 matrices σ_μ with the three Pauli matrices as well as the identity matrix as $\sigma_\mu = (\sigma_1, \sigma_2, \sigma_3, i\mathbb{1})$ and $\sigma_\mu^\dagger = (\sigma_1, \sigma_2, \sigma_3, -i\mathbb{1})$ which satisfy (6.78) and lead to a Hermitian representation for γ_μ. With this choice, we have

$$\Lambda = \gamma_1 \gamma_2 \gamma_3 \gamma_4 (= \gamma_5) = \begin{pmatrix} \mathbb{1} & 0 \\ 0 & -\mathbb{1} \end{pmatrix}, \tag{6.79}$$

which is quite useful in the study of the Dirac equation.

6.3 Clifford algebra and the $O(N)$ group

Clifford algebras have an intimate relation with the representations of the $O(N)$ group which can be seen in the following way. Let $a \in O(N)$ denote the $N \times N$ orthogonal matrix (in the defining representation) satisfying (in chapters 2 and 4 we have distinguished between a group element a from the corresponding defining matrix A by denoting $a \rightarrow A$, here we use a to stand simultaneously for both the group element and its fundamental matrix representation to avoid proliferation of symbols)

$$a^T a = a a^T = \mathbb{1}. \tag{6.80}$$

Let $\gamma_\mu, \mu = 1, 2, \cdots, N$ represent the generators of a Clifford algebra satisfying (6.1). Then writing

$$\tilde{\gamma}_\mu = \sum_{\nu=1}^N a_{\mu\nu} \gamma_\nu, \tag{6.81}$$

we determine

$$\begin{aligned}
\tilde{\gamma}_\mu \tilde{\gamma}_\nu + \tilde{\gamma}_\nu \tilde{\gamma}_\mu &= \sum_{\lambda,\rho=1}^N a_{\mu\lambda} a_{\nu\rho} \left(\gamma_\lambda \gamma_\rho + \gamma_\rho \gamma_\lambda \right) \\
&= \sum_{\lambda,\rho=1}^N a_{\mu\lambda} a_{\nu\rho} \left(2\delta_{\lambda\rho} \mathbb{1} \right) \\
&= 2 \sum_{\lambda=1}^N a_{\mu\lambda} a_{\nu\lambda} \mathbb{1} = 2 \sum_{\lambda=1}^N a_{\mu\lambda} a_{\lambda\nu}^T \mathbb{1} \\
&= 2 \left(a a^T \right)_{\mu\nu} \mathbb{1} = 2\delta_{\mu\nu} \mathbb{1},
\end{aligned} \tag{6.82}$$

so that $\tilde{\gamma}_\mu$ also satisfy the Clifford algebra. Here we have used (6.80) in the last step.

Let us assume that $N = 2M$. In this case, we know that the irreducible representation of the Clifford algebra is unique. Therefore, (6.82) implies that there exists a similarity transformation $S(a) \in O(N)$

$$S^{-1}(a) \gamma_\mu S(a) = \tilde{\gamma}_\mu = \sum_{\nu=1}^N a_{\mu\nu} \gamma_\nu, \tag{6.83}$$

which transforms between equivalent irreducible representations of the Clifford algebra. ($S(a)$ denote $2^M \times 2^M$ matrices.) For the case of $N = 2M+1$, on the other hand, this may not necessarily be correct since there exist two inequivalent irreducible representations. Let us suppose that $\widetilde{\gamma}_\mu$ belongs to the second inequivalent representation $(-\gamma_\mu)$, namely,

$$S^{-1}(a)(-\gamma_\mu)S(a) = \widetilde{\gamma}_\mu = \sum_{\nu=1}^{N} a_{\mu\nu}\gamma_\nu,$$

$$\text{or,} \quad S^{-1}(a)\gamma_\mu S(a) = -\sum_{\nu=1}^{N} a_{\mu\nu}\gamma_\nu. \tag{6.84}$$

In this case (odd N), we recall (see (6.24)) that

$$\Lambda = \gamma_1\gamma_2\cdots\gamma_N$$

$$= \frac{1}{N!} \sum_{\mu_1\mu_2\cdots\mu_N=1}^{N} \epsilon_{\mu_1\mu_2\cdots\mu_N}\gamma_{\mu_1}\gamma_{\mu_2}\cdots\gamma_{\mu_N}, \tag{6.85}$$

which follows because of the anti-commutativity of γ_μs (see (6.3)) as well as the antisymmetry of the Levi-Civita tensor $\epsilon_{\mu_1\mu_2\cdots\mu_N}$. Furthermore, it follows from (6.84) and (6.85) that

$$S^{-1}(a)\Lambda S(a) = \frac{1}{N!} \sum_{\mu_1,\mu_2,\cdots,\mu_N=1}^{N} \epsilon_{\mu_1\mu_2\cdots\mu_N}$$

$$\times (S^{-1}\gamma_{\mu_1}S)(S^{-1}\gamma_{\mu_2}S)\cdots(S^{-1}\gamma_{\mu_N}S)$$

$$= \frac{(-1)^N}{N!} \sum_{\substack{\mu_1,\mu_2,\cdots,\mu_N,\\\nu_1,\nu_2,\cdots,\nu_N=1}}^{N} \epsilon_{\mu_1\cdots\mu_N}a_{\mu_1\nu_1}\cdots a_{\mu_N\nu_N}$$

$$\times (\gamma_{\nu_1}\cdots\gamma_{\nu_N})$$

$$= -\frac{(\det a)}{N!} \sum_{\nu_1,\nu_2,\cdots,\nu_N=1}^{N} \epsilon_{\nu_1\nu_2\cdots\nu_N}\gamma_{\nu_1}\gamma_{\nu_2}\cdots\gamma_{\nu_N}$$

$$= -(\det a)\Lambda, \tag{6.86}$$

where we have used the fact that $N = 2M+1$ as well as the definition of the determinant of a matrix, namely,

$$\sum_{\mu_1,\cdots,\mu_N=1}^{N} \epsilon_{\mu_1\cdots\mu_N}a_{\mu_1\nu_1}\cdots a_{\mu_N\nu_N} = (\det a)\,\epsilon_{\nu_1\cdots\nu_N}. \tag{6.87}$$

On the other hand, we have seen in (6.24) that in odd dimensions $\Lambda = c\,\mathbb{1}$. Using this in the left hand side of (6.86) we conclude that

$$c\,S^{-1}\mathbb{1}S = c\,\mathbb{1} = \Lambda = -(\det a)\,\Lambda,$$

$$\text{or,} \quad \det a = -1. \tag{6.88}$$

Therefore, as long as we restrict ourselves to the special orthogonal group of transformations $SO(N)$ satisfying $a^T a = \mathbb{1}$ and $\det a = 1$, we must have

$$S^{-1}(a)\gamma_\mu S(a) = \sum_{\nu=1}^{N} a_{\mu\nu}\gamma_\nu, \tag{6.89}$$

even for the case $N = 2M + 1$. We note here that the mirror reflection (or parity) transformation is an example of $a \in O(N)$ satisfying $\det a = -1$. This analysis shows that for the case of even N we can consider any $a \in O(N)$ while for odd N we will choose $a \in SO(N)$ for the rest of this discussion.

Let a and b be two such matrices. Then, we can write (see (6.83))

$$S^{-1}(a)\gamma_\mu S(a) = \sum_{\nu=1}^{N} a_{\mu\nu}\gamma_\nu, \quad S^{-1}(b)\gamma_\mu S(b) = \sum_{\nu=1}^{N} b_{\mu\nu}\gamma_\nu, \tag{6.90}$$

which leads to

$$S^{-1}(b)\left(S^{-1}(a)\gamma_\mu S(a)\right)S(b) = \sum_{\nu=1}^{N} a_{\mu\nu}\left(S^{-1}(b)\gamma_\nu S(b)\right)$$

$$= \sum_{\nu,\rho=1}^{N} a_{\mu\nu}b_{\nu\rho}\gamma_\rho = \sum_{\rho=1}^{N}\left(\sum_{\nu=1}^{N} a_{\mu\nu}b_{\nu\rho}\right)\gamma_\rho$$

$$= \sum_{\rho=1}^{N}(ab)_{\mu\rho}\gamma_\rho = S^{-1}(ab)\gamma_\mu S(ab), \tag{6.91}$$

where ab denotes the matrix product of the two $N \times N$ matrices a and b (in the first term in the last step) so that $ab \in O(N)$ and satisfies (for the case of odd N) $\det(ab) = 1$ if $\det a = \det b = 1$. It follows from (6.91) that

$$S(ab)S^{-1}(b)S^{-1}(a)\gamma_\mu = \gamma_\mu S(ab)S^{-1}(b)S^{-1}(a), \tag{6.92}$$

and since $S(ab)S^{-1}(b)S^{-1}(a)$ commutes with all the generators γ_μ (and, therefore, with the set Γ_A), by Schur's lemma we have

$$S(ab)S^{-1}(b)S^{-1}(a) = \lambda \mathbb{1}_{2^{[N/2]}}, \tag{6.93}$$

for some parameter $\lambda = \lambda(a,b)$ in general depending on the group elements a, b. (Here $[N/2]$ stands for the integer part of $N/2$.) As a result, we can write

$$S(ab) = \lambda(a,b)S(a)S(b). \tag{6.94}$$

This implies that $S(a)$ defines a projective (ray) representation (see section 4.4) of $O(N)$ for N even and of $SO(N)$ for N odd.

We note that a relation of the form (6.90)

$$S^{-1}(a)\gamma_\mu S(a) = \sum_{\nu=1}^{N} a_{\mu\nu}\gamma_\nu, \tag{6.95}$$

is invariant under $S(a) \to cS(a)$ so that with a proper choice of c, we can always assume

$$\det S(a) = 1, \tag{6.96}$$

for the transformation matrices $S(a), S(b), S(ab)$. Using this in (6.94) we obtain

$$\det S(ab) = (\lambda(a,b))^{2^{[\frac{N}{2}]}} \det S(a) \det S(b),$$

$$\text{or,} \quad \lambda(a,b) = e^{2\pi i k/[\frac{N}{2}]}. \tag{6.97}$$

If we restrict ourselves to the case of $SO(N)$ for even N, we can assume $\lambda(a,b) = 1$ because of the following argument. For $a = \mathbb{1}_N$, we can choose $S(a) = S(\mathbb{1}) = \mathbb{1}_{2^{N/2}}$ since this corresponds to the trivial transformation. Therefore, starting with the point $a = b = \mathbb{1}_N$, we can set $\lambda(\mathbb{1}_N, \mathbb{1}_N) = 1$. Since $S(a)$ is supposed to be a continuous function of the group element a, as long as any other point on the group manifold can be reached continuously from the starting point (origin) $a = \mathbb{1}_N$, it follows from the argument of continuity that $\lambda(a,b) = 1$. This is true, in particular, for the group $SO(N)$. On the other hand, the orthogonal group $O(N)$ contains a disjoint mirror reflection symmetry which canot be reached continuously from the origin. In this case, for $\lambda(a,b) = 1$, we have

$$S(ab) = S(a)S(b), \tag{6.98}$$

and $S(a)$ realizes the "spinor" representation of the $SO(N)$ group. Namely, we note that both $S(a)$ and $-S(a)$ correspond to the same rotation

$$\gamma_\mu \to S^{-1}(a)\gamma_\mu S(a) = \sum_{\nu=1}^{N} a_{\mu\nu}\gamma_\nu, \tag{6.99}$$

and we have a double valued representation. This is due to the fact that the group $SO(N)$ is not simply connected and we need to define the covering group which we will not go into.

There is, however, a slight subtlety of the following nature that needs to be addressed. Even though we have assumed that γ_μ defines an irreducible representation of the Clifford algebra, this does not imply that $S(a)$ would correspond to an irreducible representation of the $SO(N)$ group. For example, let $N = 2M$ and let us consider as in (6.77)

$$\gamma_\mu = \begin{pmatrix} 0 & \sigma_\mu \\ \sigma_\mu^\dagger & 0 \end{pmatrix}. \tag{6.100}$$

In this case, we can find the block diagonal matrix $S(a)$ (see (6.90))

$$S(a) = \begin{pmatrix} S_1(a) & 0 \\ 0 & S_2(a) \end{pmatrix}, \tag{6.101}$$

with the $2^{M-1} \times 2^{M-1}$ matrices $S_i(a), i = 1, 2$ satisfying

$$S_1^{-1}(a)\sigma_\mu S_1(a) = \sum_{\nu=1}^{N} a_{\mu\nu}\sigma_\nu,$$

$$S_2^{-1}(a)\sigma_\mu^\dagger S_2(a) = \sum_{\nu=1}^{N} a_{\mu\nu}\sigma_\nu^\dagger, \tag{6.102}$$

and leading to $S_i(ab) = S_i(a)S_i(b), i = 1, 2$. Therefore, $S_1(a)$ and $S_2(a)$ define $2^{M-1} \times 2^{M-1}$ dimensional irreducible representations of $SO(N)$ while the block diagonal matrix $S(a)$ corresponds to a reducible representation. The case of $M = 2, N = 2M = 4$ is of interest since in this case, the particular representation in (6.100) is related to the Weyl spinors in relativistic theories (and is known as the Weyl representation). Furthermore, for some relativistic problems, it is of interest to consider a representation where all the generators γ_μ are real matrices (known as the Majorana representation in physics) which we will not go into.

Let us now discuss briefly about the connection between the Clifford algebra and the Lie algebra of the $SO(N)$ group. If we define

$$J_{\mu\nu} = \frac{1}{4} [\gamma_\mu, \gamma_\nu] = -J_{\nu\mu}, \tag{6.103}$$

then, it is straightforward to check that it satisfies the Lie algebra (3.24) of $so(N)$

$$[J_{\mu\nu}, J_{\alpha\beta}] = -\delta_{\mu\alpha} J_{\nu\beta} - \delta_{\nu\beta} J_{\mu\alpha} + \delta_{\mu\beta} J_{\nu\alpha} + \delta_{\nu\alpha} J_{\mu\beta}. \tag{6.104}$$

If we further assume that $\gamma_\mu^\dagger = \gamma_\mu$ (namely, Hermitian generators), it follows from (6.103) that $J_{\mu\nu}^\dagger = -J_{\mu\nu}$ corresponding to the Lie algebra structure of the compact group $SO(N, \mathbb{R}) = \mathbb{R}^N$. However, following the earlier arguments, we can readily see that the $J_{\mu\nu}$ constructed in this manner is not irreducible. In fact, considering the representation in (6.100), we can determine that

$$J_{\mu\nu} = \begin{pmatrix} J_{\mu\nu}^{(+)} & 0 \\ 0 & J_{\mu\nu}^{(-)} \end{pmatrix}, \tag{6.105}$$

where

$$J_{\mu\nu}^{(+)} = \frac{1}{4} \left(\sigma_\mu \sigma_\nu^\dagger - \sigma_\nu \sigma_\mu^\dagger \right) = -J_{\nu\mu}^{(+)},$$

$$J_{\mu\nu}^{(-)} = \frac{1}{4} \left(\sigma_\mu^\dagger \sigma_\nu - \sigma_\nu^\dagger \sigma_\mu \right) = -J_{\nu\mu}^{(-)}, \tag{6.106}$$

and which satisfy

$$[J_{\mu\nu}^{(\pm)}, J_{\alpha\beta}^{(\pm)}] = -\delta_{\mu\alpha} J_{\nu\beta}^{(\pm)} - \delta_{\nu\beta} J_{\mu\alpha}^{(\pm)} + \delta_{\mu\beta} J_{\nu\alpha}^{(\pm)} + \delta_{\nu\alpha} J_{\mu\beta}^{(\pm)}. \tag{6.107}$$

For the case of $M = 2, N = 2M = 4$, we note that $J_{\mu\nu}^{(\pm)}$ correspond to infinitesimal generators of rotation for Weyl spinors and lead to 2×2 matrix representations of $SO(N)$.

We can now construct more general spinor representations of the $SO(N)$ group. Let us consider the case of $N = 2M$ and write (the prefactor is a normalization constant, $i = \sqrt{-1}$)

$$\Lambda = \exp(\frac{i\pi M}{2}) \gamma_1 \gamma_2 \cdots \gamma_{2M}, \tag{6.108}$$

which, as we have seen in (6.13), satisfies

$$\Lambda \gamma_\mu + \gamma_\mu \Lambda = 0, \quad \mu = 1, 2, \cdots, 2M, \tag{6.109}$$

and $\Lambda^2 = \mathbb{1}$ with the particular choice of our normalization in (6.108).

Let $(f_1, f_2, \cdots, f_n)$ with $f_1 \geq f_2 \geq \cdots \geq f_n \geq 0$ represent a Young tableau for the group $SO(N)$ and $T_{\mu_1\mu_2\cdots\mu_n}$ denote the irreducible tensor corresponding to this diagram. (Here we are assuming that $n < N$ and that $f_{n+1}, f_{n+2}, \cdots, f_N = 0$.) For the 2^{M+1} dimensional spinor of $SO(N)$, we consider a general spinor-tensor given by

$$\psi_{\mu_1\mu_2\cdots\mu_n} \simeq \psi \otimes T_{\mu_1\mu_2\cdots\mu_n}, \tag{6.110}$$

and introduce the two spinors obtained from this as

$$\phi^{(\pm)}_{\mu_1\mu_2\cdots\mu_n} = \frac{1}{2}\left(1 \pm \Lambda\right)\psi_{\mu_1\mu_2\cdots\mu_n}. \tag{6.111}$$

If we choose Λ to be $(2^M \times 2^M)$ block diagonal of the form $(\Lambda^2 = \mathbb{1})$

$$\Lambda = \begin{pmatrix} \mathbb{1} & 0 \\ 0 & -\mathbb{1} \end{pmatrix}, \tag{6.112}$$

with the γ_μ matrices of the form in (6.77)

$$\gamma_\mu = \begin{pmatrix} 0 & \sigma_\mu \\ \sigma_\mu^\dagger & 0 \end{pmatrix}, \tag{6.113}$$

it is clear from the definitions in (6.111) and (6.112) that the spinors $\phi^{(\pm)}_{\mu_1\mu_2\cdots\mu_n}$ are effectively 2^M dimensional spinors which are the analogs of the 2 dimensional Weyl spinors of the Lorentz group.

Although the spinors $\phi_{\mu_1\mu_2\cdots\mu_n}$ furnish representations of the $SO(N)$ group in a natural manner, these are not irreducible in general, since it contains nontrivial subspaces such as

$$\sum_{\lambda=1}^{N} \gamma^\lambda \phi^{(\pm)}_{\lambda\mu_2\mu_3\cdots\mu_n}. \tag{6.114}$$

In order to make them irreducible, we need to impose conditions such as

$$\sum_{\lambda=1}^{N} \gamma^\lambda \phi^{(\pm)}_{\lambda\mu_2\mu_3\cdots\mu_n} = 0, \tag{6.115}$$

and so on. We note here that the condition (6.115) also automatically implies that

$$\sum_{\lambda=1}^{N} \gamma^\lambda \phi^{(\pm)}_{\mu_1\lambda\mu_3\cdots\mu_n} = 0, \tag{6.116}$$

which can be seen, for example, by considering a simple case. Let us consider a symmetric or ant-symmetric spinor-tensor satisfying $\phi_{\mu\nu}^{(\pm)} = \epsilon\phi_{\nu\mu}^{(\pm)}, \epsilon = \pm 1$. In this case,

$$\sum_{\lambda=1}^{N} \gamma^\lambda \phi_{\mu\lambda}^{(\pm)} = 0, \tag{6.117}$$

implies

$$\sum_{\lambda=1}^{N} \gamma^\lambda \phi_{\mu\lambda}^{(\pm)} = \epsilon \sum_{\lambda=1}^{N} \gamma^\lambda \phi_{\lambda\mu}^{(\pm)} = 0. \tag{6.118}$$

As a more complicated example, let us consider a spinor-tensor $\phi_{\mu\nu\lambda}^{(\pm)}$ corresponding to the Young tableau (see (5.43))

$$\phi_{\mu\nu\lambda}^{(\pm)} : \begin{array}{|c|c|} \hline \mu & \nu \\ \hline \lambda \\ \cline{1-1} \end{array}$$

which satisfies (see (5.31))

$$\phi_{\mu\nu\lambda}^{(\pm)} = \phi_{\nu\mu\lambda}^{(\pm)} = -\phi_{\lambda\nu\mu}^{(\pm)}, \quad \phi_{\mu\nu\lambda}^{(\pm)} + \phi_{\nu\lambda\mu}^{(\pm)} + \phi_{\lambda\mu\nu}^{(\pm)} = 0. \tag{6.119}$$

In this case, the relation

$$\sum_{\mu=1}^{N} \gamma^\mu \phi_{\mu\nu\lambda}^{(\pm)} = 0, \tag{6.120}$$

leads to

$$\sum_{\nu=1}^{N} \gamma^\nu \phi_{\mu\nu\lambda}^{(\pm)} = \sum_{\nu=1}^{N} \gamma^\nu \phi_{\nu\mu\lambda}^{(\pm)} = 0,$$

$$\sum_{\lambda=1}^{N} \gamma^\lambda \phi_{\mu\nu\lambda}^{(\pm)} = -\sum_{\lambda=1}^{N} \gamma^\lambda \left(\phi_{\nu\lambda\mu}^{(\pm)} + \phi_{\lambda\mu\nu}^{(\pm)} \right) = 0. \tag{6.121}$$

The representations for (odd) $N = 2M+1$ can be reduced to the case of $SO(2M)$ since (see (6.24)) $\Lambda = c_N \mathbb{1}$ (or they can be constructed in a similar manner).

In ending this section, we note that if $a_i, a_i^\dagger, i = 1, 2, \cdots, M$ denote annihilation and creation operators for fermions satisfying

$$a_i a_j + a_j a_i = 0 = a_i^\dagger a_j^\dagger + a_j^\dagger a_i^\dagger,$$

$$a_i a_j^\dagger + a_j^\dagger a_i = \delta_{ij} \mathbb{1}, \quad i, j = 1, 2, \cdots, M, \tag{6.122}$$

then, the $2M$ Hermitian operators $(i = 1, 2, \cdots, M)$

$$\gamma_i = \frac{1}{\sqrt{2}} \left(a_i + a_i^\dagger \right) = \gamma_i^\dagger, \quad \gamma_{i+M} = \frac{i}{\sqrt{2}} \left(a_i - a_i^\dagger \right) = \gamma_{i+M}^\dagger,$$

$$(6.123)$$

can be checked to satisfy the algebra

$$\gamma_\mu \gamma_\nu + \gamma_\nu \gamma_\mu = 2\delta_{\mu\nu} \, \mathbb{1}, \quad \mu, \nu = 1, 2, \cdots, 2M. \tag{6.124}$$

The underlying (Hilbert) vector space of the fermionic theory can be constructed starting with the vacuum state $|0\rangle$ satisfying

$$a_i |0\rangle = 0, \quad i = 1, 2, \cdots, M, \tag{6.125}$$

and consists of the states

$$|i\rangle = a_i^\dagger |0\rangle, \; |i_1, i_2\rangle = a_{i_1}^\dagger a_{i_2}^\dagger |0\rangle, \; |i_1, i_2, i_3\rangle = a_{i_1}^\dagger a_{i_2}^\dagger a_{i_3}^\dagger |0\rangle, \tag{6.126}$$

and so on where the states are anti-symmetric under the interchange of any pair of indices. In particular, the state with the maximum number of possible labels is given by

$$|i_1, i_2, \cdots, i_M\rangle = a_{i_1}^\dagger a_{i_2}^\dagger \cdots a_{i_M}^\dagger |0\rangle, \tag{6.127}$$

and is totally anti-symmetric in all its indices. The total dimension of the vector space can now be calculated easily

$$1 + M + \frac{1}{2!} M(M-1) + \cdots + M + 1 = (1+1)^M = 2^M, \tag{6.128}$$

which reproduces (yet another way) the dimension of the irreducible representation of the Clifford algebra for the case $N = 2M$ (see (6.41)).

6.4 References

W. Pauli, Inst. H. Poincare Ann. 6 (1936) 109.

R. H. Good, Rev. Mod. Phys. 27 (1955) 187.

S. Okubo, J. Math. Phys. 32 (1991) 1657; *ibid* 1669.

S. Okubo, Matematica Japonica 41 (1995) 59.

CHAPTER 7

Lorentz group and the Dirac equation

7.1 Lorentz group

In section 1.2.5 we had briefly introduced the concept of the Lorentz group. Since it plays an important role in the study of relativistic phenomena, we will look at this group in more detail in this chapter.

Special relativity requires consideration of the non-compact orthogonal group $O(1,3) \simeq O(3,1)$ instead of the compact group $O(4)$ more appropriate for Euclidean rotations. As we know, the quadratic form of the Minkowski space

$$c^2 t^2 - (x^2 + y^2 + z^2), \tag{7.1}$$

where c denotes the speed of light in vacuum, is invariant under transformations belonging to $O(3,1)$. It is more convenient to define

$$x^\mu = (x^0, x^1, x^2, x^3), \tag{7.2}$$

with

$$x^0 = ct, \quad x^1 = x, \quad x^2 = y, \quad x^3 = z, \tag{7.3}$$

so that we can write the quadratic form in (7.1) also as

$$(x^0)^2 - (x^1)^2 - (x^2)^2 - (x^3)^2 = \eta_{\mu\nu} x^\mu x^\nu, \tag{7.4}$$

where the repeated indices $\mu, \nu = 0, 1, 2, 3$, are summed and we have identified

$$\eta_{\mu\nu} = \eta_{\nu\mu} = \begin{cases} 0, & \text{if } \mu \neq \nu, \\ 1, & \text{if } \mu = \nu = 0, \\ -1 & \text{if } \mu = \nu = 1, 2, \text{ or } 3. \end{cases} \tag{7.5}$$

In this compact notation we can define the Lorentz transformations as

$$x^\mu \to x'^\mu = a^\mu{}_\nu x^\nu, \tag{7.6}$$

for real constants $a^\mu{}_\nu = (a^\mu{}_\nu)^*$. Invariance of the quadratic form (7.4) under these transformations now leads to

$$\eta_{\mu\nu} x'^\mu x'^\nu = \eta_{\mu\nu} x^\mu x^\nu,$$

or, $\quad \eta_{\mu\nu} \left(a^\mu{}_\lambda x^\lambda \right) \left(a^\nu{}_\rho x^\rho \right) = \eta_{\lambda\rho} x^\lambda x^\rho,$

or, $\quad \eta_{\mu\nu} a^\mu{}_\lambda a^\nu{}_\rho = \eta_{\lambda\rho}. \hfill (7.7)$

Let us next introduce the 4×4 matrices η and a as

$$(\eta)_{\mu\nu} = \eta_{\mu\nu}, \quad \text{or,} \quad \eta = \begin{pmatrix} 1 & 0 & 0 & 0 \\ 0 & -1 & 0 & 0 \\ 0 & 0 & -1 & 0 \\ 0 & 0 & 0 & -1 \end{pmatrix},$$

$$(a)_{\mu\nu} = a^\mu{}_\nu, \quad \text{or,} \quad a = \begin{pmatrix} a^0{}_0 & a^0{}_1 & a^0{}_2 & a^0{}_3 \\ a^1{}_0 & a^1{}_1 & a^1{}_2 & a^1{}_3 \\ a^2{}_0 & a^2{}_1 & a^2{}_2 & a^2{}_3 \\ a^3{}_0 & a^3{}_1 & a^3{}_2 & a^3{}_3 \end{pmatrix}, \hfill (7.8)$$

so that we can write the condition in (7.7) also as the matrix equation (note that $(a^T)_{\mu\nu} = (a)_{\nu\mu} = a^\nu{}_\mu$)

$$a^T \eta a = \eta, \hfill (7.9)$$

with the reality condition

$$a^* = a. \hfill (7.10)$$

We also note from (7.8) that

$$\eta^T = \eta, \quad \eta^* = \eta, \quad \det \eta = -1. \hfill (7.11)$$

Let us next define $\eta^{\mu\nu}$ as the tensor with the same numerical values as $\eta_{\mu\nu}$, namely,

$$\eta^{\mu\nu} = \eta^{\nu\mu} = \begin{cases} 0, & \text{if } \mu \neq \nu, \\ 1, & \text{if } \mu = \nu = 0, \\ -1 & \text{if } \mu = \nu = 1, 2, \text{ or } 3. \end{cases} \hfill (7.12)$$

It is clear that it satisfies

$$\eta^{\mu\lambda} \eta_{\lambda\nu} = \delta^\mu_\nu, \quad \mu, \nu = 0, 1, 2, 3, \hfill (7.13)$$

so that $\eta^{\mu\nu}$ really represents the matrix elements of the inverse matrix η^{-1} although numerically we have $\eta^{-1} = \eta$, namely, (see (7.6))

$$\left(\eta^{-1}\right)_{\mu\nu} = \eta^{\mu\nu}. \tag{7.14}$$

We can now derive from (7.9) that

$$a^T = \eta a^{-1}\eta^{-1}, \quad a^{-1} = \eta^{-1}a^T\eta, \tag{7.15}$$

as well as

$$a\eta^{-1}a^T = a\eta^{-1}(\eta a^{-1}\eta^{-1}) = \eta^{-1}. \tag{7.16}$$

The tensors $\eta^{\mu\nu}$ and $\eta_{\mu\nu}$ define the metric tensors of the underlying Minkowski space and using them we can now raise and lower the tensor indices as

$$x_\mu = \eta_{\mu\nu}x^\nu = (x^0, -x^1, -x^2, -x^3),$$
$$x^\mu = \eta^{\mu\nu}x_\nu = (x^0, x^1, x^2, x^3), \tag{7.17}$$

as we would expect (see (7.2)).

The set of all such matrices a (satisfying (7.9), (7.10) and (7.16)) defines the Lorentz group $\mathbb{L}_4$. In fact, if $a, b \in \mathbb{L}_4$, then we note from (7.9) and (7.10) that their matrix product satisfies

$$(ab)^* = a^*b^* = ab,$$
$$(ab)^T\eta(ab) = b^T(a^T\eta a)b = b^T\eta b = \eta, \tag{7.18}$$

so that $ab \in \mathbb{L}_4$. Moreover, $\mathbb{1}_4$ defines the identity matrix of $\mathbb{L}_4$. However, $\mathbb{L}_4$ is non-compact unlike the compact real rotation group $\mathbb{R}_4$ (see also section 1.2.5). To see this, we note that the 00 element of (7.9) (or for $\lambda = \rho = 0$ in equation (7.7)) leads to

$$\left(a^0{}_0\right)^2 - \left(a^i{}_0\right)^2 = 1,$$

$$\text{or,} \quad \left(a^0{}_0\right)^2 = 1 + \left(a^i{}_0\right)^2 = 1 + (a_{i0})^2, \quad i = 1, 2, 3. \tag{7.19}$$

Similarly, taking the 00 element of (7.16) we obtain

$$\left(a^0{}_0\right)^2 = 1 + \left(a^0{}_i\right)^2 = 1 + (a_{0i})^2. \tag{7.20}$$

Since $a^0{}_0, a^i{}_0$ and $a^0{}_i$ are real numbers, (7.19) and (7.20) show that they can take unbounded values so that the group manifold for $\mathbb{L}_4$ is unbounded and non-compact. We also note from (7.19) that

$$\left(a^0{}_0\right)^2 \geq 1,$$

$$\text{or,} \quad a^0{}_0 \geq 1, \quad \text{or,} \quad a^0{}_0 \leq -1. \tag{7.21}$$

Using (7.11) we note from (7.7) that

$$\det\left(a^T \eta a\right) = \det \eta,$$

$$\text{or,} \quad \det \eta \left(\det a\right)^2 = -\left(\det a\right)^2 = -1,$$

$$\text{or,} \quad \det a = \pm 1. \tag{7.22}$$

Lorentz transformations for which $a^0{}_0 \leq -1$ involve time reversal. For example, consider

$$x^0 \to -x^0, \quad x^i \to x^i, \tag{7.23}$$

belongs to this class and it is clear that for such a transformation $\det a = -1$. Similarly, the transformations with a space reflection (parity)

$$x^0 \to x^0, \quad x^i \to -x^i, \quad i = 1, 2, 3, \tag{7.24}$$

correspond to the class of transformations with $a^0{}_0 \geq 1$ and also satisfy $\det a = -1$. Clearly these transformations (group elements) cannot be continuously obtained from the trivial transformation (identity element). (Time reversal and space reflection are, in fact, discrete space-time symmetries. We note from (7.24) that $x^\mu \leftrightarrow x_\mu$ (see (7.17)) corresponds to the transformation of space reflection.) This leads us to define various subgroups of $\mathbb{L}_4$ depending on the values of $a^0{}_0$ and $\det a$ in the following way.

7.1.1 Proper orthochronous Lorentz group.
If we restrict the set of matrices $a \in \mathbb{L}_4$ to satisfy

$$a^0{}_0 \geq 1, \quad \det a = 1, \tag{7.25}$$

this defines the proper orthochronous Lorentz group. Clearly, (7.25) does not allow either time reversal or space reflection (parity). This contains the identity matrix and it can be checked that this defines a subgroup of $\mathbb{L}_4$ in the following manner. If $a, b \in \mathbb{L}_4$ satisfying $a^0{}_0 \geq 1$ and $b^0{}_0 \geq 1$, then

$$(ab)^0{}_0 = a^0{}_\mu b^\mu{}_0 = a^0{}_0 b^0{}_0 + a^0{}_i b^i{}_0 = a^0{}_0 b^0{}_0 - a^0{}_i b_{i0}, \tag{7.26}$$

where we have used (7.17) to lower the spatial index. On the other hand, using the Schwarz inequality (as well as (7.19) and (7.20)) we have

$$a^0{}_i b_{i0} \leq |a^0{}_i||b_{j0}| = \sqrt{(a^0{}_i)^2 (b_{j0})^2}$$

$$= \sqrt{((a^0{}_0)^2 - 1)((b^0{}_0)^2 - 1)}, \tag{7.27}$$

so that (7.26) leads to

$$(ab)^0{}_0 \geq a^0{}_0 b^0{}_0 - \sqrt{((a^0{}_0)^2 - 1)((b^0{}_0)^2 - 1)}. \tag{7.28}$$

Let us further note that

$$(a^0{}_0 b^0{}_0 - 1)^2 - ((a^0{}_0)^2 - 1)((b^0{}_0)^2 - 1)$$
$$= (a^0{}_0)^2 + (b^0{}_0)^2 - 2a^0{}_0 b^0{}_0 = (a^0{}_0 - b^0{}_0)^2 \geq 0,$$

$$\text{or,} \quad (a^0{}_0 b^0{}_0 - 1) \geq \sqrt{((a^0{}_0)^2 - 1)((b^0{}_0)^2 - 1)},$$

$$\text{or,} \quad a^0{}_0 b^0{}_0 - \sqrt{((a^0{}_0)^2 - 1)((b^0{}_0)^2 - 1)} \geq 1, \tag{7.29}$$

where we have used $a^0{}_0, b^0{}_0 \geq 1$ and, consequently (7.28) leads to

$$(ab)^0{}_0 \geq 1. \tag{7.30}$$

7.1.2 Orthochronous Lorentz group. If we restrict to group elements $a \in \mathbb{L}_4$ satisfying only $a^0{}_0 \geq 1$ and allowing for both $\det a = \pm 1$, then such elements also define a group known as the orthochronous Lorentz group. This does not allow time reversal, but includes space reflection (parity). We note that for the pure space reflection defined in (7.24), we can identify

$$a = \eta, \tag{7.31}$$

and it is clear that (7.9) and (7.10) are satisfied trivially in this case.

7.1.3 Improper Lorentz group. If $a \in \mathbb{L}_4$ satisfies $\det a = 1$, but allows for both $a^0{}_0 \geq 1$ and $a^0{}_0 \leq -1$, then this defines the improper Lorentz group. Clearly, in this case both time reversal and space reflection (parity) are allowed together.

7.2 Generalized Clifford algebra

In chapter 6, we have studied the Clifford algebra in the Euclidean space where the metric tensor was the Kronecker delta $\delta_{\mu\nu}$. In Minkowski space we have the nontrivial metric tensors given by $\eta_{\mu\nu}$ and $\eta^{\mu\nu}$ (defined in (7.5) and (7.12)). We can now generalize the Clifford algebra (6.1) to the Minkowski space as

$$\gamma_\mu \gamma_\nu + \gamma_\nu \gamma_\mu = 2\eta_{\mu\nu} \mathbb{1}, \quad \mu, \nu = 0, 1, 2, 3, \tag{7.32}$$

in order to study the Lorentz group. The representation theory of this generalized Clifford algebra, however, remains unchanged. In fact, if we redefine

$$\tilde{\gamma}_0 = \gamma_0, \quad \tilde{\gamma}_i = \pm i\gamma_i, \tag{7.33}$$

the algebra (7.32) takes the form of (6.1)

$$\tilde{\gamma}_\mu \tilde{\gamma}_\nu + \tilde{\gamma}_\nu \tilde{\gamma}_\mu = 2\delta_{\mu\nu}\, \mathbb{1}, \tag{7.34}$$

so that it has a unique four dimensional irreducible representation.

The standard representation of the algebra (7.32) is known as the Dirac representation and has the form ($\gamma^\mu = \eta^{\mu\nu}\gamma_\nu$)

$$\gamma_D^0 = \gamma_{0D} = \begin{pmatrix} \mathbb{1}_2 & 0 \\ 0 & -\mathbb{1}_2 \end{pmatrix}, \gamma_D^i = -\gamma_{iD} = \begin{pmatrix} 0 & \sigma_i \\ -\sigma_i & 0 \end{pmatrix}, i = 1, 2, 3, \tag{7.35}$$

with the charge conjugation matrix C given by

$$C_D = \gamma_D^0 \gamma_D^2 = \begin{pmatrix} 0 & \sigma_2 \\ \sigma_2 & 0 \end{pmatrix}. \tag{7.36}$$

Here $\sigma_i, i = 1, 2, 3$ denote the three 2×2 (Hermitian) Pauli matrices defined in (6.55). We note from (7.35) that

$$(\gamma_D^\mu)^\dagger = \gamma_D^0 \gamma_D^\mu \gamma_D^0, \quad \mu = 0, 1, 2, 3. \tag{7.37}$$

Another commonly used representation (which is, of course, related by a similarity transformation and, therefore, equivalent) is known as the Weyl representation where the generators of the Clifford algebra take the form

$$\gamma_W^\mu = \begin{pmatrix} 0 & \sigma_\mu \\ \tilde{\sigma}_\mu & 0 \end{pmatrix}, \tag{7.38}$$

where

$$\sigma_\mu = (\mathbb{1}_2, \sigma_i), \quad \tilde{\sigma}_\mu = (\mathbb{1}_2, -\sigma_i) \quad (= \sigma^\mu). \tag{7.39}$$

The σ_μ matrices satisfy

$$\sigma_2 \sigma_\mu \sigma_2 = \tilde{\sigma}_\mu^T, \quad \sigma_2 \tilde{\sigma}_\mu \sigma_2 = \sigma_\mu^T, \tag{7.40}$$

so that, in this representation, the charge conjugation matrix can be identified with

$$C_W = \gamma_W^0 \gamma_W^2 = \begin{pmatrix} -\sigma_2 & 0 \\ 0 & \sigma_2 \end{pmatrix}. \tag{7.41}$$

Note that the γ_W^μ matrices in the Weyl representation also satisfy (7.37) which is independent of (equivalent) representations. There is yet another useful representation known as the Majorana representation where all the generators are purely imaginary and have the forms

$$\gamma_M^0 = \begin{pmatrix} 0 & \sigma_2 \\ \sigma_2 & 0 \end{pmatrix}, \quad \gamma_M^1 = \begin{pmatrix} i\sigma_3 & 0 \\ 0 & i\sigma_3 \end{pmatrix},$$

$$\gamma_M^2 = \begin{pmatrix} 0 & -\sigma_2 \\ \sigma_2 & 0 \end{pmatrix}, \quad \gamma_M^3 = \begin{pmatrix} -i\sigma_1 & 0 \\ 0 & -i\sigma_1 \end{pmatrix}. \tag{7.42}$$

These matrices still satisfy $(\gamma_M^\mu)^\dagger = \gamma_M^0 \gamma_M^\mu \gamma_M^0$ and since all the γ_M^μ matrices are purely imaginary, namely, $(\gamma_M^\mu)^* = -\gamma_M^\mu$, it follows that

$$\gamma_M^0 \gamma_M^\mu \gamma_M^0 = (\gamma_M^\mu)^\dagger = -(\gamma_M^\mu)^T, \tag{7.43}$$

so that we can identify the charge conjugation matrix with (the negative sign is a phase convention and the reason behind this will become clear at the end of the next section)

$$C_M = -\gamma_M^0 = \begin{pmatrix} 0 & -\sigma_2 \\ -\sigma_2 & 0 \end{pmatrix}. \tag{7.44}$$

7.3 Dirac equation

Let us next consider the motion of a charged fermion with mass m and charge e interacting with an external electromagnetic field described by the Dirac equation (repeated indices are summed)

$$\left(\gamma^\mu \left(i\partial_\mu - \frac{e}{\hbar c} A_\mu \right) - \frac{mc}{\hbar} \right) \psi(x) = 0, \tag{7.45}$$

where x^μ denotes the space-time coordinates defined in (7.2) and $\psi(x)$ is a four component spinor

$$\psi(x) = \begin{pmatrix} \psi_1(x) \\ \psi_2(x) \\ \psi_3(x) \\ \psi_4(x) \end{pmatrix}, \tag{7.46}$$

on which the 4×4 matrices, γ^μ, operate (note that $\gamma^\mu = \eta^{\mu\nu}\gamma_\nu$). We also note that c and $\hbar$ denote respectively the speed of light in vacuum and the Planck's constant while A_μ represents the vector potential for the electromagnetic field which is assumed to be Hermitian, namely, $A_\mu^\dagger(x) = A_\mu(x)$. Furthermore, to simplify our discussions we will choose units in which $\hbar = c = 1$ so that (7.45) takes the form ($\partial_\mu = \frac{\partial}{\partial x^\mu}$)

$$\left(\gamma^\mu \left(i\partial_\mu - eA_\mu(x)\right) - m\right)\psi(x) = 0. \tag{7.47}$$

The Dirac equation is a relativistic equation and we will next show that it transforms covariantly under a Lorentz transformation (7.6)

$$x^\mu \to x'^\mu = a^\mu{}_\nu x^\nu, \quad a \in \mathbb{L}_4. \tag{7.48}$$

To this end let us define

$$\widetilde{\gamma}^\mu = a^\mu{}_\nu \gamma^\nu, \tag{7.49}$$

which leads to

$$
\begin{aligned}
\widetilde{\gamma}^\mu \widetilde{\gamma}^\nu + \widetilde{\gamma}^\nu \widetilde{\gamma}^\mu &= a^\mu{}_\lambda a^\nu{}_\rho \left(\gamma^\lambda \gamma^\rho + \gamma^\rho \gamma^\lambda\right) \\
&= a^\mu{}_\lambda a^\nu{}_\rho \left(2\eta^{\lambda\rho}\, \mathbb{1}\right) = 2\left(a\eta^{-1}a^T\right)_{\mu\nu} \mathbb{1} \\
&= 2\left(\eta^{-1}\right)_{\mu\nu} \mathbb{1} = 2\eta^{\mu\nu}\, \mathbb{1},
\end{aligned}
\tag{7.50}
$$

where we have made use of (7.16). From the uniqueness of the irreducible representation of the Clifford algebra, we conclude that there exists a 4×4 matrix similarity transformation $S(a)$ such that (see also (6.99) for the Euclidean case)

$$\widetilde{\gamma}^\mu = S^{-1}(a)\gamma^\mu S(a), \tag{7.51}$$

such that we can write (see (7.49) and (7.51), summation over repeated indices is understood)

$$S^{-1}(a)\gamma^\mu S(a) = a^\mu{}_\nu \gamma^\nu, \tag{7.52}$$

for any Lorentz transformation $a \in \mathbb{L}_4$.

We next note (using the chain rule) that the cogradient vector can be written as

$$\partial_\mu = \frac{\partial}{\partial x^\mu} = \frac{\partial x'^\nu}{\partial x^\mu} \frac{\partial}{\partial x'^\nu} = a^\nu{}_\mu \partial'_\nu, \tag{7.53}$$

where we have used (7.6). This can be inverted (using (7.15)) to write

$$\partial'_\mu = a_\mu{}^\nu \, \partial_\nu. \tag{7.54}$$

Under a Lorentz transformation the electromagnetic potential $A_\mu(x)$ transforms like the vector x_μ or the vector ∂_μ so that we can write (see (7.53) and (7.54))

$$A_\mu(x) \to A'_\mu(x') = a_\mu{}^\nu \, A_\nu(x),$$

$$\text{or,} \quad A_\mu(x) = a^\nu{}_\mu \, A'_\nu(x'). \tag{7.55}$$

Therefore, using (7.52)-(7.55) we can write

$$\gamma^\mu \left(i\partial_\mu - eA_\mu(x) \right) = \gamma^\mu a^\nu{}_\mu \left(i\partial'_\nu - eA'_\nu(x') \right)$$

$$= \left(\sum_{\mu=0}^3 a^\nu{}_\mu \gamma^\mu \right) \left(i\partial'_\nu - eA'_\nu(x') \right)$$

$$= S^{-1}(a)\gamma^\nu S(a) \left(i\partial'_\nu - eA'_\nu(x') \right)$$

$$= S^{-1}(a) \left(\gamma^\nu \left(i\partial'_\nu - eA'_\nu(x') \right) \right) S(a). \tag{7.56}$$

Substituting this into the Dirac equation (7.47) we obtain

$$\left[S^{-1}(a) \left(\gamma^\nu \left(i\partial'_\nu - eA'_\nu(x') \right) \right) S(a) - m \right] \psi(x)$$

$$= S^{-1}(a) \left[\gamma^\nu \left(i\partial'_\nu - eA'_\nu(x') \right) - m \right] S(a)\psi(x) = 0,$$

$$\text{or,} \quad \left[\gamma^\nu \left(i\partial'_\nu - eA'_\nu(x') \right) - m \right] \psi'(x') = 0, \tag{7.57}$$

where we have identified

$$\psi'(x') = S(a)\psi(x). \tag{7.58}$$

We note that (7.57) has exactly the same form as the original Dirac equation (7.47).

Let us next consider two Lorentz transformations, $a, b \in \mathbb{L}_4$ and note that (see (7.52))

$$S^{-1}(b) \left(S^{-1}(a)\gamma^\mu S(a) \right) S(b) = S^{-1}(b) \left(a^\mu{}_\nu \gamma^\nu \right) S(b)$$

$$= a^\mu{}_\nu S^{-1}(b)\gamma^\nu S(b) = a^\mu{}_\nu b^\nu{}_\lambda \gamma^\lambda$$

$$= (ab)^\mu{}_\lambda \gamma^\lambda = S^{-1}(ab)\gamma^\mu S(ab), \tag{7.59}$$

so that

$$[\gamma^\mu, S(ab)S^{-1}(b)S^{-1}(a)] = 0, \tag{7.60}$$

which lets us identify (by Schur's lemma)

$$S(ab)S^{-1}(b)S^{-1}(a) = \lambda \mathbb{1},$$

$$\text{or,} \quad S(ab) = \lambda S(a)S(b), \tag{7.61}$$

with λ a constant. If we impose the condition $\det S(a) = 1$ as in the previous section and note that $S(a)$ is a 4×4 matrix (in the Dirac spinor space), (7.61) determines

$$\lambda^4 = 1, \quad \text{or,} \quad \lambda = \pm 1, \pm i. \tag{7.62}$$

The continuity of the group elements (as we have argued earlier) selects

$$\lambda = 1, \tag{7.63}$$

if we are restricting to proper orthochronous Lorentz transformations and in this case the similarity transformation (matrix) defines a 4-dimensional representation of the proper orthochronous group. (In general, however, $S(a)$ can lead to a ray representation of the Lorentz group.) With this we can identify $\psi'(x')$ in (7.58) as the Lorentz transformed Dirac spinor and (7.57) as showing the covariance of the Dirac equation under a Lorentz transformation.

Taking the Hermitian conjugate of (7.52) and recalling that $a^\mu_{\ \nu}$ are real for Lorentz transformations, we obtain

$$\left(S^{-1}(a)\gamma^\mu S(a)\right)^\dagger = a^\mu_{\ \nu} (\gamma^\nu)^\dagger,$$

$$\text{or,} \quad \gamma^0\left(S^\dagger(a)(\gamma^\mu)^\dagger(S^{-1})^\dagger(a)\right)\gamma^0 = a^\mu_{\ \nu}\,\gamma^0(\gamma^\nu)^\dagger\gamma^0 = a^\mu_{\ \nu}\gamma^\nu,$$

$$\text{or,} \quad \left(\gamma^0 S^\dagger(a)\gamma^0\right)\gamma^\mu\left(\gamma^0 S^\dagger(a)\gamma^0\right)^{-1} = S^{-1}(a)\gamma^\mu S(a),$$

$$\text{or,} \quad \left[S(a)\gamma^0 S^\dagger(a)\gamma^0, \gamma^\mu\right] = 0, \tag{7.64}$$

where we have used (7.37) as well as the properties $\gamma^0 = (\gamma^0)^\dagger = (\gamma^0)^{-1}$ in the Dirac representation. Therefore, we can write (by Schur's lemma)

$$S(a)\gamma^0 S^\dagger(a)\gamma^0 = b\mathbb{1}, \tag{7.65}$$

with b a constant. Taking the Hermitian conjugate of this relation (and using $\gamma^0 = (\gamma^0)^\dagger, (\gamma^0)^2 = \mathbb{1}$) we obtain

$$\gamma^0 S(a)\gamma^0 S^\dagger(a) = b^*\mathbb{1},$$

$$\text{or,} \quad S(a)\gamma^0 S^\dagger(a)\gamma^0 = b^*\mathbb{1}. \tag{7.66}$$

Comparing (7.65) and (7.66) we determine

$$b = b^* = \text{real}. \tag{7.67}$$

Taking the determinant of both sides in (7.65) and using the fact that $\det S(a) = \pm 1$, we conclude that

$$b^4 = 1, \quad \text{or,} \quad b = \pm 1, \pm i, \tag{7.68}$$

and the two real roots of this equation are given by $b = \pm 1$. Once again, the continuity of the group elements for the proper orthochronous Lorentz group with $\det S(a) = 1$ determines $b = 1$. In fact, in this case, the argument can be extended to the entire Lorentz group in the following manner. Let us consider the reflection (parity) operation (see (7.24)) $(x^0, x^i) \rightarrow (x^0, -x^i)$ corresponding to the choice $a = \eta$ (see (7.31)). A simple analysis of the Dirac equation shows that it is invariant under (up to a phase)

$$\psi(x^0, x^i) \rightarrow \psi'(x^0, -x^i) = \gamma^0\, \psi(x^0, x^i), \tag{7.69}$$

which implies that we can identify $S(\eta) = \gamma^0$. On the other hand, this directly leads to (the phase simply drops out)

$$S(\eta)\gamma^0 S^\dagger(\eta)\gamma^0 = (\gamma^0)^4 = \mathbb{1}, \tag{7.70}$$

and comparing with (7.65), we conclude that $b = 1$.

Although the finite dimensional matrices $S(a)$ provide a representation of the Lorentz group, the representation is not unitary ($S^\dagger(a)S(a) \neq \mathbb{1}$) in conformity with the fact that the Lorentz group is non-compact (see, for example, the discussion in section 1.2.5). However, we will show next that $\psi(x)$ provides an infinite dimensional unitary representation of the Lorentz group. To see this, let us take the Hermitian conjugate of (7.47) to obtain

$$\psi^\dagger(x)\left((\gamma^\mu)^\dagger\left(-i\overleftarrow{\partial_\mu} - eA_\mu(x)\right) - m\right) = 0, \tag{7.71}$$

where we have used the fact that $A^\dagger_\mu(x) = A_\mu(x)$. Using $(\gamma^\mu)^\dagger = \gamma^0 \gamma^\mu \gamma^0$ (see (7.37)) as well as $(\gamma^0)^2 = \mathbb{1}$, we can write (7.71) also as

$$\psi^\dagger(x)\gamma^0 \left(\gamma^\mu \left(-i\overleftarrow{\partial}_\mu - eA_\mu(x)\right) - m\right)\gamma^0 = 0,$$

$$\text{or,} \quad \overline{\psi}(x)\left(\gamma^\mu \left(i\overleftarrow{\partial}_\mu + eA_\mu(x)\right) + m\right) = 0, \tag{7.72}$$

where we have identified ($\overline{\psi}$ is known as the adjoint spinor)

$$\overline{\psi}(x) = \psi^\dagger(x)\gamma^0. \tag{7.73}$$

As a result, from (7.47) and (7.72) we can deduce the conservation law (continuity equation) for the electric current in the standard manner

$$\partial_\mu \left(\overline{\psi}(x)\gamma^\mu \psi(x)\right) = 0. \tag{7.74}$$

Under a general Lorentz transformation $\overline{\psi}(x)$ transforms as

$$\overline{\psi}(x) \to \overline{\psi}'(x') = (\psi'(x'))^\dagger \gamma^0 = (S(a)\psi(x))^\dagger \gamma^0 = \psi^\dagger(x)S^\dagger(a)\gamma^0$$

$$= \psi^\dagger(x)\gamma^0 \gamma^0 S^\dagger(a)\gamma^0 = \overline{\psi}(x)S^{-1}(a), \tag{7.75}$$

where we have used (7.65) (with $b = 1$). This shows that the adjoint spinor $\overline{\psi}(x)$ transforms inversely as $\psi(x)$ under a Lorentz transformation (which is why $\overline{\psi}$ is so important in studying relativistic Dirac systems). It follows now that

$$\overline{\psi}'(x')\psi'(x') = \overline{\psi}(x)S^{-1}(a)S(a)\psi(x) = \overline{\psi}(x)\psi(x), \quad \text{(scalar)},$$

$$\overline{\psi}'(x')\gamma^\mu \psi'(x') = \overline{\psi}(x)S^{-1}\gamma^\mu S(a)\psi(x)$$

$$= a^\mu_{\ \nu}\overline{\psi}(x)\gamma^\nu \psi(x), \quad \text{(vector)},$$

$$\overline{\psi}'(x')[\gamma^\mu, \gamma^\nu]\psi'(x') = \overline{\psi}(x)S^{-1}(a)[\gamma^\mu, \gamma^\nu]S(a)\psi(x)$$

$$= a^\mu_{\ \lambda}a^\nu_{\ \rho}\overline{\psi}(x)[\gamma^\lambda, \gamma^\rho]\psi(x), \quad \text{(tensor)}. \tag{7.76}$$

Similarly, defining (see, for example, (6.12) and the factor of i is included in the definition to make the matrix Hermitian)

$$\Lambda = \gamma_5 = i\gamma^0 \gamma^1 \gamma^2 \gamma^3 = -\frac{i}{4!}\,\epsilon_{\mu\nu\lambda\rho}\gamma^\mu \gamma^\nu \gamma^\lambda \gamma^\rho, \tag{7.77}$$

where $\epsilon_{\mu\nu\lambda\rho}$ is the completely anti-symmetric Levi-Civita tensor with $\epsilon_{0123} = -1$, we can show that

$$\overline{\psi}'(x')\gamma_5\psi'(x') = (\det a)\,\overline{\psi}(x)\gamma_5\psi(x), \quad \text{(pseudo-scalar)},$$

$$\overline{\psi}'(x')\gamma_5\gamma^\mu \psi'(x') = (\det a)\,a^\mu_{\ \nu}\overline{\psi}(x)\gamma_5\gamma^\nu \psi(x), \quad \text{(axial vector)}. \tag{7.78}$$

Let us consider the inner product for the Dirac field (spinor) defined as

$$\langle \psi | \psi \rangle = \int \mathrm{d}^3 x \, \psi^\dagger(x^0, \mathbf{x}) \psi(x^0, \mathbf{x}), \qquad (7.79)$$

which is clearly positive definite. We also note that ($\mu = 0, 1, 2, 3$ and $i = 1, 2, 3$)

$$\frac{\mathrm{d}}{\mathrm{d}x^0} \int \mathrm{d}^3 x \, \psi^\dagger(x^0, \mathbf{x}) \psi(x^0, \mathbf{x}) = \int \mathrm{d}^3 x \, \partial_0 \left(\psi^\dagger(x^0, \mathbf{x}) \psi(x^0, \mathbf{x}) \right)$$

$$= \int \mathrm{d}^3 x \, \partial_0 \left(\overline{\psi}(x^0, \mathbf{x}) \gamma^0 \psi(x^0, \mathbf{x}) \right) = \int \mathrm{d}^3 x \, \partial_0 \left(\overline{\psi}(x) \gamma^0 \psi(x) \right)$$

$$= \int \mathrm{d}^3 x \left[\partial_\mu \left(\overline{\psi}(x) \gamma^\mu \psi(x) \right) - \partial_i \left(\overline{\psi}(x) \gamma^i \psi(x) \right) \right]$$

$$= - \int \mathrm{d}^3 x \, \partial_i \left(\overline{\psi}(x) \gamma^i \psi(x) \right) = 0, \qquad (7.80)$$

where we have used (7.74) in the intermediate step and the last step follows if the fields fall off asymptotically, namely, $\psi(x^0, \mathbf{x}) \to 0$ for $|\mathbf{x}| \to \infty$. Thus, we see that for such fields, $\langle \psi | \psi \rangle$ is independent of x^0.

In order to prove that the inner product $\langle \psi | \psi \rangle$ is also Lorentz invariant, we use the following argument. Let σ denote an arbitrary space-like surface, namely, any two distinct points x^μ, y^μ on the surface σ satisfy

$$\eta_{\mu\nu}(x^\mu - y^\mu)(x^\nu - y^\nu) = (x^0 - y^0)^2 - (x^i - y^i)^2 < 0. \qquad (7.81)$$

Such a surface is manifestly Lorentz invariant. The most familiar of such a surface is, of course, a time slice defined by $x^0 = $ constant (equal-time surface). The infinitesimal volume vector at any point on such a surface is given by

$$\mathrm{d}\sigma_\mu(x) = (\mathrm{d}x^1 \mathrm{d}x^2 \mathrm{d}x^3, \mathrm{d}x^0 \mathrm{d}x^2 \mathrm{d}x^3, \mathrm{d}x^0 \mathrm{d}x^1 \mathrm{d}x^3, \mathrm{d}x^0 \mathrm{d}x^1 \mathrm{d}x^2)$$

$$= \frac{(\mathrm{d}x^0 \mathrm{d}x^1 \mathrm{d}x^2 \mathrm{d}x^3)}{\mathrm{d}x^\mu}, \qquad (7.82)$$

and this transforms like the vector x_μ under a Lorentz transformation. Using this we can write the inner product as

$$\langle \psi | \psi \rangle_\sigma = \int_\sigma \mathrm{d}\sigma_\mu(x) \, \overline{\psi}(x) \gamma^\mu \psi(x), \qquad (7.83)$$

which reduces to (7.79) for the surface $x^0 = $ constant. Under a Lorentz transformation we have

$$\langle\psi|\psi\rangle_\sigma \rightarrow \langle\psi'|\psi'\rangle_{\sigma'} = \int_{\sigma'} d\sigma'_\mu(x)\,\overline{\psi}'(x')\gamma^\mu\psi'(x)$$

$$= \int_{\sigma'} d\sigma_\mu(x)\,\overline{\psi}(x)\gamma^\mu\psi(x) = \int_\sigma d\sigma_\mu(x)\,\overline{\psi}(x)\gamma^\mu\psi(x)$$

$$= \langle\psi|\psi\rangle_\sigma, \tag{7.84}$$

so that the inner product is invariant under a Lorentz transformation. Here we have used the fact that $d\sigma_\mu$ and $\overline{\psi}\gamma^\mu\psi$ transform inversely under a Lorentz transformation (one is like x_μ while the other is like x^μ). We have also used in the intermediate step our earlier observation that the surface σ is Lorentz invariant.

The inner product $\langle\psi|\psi\rangle_\sigma$ is actually independent of the surface σ on which it is defined and, therefore, we can identify this with the definition in (7.79). To demonstarte this, let us consider the differential

$$\frac{\delta F(\sigma)}{\delta\sigma_\mu(y)} = \lim_{\Delta V \to 0} \frac{1}{\Delta V}\left(F(\sigma') - F(\sigma)\right), \tag{7.85}$$

where

$$F(\sigma) = \int_\sigma d\sigma_\mu(x)\,\overline{\psi}(x)\gamma^\mu\psi(x), \tag{7.86}$$

and we assume that the space-like surfaces σ and σ' differ only in an infinitesimal volume ΔV around the point y on σ (see Fig. 7.1). In

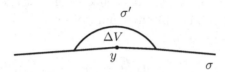

Figure 7.1: The surfaces σ and σ' differ only around a point y in an infinitesimal volume ΔV.

this case, using Gauss' theorem we obtain

$$\frac{\delta}{\delta\sigma(y)}\int_\sigma d\sigma_\mu(x)\,\overline{\psi}(x)\gamma^\mu\psi(x) = \frac{\partial}{\partial y^\mu}\left(\overline{\psi}(y)\gamma^\mu\psi(y)\right) = 0, \tag{7.87}$$

which follows from (7.74). Although we derived this result by considering two space-like surfaces σ and σ' differing only in an infinitesimal region, this result can be extended to other cases and proves that $\langle \psi | \psi \rangle_\sigma$ is independent of the surface σ. As a result, we conclude that the inner product defined in (7.79) is invariant under any Lorentz transformation

$$\langle \psi' | \psi' \rangle = \langle \psi | \psi \rangle, \tag{7.88}$$

so that the transformation

$$\psi(x) \to \psi'(x') = S(a)\psi(x), \tag{7.89}$$

is unitary with respect to this inner product. We would like to note here that the Hilbert space under consideration here is that of the first quantized Dirac field. A more physically interesting case is that for the second quantized Dirac field which we will not go into.

7.3.1 Charge conjugation. Let us now return to equation (7.72) satisfied by the adjoint spinor and take its matrix transpose to obtain

$$((\gamma^\mu)^T (i\partial_\mu + eA_\mu(x)) + m) \overline{\psi}^T(x) = 0. \tag{7.90}$$

Furthermore, recalling that (see (6.58)) the charge conjugation matrix relates

$$(\gamma^\mu)^T = -C^{-1}\gamma^\mu C. \tag{7.91}$$

we can rewrite (7.90) also as

$$C^{-1} (\gamma^\mu (i\partial_\mu + eA_\mu(x)) - m) C\overline{\psi}^T(x) = 0,$$

$$\text{or,} \quad (\gamma^\mu (i\partial_\mu + eA_\mu(x)) - m) \psi_C(x) = 0, \tag{7.92}$$

where we have identified

$$\psi_C(x) = C\overline{\psi}^T(x). \tag{7.93}$$

Therefore, we see that the spinor $\psi_C(x)$ satisfies the same equation as $\psi(x)$ with the same mass m and opposite electric charge ($-e$ instead of e). Consequently, we can interpret $\psi_C(x)$ to represent the antiparticle state of the Dirac particle represented by $\psi(x)$ (and is known as the charge conjugate spinor which is the reason for the subscript).

Under a general Lorentz transformation, this spinor transforms as (see (7.75))

$$\psi_C'(x') = C\overline{\psi'}^T(x') = C\left(\overline{\psi}(x)S^{-1}(a)\right)^T = C\left(\overline{\psi}(x)S(a^{-1})\right)^T$$
$$= CS^T(a^{-1})\overline{\psi}^T(x) = CS^T(a^{-1})C^{-1}C\overline{\psi}^T(x),$$

or, $\quad \psi_C'(x') = \left(CS^T(a^{-1})C^{-1}\right)\psi_C(x).$ \hfill (7.94)

In particular, for space reflection (parity), we have already seen in (7.69) that

$$\psi(x^0, -x^i) = \gamma^0\psi(x^0, x^i),$$
$$S(a) = \gamma^0, \quad S(a^{-1}) = S^{-1}(a) = \gamma^0.$$ \hfill (7.95)

For such a transformation

$$CS^T(a^{-1})C^{-1} = C(\gamma^0)^T C^{-1} = -\gamma^0,$$ \hfill (7.96)

so that we see from (7.94) that under a parity transformation, the charge conjugate spinor will transform as

$$\psi_C(x^0, -x^i) = -\gamma^0\psi_C(x^0, x^i).$$ \hfill (7.97)

This should be contrasted with the transformation of $\psi(x)$ in (7.95) and implies that the parity quantum number of the anti-particle is opposite of that for the particle.

On the other hand, for a proper orthochronous Lorentz transformations, we can show that

$$C(S(a^{-1}))^T C^{-1} = S(a),$$ \hfill (7.98)

in the following way. Let us recall from (7.52) that

$$S^{-1}(a)\gamma^\mu S(a) = a^\mu{}_\nu \gamma^\nu.$$ \hfill (7.99)

Taking the transpose of this equation, we obtain

$$(S(a))^T(\gamma^\mu)^T(S^{-1}(a))^T = a^\mu{}_\nu(\gamma^\nu)^T,$$

or, $\quad (S(a))^T(-C^{-1}\gamma^\mu C)(S^{-1}(a))^T = a^\mu{}_\nu(-C\gamma^\nu C),$

or, $\quad (S(a))^T C^{-1}\gamma^\mu C(S^{-1}(a))^T = C^{-1}(a^\mu{}_\nu\gamma^\nu)C,$

or, $\quad (S(a))^T C^{-1}\gamma^\mu C(S^{-1}(a))^T = C^{-1}S^{-1}(a)\gamma^\mu S(a)C,$

or, $\quad \gamma^\mu C(S^{-1}(a))^T C^{-1}S^{-1}(a) = C(S^{-1}(a))^T C^{-1}S^{-1}(a)\gamma^\mu,$

or, $\quad [C(S^{-1}(a))^T C^{-1}S^{-1}(a), \gamma^\mu] = 0,$ \hfill (7.100) where we have used (7.91) in the second line of (7.100). Therefore, by Schur's lemma, we can write

$$C(S^{-1}(a))^T C^{-1} S^{-1}(a) = \lambda \mathbb{1}. \tag{7.101}$$

Taking the determinant of both sides and recalling that $\det S(a) = 1$ for a proper orthochronous Lorentz transformation, as in (7.62), we determine that $\lambda^4 = 1$ which leads to $\lambda = \pm 1, \pm i$. The argument of continuity then selects $\lambda = 1$ and with this (7.101) leads to

$$C(S^{-1}(a))^T C^{-1} S^{-1}(a) = C(S(a^{-1}))^T C^{-1} S^{-1}(a) = \mathbb{1},$$

$$\text{or,} \quad C(S(a^{-1}))^T C^{-1} = S(a), \tag{7.102}$$

which proves (7.98). As a result, we conclude from (7.94) that $\psi'_C(x') = S(a)\psi_C(x)$ and $\psi_C(x)$ behaves much like the particle $\psi(x)$ under a proper orthochronous Lorentz transformation.

7.3.2 Weyl and Majorana particles. In ending this section, let us discuss briefly the concept of Weyl particles as well as Majorana particles. Let us first consider the Weyl representation of the γ^μ matrices (see (7.38)) given by

$$\gamma^\mu = \begin{pmatrix} 0 & \sigma^\mu \\ \tilde{\sigma}^\mu & 0 \end{pmatrix}, \tag{7.103}$$

and we write the four component spinor in terms of two component spinors as

$$\psi(x) = \begin{pmatrix} \phi(x) \\ \xi(x) \end{pmatrix}, \tag{7.104}$$

where the two component spinors can be written as

$$\phi(x) = \begin{pmatrix} \psi_1(x) \\ \psi_2(x) \end{pmatrix}, \quad \xi(x) = \begin{pmatrix} \psi_3(x) \\ \psi_4(x) \end{pmatrix}. \tag{7.105}$$

In terms of the two component spinors in (7.104), the Dirac equation (7.47) decomposes into two coupled equations

$$\sigma^\mu \left(i\partial_\mu - eA_\mu(x) \right) \xi(x) = m\phi(x),$$

$$\tilde{\sigma}^\mu \left(i\partial_\mu - eA_\mu(x) \right) \phi(x) = m\xi(x). \tag{7.106}$$

We note that under a parity transformation $\psi(x^0, x^i) \to \gamma^0 \psi(x^0, x^i)$ (see (7.69)) which leads to the fact that under a parity transformation (mirror reflection)

$$\phi(x^0, x^i) \leftrightarrow \xi(x^0, x^i). \tag{7.107}$$

This is consistent with the fact that under a Lorentz transformation the spinor components $\phi(x)$ and $\xi(x)$ transform among themselves.

If the particle is massless ($m = 0$), then from (7.106) we see that the system of equations decouples

$$\sigma^\mu \left(i\partial_\mu - eA_\mu(x) \right) \xi(x) = 0,$$

$$\tilde{\sigma}^\mu \left(i\partial_\mu - eA_\mu(x) \right) \phi(x) = 0, \tag{7.108}$$

and we can set either $\phi(x) = 0$ or $\xi(x) = 0$ to have a consistent (two component) equation known as the Weyl equation (this is possible only if $m = 0$). The two component spinors $\phi(x)$ and $\xi(x)$ are known as Weyl spinors and each of the dynamical equations in (7.108) is invariant under a proper orthochronous Lorentz transformation. However, it follows from (7.107) that they are not invariant under a parity transformation.

Let us next consider the Majorana representation of the γ^μ matrices given in (7.42). In this case, the γ^μ matrices are purely imaginary, $(\gamma^0)^T = -\gamma^0$ and we can identify the charge conjugation matrix with $C = -\gamma^0$ (see (7.44)). As a result, the charge conjugate spinor can be written as

$$\psi_C(x) = C\overline{\psi}^T(x) = (-\gamma^0) \left(\psi^\dagger(x)\gamma^0 \right)^T$$

$$= (-\gamma^0)(\gamma^0)^T \psi^*(x) = (-\gamma^0)(-\gamma^0)\psi^*(x)$$

$$= \psi^*(x), \tag{7.109}$$

where we have used $(\gamma^0)^T = -\gamma^0$ in the Majorana representation. Namely, the charge conjugate spinor (describing the anti-particle) can be identified with the complex conjugate of the original spinor itself. (This is the reason for the negative sign in the definition of the charge conjugation matrix in (7.44).) From (7.92) we see that in the absence of the electromagnetic coupling ($e = 0$), the equation satisfied by the charge conjugate spinor is given by (see (7.92) as well as (7.109))

$$(i\gamma^\mu \partial_\mu - m) \psi^*(x) = 0, \tag{7.110}$$

which has the same form as the original equation (7.47) (for $e = 0$). Therefore, we can choose a real spinor $\psi^*(x) = \psi(x)$ or $\psi_C(x) = \psi(x)$ so that in this case the particle is identical to its anti-particle (self charge conjugate). Such a particle is known as a Majorana particle and in this case, the phase of the parity transformation needs to be chosen as

$$\psi(x^0, x^i) \to \psi'(x^0, -x^i) = i\gamma^0 \psi(x^0, x^i), \tag{7.111}$$

so that the reality condition for the spinor is maintained under the transformation (recall that γ^0 is purely imaginary in this representation), namely,

$$(\psi'(x^0, -x^i))^* = \psi'(x^0, -x^i). \tag{7.112}$$

We conclude simply by noting that there is no evidence so far for the existence of a Majorana fermion except possibly for a neutrino.

7.4 References

A. Das, *Lectures on Quantum Field Theory*, World Scientific, Singapore (2008).

CHAPTER 8

Yang-Mills gauge theory

In physics gauge theories play an important role since the basic fundamental forces are described by such theories. We are familiar with the Maxwell gauge theory from our study of electrodynamics. This is the simplest of gauge theories based on the symmetry group $U(1)$. However, there are other gauge theories based on larger symmetry groups that, for example, are necessary to describe such fundamental forces as the weak force or the strong force. In this chapter we will discuss how such dynamical theories are constructed based on larger symmetry groups.

8.1 Gauge field dynamics

Let $t_a, a = 1, 2, \cdots, N$ define a basis of a Lie algebra L with commutation relations (here we are trying to use the notation used in discussions in physics as opposed to the notation in section 3.3)

$$[t_a, t_b] = i f^c_{ab} t_c, \quad a, b, c = 1, 2, \cdots, N, \tag{8.1}$$

where

$$f^c_{ab} = -f^c_{ba}, \tag{8.2}$$

denote the (real) structure constants of the group and repeated indices are assumed to be summed (in chapter 3, the structure constants were called $C^\lambda_{\mu\nu}$ without the factor of "i", see (3.93)). The generators t_a are assumed to be Hermitian, $t^\dagger_a = t_a$. Let $A^a_\mu(x), a = 1, 2, \cdots, N$ represent the gauge field which has a Lorentz vector index $\mu = 0, 1, 2, 3$ as well as a group index a and is a real function of the space-time coordinates $x = (x^0, x^1, x^2, x^3)$ in the 4-dimensional Minkowski space. We next introduce a L-valued gauge field $A_\mu(x)$ (also known as the connection) by (the group index is being summed)

$$A_\mu(x) = A^a_\mu(x) t_a, \tag{8.3}$$

as well as the field strength tensor (or the curvature) tensor $F_{\mu\nu}(x)$ associated with the gauge field by (we are setting the coupling constant for gauge interaction to unity as well as $\hbar = c = 1$)

$$F_{\mu\nu}(x) = -F_{\nu\mu}(x) = \partial_\mu A_\nu(x) - \partial_\nu A_\mu(x) + i[A_\mu(x), A_\nu(x)]$$

$$= F^a_{\mu\nu}(x) t_a, \tag{8.4}$$

where the components of the field strength tensor are given by

$$F^a_{\mu\nu}(x) = \partial_\mu A^a_\nu(x) - \partial_\nu A^a_\mu(x) - f^a_{bc} A^b_\mu(x) A^c_\nu(x) = -F^{(a)}_{\nu\mu}(x), \tag{8.5}$$

and the anti-symmetry is manifest because of (8.2). Here we have used the result that (see (8.1))

$$i[A_\mu(x), A_\nu(x)] = i \left[A^b_\mu(x) t_b, A^c_\nu(x) t_c \right]$$

$$= i A^b_\mu(x) A^c_\nu(x) [t_b, t_c] = i A^b_\mu(x) A^c_\nu \left(i f^a_{bc} t_a \right)$$

$$= -\left(f^a_{bc} A^b_\mu(x) A^c_\nu(x) \right) t_a. \tag{8.6}$$

This follows because $A^b_\mu(x)$, $A^c_\nu(x)$ are classical functions of the space-time coordinates and we have used (8.1). (Note that if t^as correspond to a matrix representation of the symmetry group, A_μ as well as $F_{\mu\nu}$ will also be matrices in that particular representation.)

Let us next define a space-time coordinate dependent group element

$$g(x) = e^{-i\xi^a(x) t_a}, \tag{8.7}$$

of the corresponding Lie group G where $\xi^a(x)$ are N real functions of the space-time coordinates x. To get a feeling for $g(x)$, we note that it cannot clearly be an element of the Lie algebra since the Lie algebra does not contain quadratic and higher order terms in the generators (basis elements). On the other hand, if we are considering a $n \times n$ matrix representation $\rho(t_a)$ of the Lie algebra, the generators will be $n \times n$ matrices. The exponential in such a representation, $\exp(-i\xi^a(x)\rho(t_a))$, is also a $n \times n$ unitary matrix which is quite simple to manipulate with. We will use such a matrix representation for t_a in this chapter (without explicitly writing $\rho(t_a)$).

Let us now consider a local gauge transformation of the form (it is called local because the matrix of transformation is space-time coordinate dependent, space-time independent transformations are called global)

$$A_\mu(x) \rightarrow A'_\mu(x) = g(x) A_\mu(x) g^{-1}(x) + i(\partial_\mu g(x)) g^{-1}(x). \tag{8.8}$$

We note here that $(A'_\mu(x))^\dagger = A'_\mu(x)$ is Hermitian if we recall that (see also (8.13))

$$\partial_\mu(g(x))^\dagger = \partial_\mu(g(x))^{-1} = -g^{-1}(x)(\partial_\mu g(x))g^{-1}(x). \qquad (8.9)$$

We will verify shortly that $A'_\mu(x)$ can also be written as

$$A'_\mu(x) = (A')^a_\mu(x)t_a, \qquad (8.10)$$

namely, $A'_\mu(x)$ also belongs to the Lie algebra, but before doing that let us show that the field strength tensor $F_{\mu\nu}(x)$ transforms covariantly under the gauge transformation (8.8), namely,

$$F_{\mu\nu}(x) \to F'_{\mu\nu}(x) = \partial_\mu A'_\nu(x) - \partial_\nu A'_\mu(x) + i[A'_\mu(x), A'_\nu(x)]$$

$$= g(x)F_{\mu\nu}(x)g^{-1}(x). \qquad (8.11)$$

We note from (8.8) that

$$\partial_\mu A'_\nu(x) = \partial_\mu \left(g(x)A_\nu(x)g^{-1}(x) + i(\partial_\nu g(x))g^{-1}(x)\right)$$

$$= g(x)(\partial_\mu A_\nu(x))g^{-1}(x) + (\partial_\mu g(x))A_\nu(x)g^{-1}(x)$$

$$+ g(x)A_\nu(x)(\partial_\mu g^{-1}(x)) + i(\partial_\mu \partial_\nu g(x))g^{-1}(x)$$

$$+ i(\partial_\nu g(x))(\partial_\mu g^{-1}(x)). \qquad (8.12)$$

We recall that $g(x)g^{-1}(x) = \mathbb{1} = g^{-1}(x)g(x)$, which leads to

$$(\partial_\mu g^{-1}(x)) = -g^{-1}(x)(\partial_\mu g(x))g^{-1}(x). \qquad (8.13)$$

Using this result in (8.12) we obtain

$$\partial_\mu A'_\nu(x) = g(x)(\partial_\mu A_\nu(x))g^{-1}(x) + (\partial_\mu g(x))g^{-1}(x)g(x)A_\nu g^{-1}(x)$$

$$- g(x)A_\nu(x)g^{-1}(x)(\partial_\mu g(x))g^{-1}(x) + i(\partial_\mu \partial_\nu g(x))g^{-1}(x)$$

$$- i(\partial_\nu g(x))g^{-1}(x)(\partial_\mu g(x))g^{-1}(x)$$

$$= g(x)(\partial_\mu A_\nu(x))g^{-1}(x) + [(\partial_\mu g(x))g^{-1}(x), g(x)A_\nu(x)g^{-1}(x)]$$

$$+ i(\partial_\mu \partial_\nu g(x))g^{-1}(x) - i(\partial_\nu g(x))(\partial_\mu g(x))g^{-1}(x). \qquad (8.14)$$

As a result, we can write

$$\partial_\mu A'_\nu(x) - \partial_\nu A'_\mu(x) = g(x)\left(\partial_\mu A_\nu(x) - \partial_\nu A_\mu(x)\right)g^{-1}(x)$$

$$+ [(\partial_\mu g(x))g^{-1}(x), g(x)A_\nu(x)g^{-1}(x)]$$

$$+ [g(x)A_\mu(x)g^{-1}(x), (\partial_\nu g(x))g^{-1}(x)]$$

$$+ i[(\partial_\mu g(x))g^{-1}(x), (\partial_\nu g(x))g^{-1}(x)]. \qquad (8.15)$$

Similarly, we obtain in a straightforward manner

$$i[A'_\mu(x), A'_\nu(x)] = ig(x)[A_\mu, A_\nu(x)]g^{-1}(x)$$
$$- [(\partial_\mu g(x))g^{-1}(x), g(x)A_\nu(x)g^{-1}(x)]$$
$$- [g(x)A_\mu(x)g^{-1}(x), (\partial_\nu g(x))g^{-1}(x)]$$
$$- i[(\partial_\mu g(x))g^{-1}(x), (\partial_\nu g(x))g^{-1}(x)]. \tag{8.16}$$

It follows now from (8.15) and (8.16) that

$$F'_{\mu\nu}(x) = \partial_\mu A'_\nu(x) - \partial_\nu A'_\mu(x) + i[A'_\mu(x), A'_\nu(x)]$$
$$= g(x)\left(\partial_\mu A_\nu(x) - \partial_\nu A_\mu(x) + i[A_\mu(x), A_\nu(x)]\right)g^{-1}(x)$$
$$= g(x)F_{\mu\nu}(x)g^{-1}(x). \tag{8.17}$$

showing that the field strength tensor does indeed transform covariantly under the gauge transformation (8.8).

Let us next come back to the question of whether A'_μ is a Lie algebra valued connection (gauge field) as stated in (8.10). We note that if X and Y denote any two linear operators (matrices), then we can define a one parameter dependent family of matrices as

$$f(\lambda) = e^{\lambda X} Y e^{-\lambda X}, \tag{8.18}$$

where λ is a constant parameter. Taking the derivative of this function with respect to the parameter λ we obtain

$$f^{(1)}(\lambda) = \frac{df(\lambda)}{d\lambda} = e^{\lambda X}\left(XY - YX\right)e^{-X}$$
$$= e^{\lambda X}[X, Y]e^{-\lambda X} = e^{\lambda X}\left((ad\,X)Y\right)e^{-\lambda X},$$
$$f^{(2)}(\lambda) = \frac{df^{(1)}(\lambda)}{d\lambda} = e^{\lambda X}\left((ad\,X)^2 Y\right)e^{-\lambda X}, \tag{8.19}$$

and so on where we have used the definition of the adjoint operation (see (3.103))

$$(ad\,X)Y = [X, Y]. \tag{8.20}$$

Using these we can Taylor expand the function around $\lambda = 0$ to obtain

$$f(\lambda) = \sum_{n=0}^{\infty} \frac{\lambda^n}{n!} f^{(n)}(0) = \sum_{n=0}^{\infty} \frac{\lambda^n}{n!} (ad\,X)^n Y = e^{\lambda(ad\,X)} Y. \tag{8.21}$$

Now setting $\lambda = 1$ this leads to

$$e^X Y e^{-X} = e^{(ad\,X)} Y = \sum_{n=0}^{\infty} \frac{(ad\,X)^n}{n!} Y$$

$$= Y + [X, Y] + \frac{1}{2!}[X, [X, Y]] + \frac{1}{3!}[X, [X, [X, Y]]] + \cdots . \quad (8.22)$$

Similarly, if we define another one parameter family of matrices as

$$F(\lambda) = \left(\partial_\mu e^{\lambda X}\right) e^{-\lambda X}, \quad (8.23)$$

where λ is a constant parameter and the matrix X depends on space-time coordinates, we obtain

$$\frac{\mathrm{d}F(\lambda)}{\mathrm{d}\lambda} = \left(\partial_\mu(e^{\lambda X} X)\right) e^{-\lambda X} + (\partial_\mu e^{\lambda X})\left(-X e^{-\lambda X}\right)$$

$$= e^{\lambda X}(\partial_\mu X) e^{-\lambda X} = \sum_{n=0}^{\infty} \frac{\lambda^n}{n!}(ad\,X)^n(\partial_\mu X), \quad (8.24)$$

where we have used (8.21) (with the identification $Y = (\partial_\mu X)$). Integrating this with respect to λ (between $(0, \lambda)$), we obtain

$$F(\lambda) = \sum_{n=0}^{\infty} \frac{\lambda^{n+1}}{(n+1)!}(ad\,X)^n(\partial_\mu X). \quad (8.25)$$

We note that $F(0) = 0$ and setting $\lambda = 1$ in (8.25) we obtain

$$F(1) = \left(\partial_\mu e^X\right) e^{-X}$$

$$= (\partial_\mu X) + \frac{1}{2!}[X, (\partial_\mu X)] + \frac{1}{3!}[X, [X, (\partial_\mu X)]] + \cdots$$

$$= \sum_{n=0}^{\infty} \frac{1}{(n+1)!}(ad\,X)^n(\partial_\mu X)$$

$$= \sum_{n=0}^{\infty} \frac{1}{(n+1)!}\frac{(ad\,X)^{n+1}}{ad\,X}(\partial_\mu X)$$

$$= \left(\frac{e^{ad\,X} - 1}{ad\,X}\right)(\partial_\mu X). \quad (8.26)$$

Let us now identify

$$X = -i\xi^a(x)t_a \in L, \quad (8.27)$$

so that we can write (see (8.7))

$$g(x) = e^{-i\xi^a(x)t_a} = e^X.$$ (8.28)

Then it follows from (8.22) that

$$g(x)A_\mu(x)g^{-1}(x) = g(x)t_a g^{-1}(x)A_\mu^a(x) = \left(e^X t_a e^{-X}\right) A_\mu^a(x)$$

$$= \left(e^{ad\,X} t_a\right) A_\mu^a(x)$$

$$= \left(t_a + [X, t_a] + \frac{1}{2!}[X, [X, t_a]] + \cdots\right) A_\mu^a(x).$$ (8.29)

We note that we can write

$$[X, t_a] = -i\left[\xi^b(x)t_b, t_a\right] = -i\xi^b(x)[t_b, t_a]$$

$$= -i\xi^b(x)\left(if_{ba}^c t_c\right) = -i\xi_a{}^c(x)t_c \in L,$$ (8.30)

where we have identified (see (3.99) and (3.108))

$$\xi_a{}^c(x) = \xi^b(x)(if_{ba}^c) = -\xi^b(x)\left(ad\,t_b\right)_a{}^c.$$ (8.31)

Note that the $N \times N$ matrix $\xi(x)$ defined by

$$\xi_a{}^c(x) = (\xi(x))_{ac},$$ (8.32)

is an element of the adjoint representation of L. Therefore we can write, in this notation,

$$[X, t_a] = (\xi(x))_{ac}t_c = -i(\xi(x)t)_a.$$ (8.33)

Using this notation, we can now calculate

$$[X, [X, t_a]] = [X, -i(\xi(x)t)_a] = (-i)^2(\xi(x)(\xi(x)t))_a$$

$$= ((-i\xi(x))^2 t)_a,$$ (8.34)

where we have used (8.33) in the intermediate step. Higher order commutators can also be similarly calculated and substituting these into (8.29) we obtain

$$g(x)A_\mu g^{-1}(x) = \left(t_a + (-i\xi(x)t)_a + \frac{1}{2!}((-i\xi(x))^2 t)_a + \cdots\right) A_\mu^a(x)$$

$$= \left(\delta_a{}^c + (-i\xi(x))_a{}^c + \frac{1}{2!}((-i\xi(x))^2)_a{}^c + \cdots\right) t_c A_\mu^a(x)$$

$$= \left(e^{-i\xi(x)}\right)_a{}^c t_c A_\mu^a(x),$$ (8.35)

which is indeed an element of the Lie algebra L.

Similarly we can calculate (see (8.27))

$$\partial_\mu X = (-i\partial_\mu \xi^a(x))t_a,$$
$$[X, (\partial_\mu X)] = [X, t_a](-i\partial_\mu \xi^a(x)) = (-i\xi(x)t)_a(-i\partial_\mu \xi^a(x)), \quad (8.36)$$

and so on. Here we have used (8.33) in the last step. Using these in (8.26) we obtain

$$
\begin{aligned}
(\partial_\mu g(x))g^{-1}(x) &= (\partial_\mu X) + \frac{1}{2!}[X, (\partial_\mu X)] + \frac{1}{3!}[X, [X, (\partial_\mu X)]] + \cdots \\
&= \left(t_a + \frac{1}{2!}(-i\xi(x)t)_a + \frac{1}{3!}((-i\xi(x))^2 t)_a + \cdots\right)(-i\partial_\mu \xi^a(x)) \\
&= \left(\frac{e^{-i\xi(x)} - 1}{-i\xi(x)}\right)_a^c t_c(-i\partial_\mu \xi^a(x)) \\
&= \Lambda_a{}^c t_c(-i\partial_\mu \xi^a(x)) \in L, \quad (8.37)
\end{aligned}
$$

where we have denoted

$$\Lambda(x) = \sum_{n=0}^\infty \frac{(-i\xi(x))^n}{(n+1)!} = \frac{e^{-i\xi(x)} - 1}{-i\xi(x)}. \quad (8.38)$$

We conclude from (8.35) and (8.37) that

$$
\begin{aligned}
A_\mu(x) \to A'_\mu(x) &= g(x)A_\mu(x)g^{-1}(x) + i(\partial_\mu g(x))g^{-1}(x) \\
&= \left(\left(e^{-i\xi(x)}\right)_a^c A_\mu^a(x) + (\Lambda(x))_a{}^c(\partial_\mu \xi^a(x))\right)t_c, \quad (8.39)
\end{aligned}
$$

which belongs to the Lie algebra L and we can write the transformation laws for the component fields as

$$A_\mu^a(x) \to A_\mu'^a(x) = \left(\left(e^{-i\xi(x)}\right)_c{}^a A_\mu^c + (\Lambda(x))_c{}^a(\partial_\mu \xi^c(x))\right). \quad (8.40)$$

From (8.17) we now see using (8.35) that the components of the field strength tensor (curvature) transform under a gauge transformation covariantly as

$$F_{\mu\nu}^a(x) \to F_{\mu\nu}'^a(x) = \left(e^{-i\xi(x)}\right)_c{}^a F_{\mu\nu}^c(x), \quad (8.41)$$

without the inhomogeneous term proportional to $\Lambda(x)$ present in the transformation of the gauge field (potential) in (8.40). Since the $N \times N$ matrix $U(g)$ given by

$$(U(g))_{ab} = \left(e^{-i\xi(x)}\right)_a^b, \tag{8.42}$$

gives the adjoint representation of the group G (see (8.32)), we conclude that the field strength tensor $F_{\mu\nu}^a(x), a = 1, 2, \cdots, N$ belongs to the adjoint representation of the group G. This is independent of any particular underlying $n \times n$ matrix representation $\lambda_a = \rho(t_a)$ of the Lie algebra L used.

▶ **Example (Abelian theory).** As an example, we note that the special case of $N = 1$ is of some interest. It corresponds to an Abelian group since the only generator of the group commutes with itself, namely, $[t_1, t_1] = 0$ and we can identify the matrix $\lambda = \rho(t_1) = \mathbb{1}$. As a result, the group element (8.7)

$$g(x) = e^{-i\xi^1(x)}, \tag{8.43}$$

also becomes a commuting ordinary function. As a result, the gauge field (potential) $B_\mu(x) = A_\mu^1(x)$ transforms under a gauge transformation as

$$B_\mu(x) \rightarrow B_\mu'(x) = g(x)B_\mu(x)g^{-1}(x) + i(\partial_\mu g(x))g^{-1}(x)$$

$$= B_\mu(x) + (\partial_\mu \xi^1(x)), \tag{8.44}$$

while the field strength (curvature) tensor (the commutator in (8.5) vanishes for the Abelian case)

$$G_{\mu\nu}(x)(\equiv F_{\mu\nu}^1(x)) = \partial_\mu B_\nu(x) - \partial_\nu B_\mu(x), \tag{8.45}$$

remains invariant under the transformation

$$G_{\mu\nu}(x) \rightarrow G_{\mu\nu}'(x) = g(x)G_{\mu\nu}(x)g^{-1}(x) = G_{\mu\nu}(x). \tag{8.46}$$

These reproduce the well known classical gauge transformations of the electromagnetic potential (for example, with the identification $\xi^1(x) = \alpha(x)$). ◀

Let us next return to the general case and note that since the field strength tensor transforms covariantly under a gauge transformation (see (8.11))

$$\text{Tr}\left(F'^{\mu\nu}(x)F_{\mu\nu}'(x)\right)$$

$$= \text{Tr}\left(g(x)F^{\mu\nu}(x)g^{-1}(x)g(x)F_{\mu\nu}(x)g^{-1}(x)\right)$$

$$= \text{Tr}\left(g(x)F^{\mu\nu}(x)F_{\mu\nu}(x)g^{-1}(x)\right)$$

$$= \text{Tr}\left(F^{\mu\nu}(x)F_{\mu\nu}(x)g^{-1}(x)g(x)\right)$$

$$= \text{Tr}\left(F^{\mu\nu}(x)F_{\mu\nu}(x)\right), \tag{8.47}$$

where we have used the property of cyclicity under a trace. Therefore, we see that this quadratic combination of the fielod strength tensor is not only Lorentz invariant (all the Lorentz indices μ, ν have been contracted), but is also invariant under a gauge transformation (8.8). Furthermore, since $F_{\mu\nu} = F^a_{\mu\nu} t_a$ (see (8.4)), if we denote (see (3.111))

$$g_{ab} = \text{Tr}\,(t_a t_b),\tag{8.48}$$

we can write the invariant quadratic combination in (8.47) in terms of components as (recall that repeated indices are summed)

$$\text{Tr}\,(F^{\mu\nu}(x)F_{\mu\nu}(x)) = \text{Tr}\,(t_a t_b)\,F^{\mu\nu\,a}(x)F^b_{\mu\nu}(x)$$

$$= g_{ab}F^{\mu\nu\,a}(x)F^b_{\mu\nu}(x).\tag{8.49}$$

In general, we can choose the matrix representation $\rho(t_a)$ such that it satisfies $g_{ab} = \frac{1}{2}\delta_{ab}$ in the fundamental representation (to which the fermions generally belong) for physics problems as we will see shortly. As a result, with a proper normalization the Lagrangian density describing gauge field dynamics can be written as

$$\mathcal{L}_G = -\frac{1}{2}\,\text{Tr}\,F_{\mu\nu}F^{\mu\nu} = -\frac{1}{4}\,F^a_{\mu\nu}F^{\mu\nu a}.\tag{8.50}$$

8.2 Fermion dynamics

Let us now consider the Dirac field $\psi_i(x), i = 1, 2, \cdots, n$ in four space-time dimensions belonging to the n-dimensional representation of the group G, where for each index i, the field $\psi_i(x)$ represents a four component Dirac spinor, namely,

$$\psi_i(x) = \begin{pmatrix} \phi_{i,1}(x) \\ \phi_{i,2}(x) \\ \phi_{i,3}(x) \\ \phi_{i,4}(x) \end{pmatrix}, \quad i = 1, 2, \cdots, n.\tag{8.51}$$

Let us assume that the Dirac field satisfies the equation (see (7.47), we have set $\hbar = c = 1$ as well as the coupling constant to unity)

$$(\gamma^\mu(i\partial_\mu - A_\mu(x)) - m)\,\psi(x) = 0,\tag{8.52}$$

where we have identified $\psi(x)$ with the column matrix

$$\psi(x) = \begin{pmatrix} \psi_1(x) \\ \psi_2(x) \\ \vdots \\ \psi_n(x) \end{pmatrix},\tag{8.53}$$

on which the generators t_a (we identify t_a with its $n \times n$ matrix representation) in $A_\mu(x) = A_\mu^a(x)t_a$ act, namely,

$$(t_a\psi(x))_i = (t_a)_{ij}\psi_j(x), \tag{8.54}$$

for $i, j = 1, 2, \cdots, n$ and $a = 1, 2, \cdots, N$. (Of course, every component of the fermion field corresponds to a four component spinor as in (8.51).)

If we define the gauge transformation for the Dirac field to be

$$\psi(x) \rightarrow \psi'(x) = g(x)\psi(x), \tag{8.55}$$

where $g(x)$ is defined in (8.7) for any representation, then, it is easy to see that the Dirac equation (8.52) is invariant under the combined transformations of (8.8) and (8.55). In fact, let us note that

$$\left(\gamma^\mu(i\partial_\mu - A_\mu'(x)) - m\right)\psi'(x)$$

$$= \left(\gamma^\mu(i\partial_\mu - (g(x)A_\mu g^{-1}(x) + i(\partial_\mu g(x))g^{-1}(x)))\right.$$

$$\left. - m\right)g(x)\psi(x)$$

$$= \left(\gamma^\mu((i\partial_\mu g(x)) + g(x)i\partial_\mu - g(x)A_\mu(x) - i(\partial_\mu g(x)))\right.$$

$$\left. - mg(x)\right)\psi(x)$$

$$= g(x)\left(\gamma^\mu(i\partial_\mu - A_\mu(x)) - m\right)\psi(x) = 0. \tag{8.56}$$

For the transformation of the adjoint equation (7.72), we note from (8.55) that the adjoint spinor $\overline{\psi}(x) = \psi^\dagger(x)\gamma^0$ transforms under a gauge transformation in the form

$$\overline{\psi}(x) \rightarrow \overline{\psi}'(x) = \overline{\psi}(x)g^\dagger(x). \tag{8.57}$$

In order to make the bilinear product such as $\overline{\psi}(x)\psi(x)$ invariant under a gauge transformation, it is clear that we must assume that the $n \times n$ matrix $g(x)$ is unitary, namely,

$$g^\dagger(x)g(x) = \mathbb{1}_n, \quad g^\dagger(x) = g^{-1}(x). \tag{8.58}$$

Furthermore, in analogy to the electromagnetic case of $N = 1$ (see discussion around (8.44)) where $B_\mu^\dagger(x) = (A_\mu^1)^\dagger(x) = A_\mu^1(x) = B_\mu(x)$, we assume that

$$(A_\mu^a)^\dagger(x) = A_\mu^a(x), \qquad a = 1, 2, \cdots, N,$$

$$A_\mu^\dagger(x) = A_\mu(x), \qquad F_{\mu\nu}^\dagger(x) = F_{\mu\nu}(x). \tag{8.59}$$

(which is consistent with our choice of Hermitian generators t_a) so that the adjoint equation has the form as in (7.72)

$$\overline{\psi}(x) \left(\gamma^\mu \left(i \overleftarrow{\partial_\mu} - A_\mu(x) \right) + m \right) = 0, \tag{8.60}$$

and which can also be seen as discussed above to be covariant under a gauge transformation (using (8.58)). The continuity equation (conservation law) that follows from (8.52) and (8.60) continues to have the form (7.74)

$$\partial_\mu \left(\overline{\psi}(x) \gamma^\mu \psi(x) \right) = 0. \tag{8.61}$$

The Lagrangian density which leads to the dynamical equations (8.52) and (8.60) has the form

$$\mathcal{L}_f = \overline{\psi} \left(\gamma^\mu \left(i \partial_\mu - A_\mu \right) - m \right) \psi. \tag{8.62}$$

The total Lagrangian density describing the interaction of fermions with Yang-Mills (gauge) fields is given by (see (8.50) and (8.62))

$$\mathcal{L} = \mathcal{L}_G + \mathcal{L}_f$$

$$= -\frac{1}{2} \mathrm{Tr} \left(F^{\mu\nu}(x) F_{\mu\nu}(x) \right)$$

$$+ \overline{\psi}(x) \left(\gamma^\mu \left(i \partial_\mu - A_\mu(x) \right) - m \right) \psi(x), \tag{8.63}$$

where we assume that the generators in the gauge field Lagrangian density are in the fundamental representation as discussed earlier. The Lagrangian density in (8.63) is invariant under the combined local gauge transformations (see (8.8), (8.55), (8.57) as well as (8.58))

$$\psi(x) \to \psi'(x) = g(x)\psi(x),$$

$$\overline{\psi}(x) \to \overline{\psi}'(x) = \overline{\psi}(x) g^{-1}(x),$$

$$A_\mu(x) \to A'_\mu(x) = g(x) A_\mu(x) g^{-1}(x) + i(\partial_\mu g(x)) g^{-1}(x). \tag{8.64}$$

The invariance of the first term in (8.63) has already been shown in (8.47) and, therefore, we only need to check the invariance of the Lagrangian density involving the Dirac fields. If we use the covariant transformation derived in (8.56) as well as the transformation of $\overline{\psi}(x)$ from (8.64), we obtain

$$\overline{\psi}'(x) \left(\gamma^\mu \left(i \partial_\mu - A'_\mu(x) \right) - m \right) \psi'(x)$$

$$= \overline{\psi}(x) g^{-1}(x) g(x) \left(\gamma^\mu \left(i \partial_\mu - A_\mu(x) \right) - m \right) \psi(x)$$

$$= \overline{\psi}(x) \left(\gamma^\mu \left(i \partial_\mu - A_\mu(x) \right) - m \right) \psi(x), \tag{8.65}$$

which proves the invariance of the fermion part of the Lagrangian density as well so that the total Lagrangian density in (8.63) is invariant under the local gauge transformations in (8.64). This Lagrangian density is the starting point in the study of quantum chromodynamics (QCD) which describes strong interactions between fundamental hadronic particles.

8.3 Quantum chromodynamics

We have already noted (see (8.58)) that it is important to assume the group element $g(x)$ to be unitary for applications in physics. Furthermore, since we also know that only compact groups can have nontrivial, finite dimensional unitary representation (see Peter-Weyl theorem in section 2.4), we must consider compact Lie groups for G. In this case, the corresponding Lie algebra is known to be reductive, namely, it must be at the most a direct sum of Abelian and semi-simple Lie algebras. Moreover in such a case (see, for example, (3.111)) there exists a representation $\rho(t_a)$ such that the metric

$$g_{ab} = \text{Tr}\left(\rho(t_a)\rho(t_b)\right) = g_{ba}, \tag{8.66}$$

is non-degenerate and its inverse g^{ab} exists. To see this in some more detail, let us write any $g(x) \in G$ in the exponential form (see (8.7))

$$g(x) = e^{-i\xi^a(x)\rho(t_a)}, \tag{8.67}$$

for real local parameters $\xi^a(x)$. Since we would like $g(x)$ to be unitary (see (8.58)), this determines the matrix generators to be Hermitian (which we have already assumed)

$$(\rho(t_a))^\dagger = \rho(t_a), \tag{8.68}$$

As a result, the metric

$$g_{ab} = \text{Tr}\left(\rho(t_a)\rho(t_b)\right) = \text{Tr}\left((\rho(t_a)^\dagger\rho(t_b)\right) > 0, \tag{8.69}$$

is a positive definite second rank tensor in the sense that for any real constants $c^a, a = 1, 2, \cdots, N$ (repeated indices are summed)

$$g_{ab}c^a c^b \geq 0, \tag{8.70}$$

where equality in (8.70) holds only for $c^a = 0$ for all $a = 1, 2, \cdots, N$. Since g_{ab} is a Hermitian (in the space ab) and positive definite matrix, it can always be diagonalized through a proper choice of the basis t_a

to be proportional to the identity matrix (this takes some simple algebra to show)

$$g_{ab} = C \, \delta_{ab}, \qquad\qquad (8.71)$$

where C is a positive constant. In particular, this constant is chosen to be $C = \frac{1}{2}$ for the fundamental representation so that in this representation $g_{ab} = \frac{1}{2} \delta_{ab}$ as we have mentioned earlier (see discussion following (8.49)). Let us point out here that if L is an Abelian algebra, then clearly the generator can be chosen to be a multiple of the identity matrix leading to a non-singular g_{ab}.

With this background, let us next discuss the basics of quantum chromodynamics (QCD). It is an interacting gauge theory based on the compact gauge group $G = SU(3)$. However, in order to avoid any possible confusion with the flavor $SU(3)$ group (of fermions/quarks) to be considered in the next chapter, we will refer to this group as $G = SU_c(3)$ and denote the flavor symmetry group as $SU_F(3)$. Here the subscript c refers to the three colors - red, blue and green - that the fundamental constituent fermions, known as quarks, can have. The quarks belong to the fundamental representation of $SU_c(3)$ (as we have mentioned before, fermions belong to the fundamental representation) which is three dimensional. On the other hand, as we have seen, gauge fields belong to the adjoint representation which, for $SU_c(3)$ (in general, for any $SU(3)$), is eight dimensional (the adjoint representation of $SU(n)$ is $(n^2 - 1)$ dimensional). Correspondingly, the gauge fields (potentials) are written as $A_\mu^a(x), a = 1, 2, \cdots, 8$ and are known as gluon fields.

The three colored quarks are labelled by the index $i = 1, 2, 3$ (1 for red, 2 for blue etc.) and are known to come in six different flavors: u (up), d (down), c (charm), s (strange), t (top) and b (bottom) and, of course, each flavor of quark comes in three colors so that we can label the quark fields correspondingly as $u_i(x), d_i(x)$ and so on. In addition to the color quantum number, quarks also carry electric charge. In units of the magnitude of the charge of the electron, the charge carried by the $u_i(x), c_i(x)$ and the $t_i(x)$ quarks is $\frac{2}{3}$ while that carried by the $d_i(x), s_i(x)$ and the $b_i(x)$ quarks is $-\frac{1}{3}$. Quarks are the basic constituent of all hadronic matter and are different from leptons in that they interact strongly. (Leptons can have only weak and electromagnetic interactions while quarks participate in weak, electromagnetic as well as strong interactions.) If we denote the quarks collectively as $q_{\alpha,i}(x), \alpha = 1, 2, \cdots, 6$ where $q_{1,i}(x) = u_i(x), q_{2,i}(x) = d_i(x)$ and so on. Then we can write the general Lagrangian density for the six quarks interacting with gluons (color gauge fields) as (repeated indices are summed)

$$\mathcal{L}_q = \overline{q}_{\alpha,i}(x)\left(\gamma^\mu\left(i\partial_\mu\delta_{ij} - (A_\mu)_{ij}(x)\right) - m_\alpha\delta_{ij}\right)q_{\alpha,j}(x), \qquad (8.72)$$

where we have identified the gauge field in the representation of the quarks (fundamental representation) as

$$(A_\mu)_{ij}(x) = A_\mu^a(x)\,(t_a)_{ij}, \qquad\qquad (8.73)$$

with $(t_a)_{ij}$ representing the generators of $SU_c(3)$ in the fundamental representation. Furthermore, $m_\alpha, \alpha = 1, 2, \cdots, 6$ in (8.72) denote the bare masses (also known as the current quark masses) of the different flavors of quarks.

The peculiarity of the strongly interacting theory called QCD is that the building blocks of matter, namely, the quarks, are confined in space. In other words, we cannot experimentally observe isolated quarks. Only their bound states belonging to the trivial representation (color neutral) of the color group $SU_c(3)$ manifest as hadrons such as proton p, neutron n etc. which are observed in experiments. This is the reason why the three colors of quarks are termed red, blue and green since they can naturally combine to give white color which symbolically represents a color neutral state (trivial representation). For example, the three pi meson states (π^+, π^0, π^-) with spin-parity $J^P = 0^-$ (J, P correspond to spin and parity respectively) and iso-topic spin $I = 1$ can be described symbolically as color singlet bound states of the u_i, d_i quarks and their anti-particle states ($\overline{u}_i, \overline{d}_i$) of the forms

$$\pi^+ = u_i\overline{d}_i, \quad \pi^0 = \frac{1}{\sqrt{2}}(u_i\overline{u}_i + d_i\overline{d}_i), \quad \pi^- = d_i\overline{u}_i. \qquad (8.74)$$

The spins of the constituent quarks must be anti-parallel in order for the bound state to have spin 0. This can, of course, be made more quantitative through the introduction of creation and annihilation operators for the quarks as well as states in a Hilbert space in a second quantized theory. But, we can avoid all these technicalities here and conventionally one says that $\pi^+ = u\overline{d}$, just as we say that the deuterium is a bound state of a proton and neutron $D = pn$ (we use D for deuterium to avoid confusion with the d quark). Similarly, the proton and the neutron can also be written as bound states of u (up) and d (down) quarks of the forms

$$p = \epsilon_{ijk}u_iu_jd_k, \quad \text{or, simply} \quad p = uud,$$

$$n = \epsilon_{ijk}u_id_jd_k, \quad \text{or, simply} \quad n = udd, \qquad (8.75)$$

which are clearly color singlet states. Here $\epsilon_{ijk}, i, j, k = 1, 2, 3$ denotes the completely anti-symmetric Levi-Civita tensor in the three dimensional color space.

The values of the bare masses $m_\alpha, \alpha = 1, 2, \cdots, 6$ of quarks q_α are deduced from experiments to be around 3-5 MeV for the u (up) and the d (down) quarks while they are about 150 MeV for the s (strange) quark, 500 MeV for the c (charm) quark, 5 GeV ($= 5000$ MeV) for the b (bottom) quark and 150 GeV for the t (top) quark. The reason underlying the large spread in the values of the masses is not yet well understood theoretically. As a result, it is important to study different energy regimes (scales) separately. For very low energy experiments, say below 50 MeV, the contributions due to the s, c, b and t quarks are essentially negligible because their masses are much heavier. In this case, it is sufficient restrict the theory in (8.72) to the sector containing only the u, d quarks. Furthermore, ignoring the small masses of these two quarks, it is easy to see that the theory has a $U(2)$ flavor symmetry (also written as $U_F(2)$), namely, the Lagrangian density in (8.72) (with only the u, d quarks and masses vanishing) is invariant under a transformation of the form

$$\begin{pmatrix} u_i \\ d_i \end{pmatrix} \rightarrow \begin{pmatrix} u'_i \\ d'_i \end{pmatrix} = U \begin{pmatrix} u_i \\ d_i \end{pmatrix}, \tag{8.76}$$

where U denotes an arbitrary, global (space-time independent) 2×2 unitary matrix. However, since the $U(1)$ subgroup of $U_F(2)$ is related to baryon number conservation, we restrict the matrix U in reality to $SU_F(2)$ which is known as the isotopic spin group. It defines the isotopic spin I already introduced above and takes values $I = 0, \frac{1}{2}, 1, \frac{3}{2}, \cdots$ just as ordinary spin. In this sector (of u, d quarks), the only baryons are p (proton) and n (neutron) as we have already seen in (8.75) and they have $I = \frac{1}{2}$ just like the u, d quarks. Nuclear matter is described as bound states of these nucleons p and n. For example, we can think of He^3 and H^3 as corresponding to the bound states $He^3 = ppn$ and $H^3 = pnn$ analogous to $p = uud, n = udd$.

Next let us suppose that we are considering experiments at energies below 500 MeV. In this case we may neglect the effects of the c, b and t quarks thereby dealing with only the u, d and s quarks. Restricting the Lagrangian density (8.72) to only these three flavors of quarks and ignoring their mass differences, it is easy to check that the theory is invariant under transformations belonging to the $SU_F(3)$ group (actually, it is invariant under $U(3)$ group, but we ignore the

$U(1)$ subgroup as explained earlier),

$$
\begin{pmatrix} u_i \\ d_i \\ s_i \end{pmatrix} \rightarrow \begin{pmatrix} u_i' \\ d_i' \\ s_i' \end{pmatrix} = U \begin{pmatrix} u_i \\ d_i \\ s_i \end{pmatrix}, \tag{8.77}
$$

where U is an arbitrary global 3×3 matrix belonging to the group $SU_F(3)$. This $SU(3)$ group, which we will refer to as hereafter as the flavor $SU_F(3)$ group to distinguish it from the color $SU_c(3)$ group, is in reality violated by the mass differences between m_u, m_d and m_s. Finally, we note that if we are considering extremely high energy experiments, say in the region of energy 1 TeV or larger, then we need to take all the six quarks into consideration. Furthermore, if we ignore all the masses (or mass differences), then the Lagrangian density (8.72) is easily seen to be invariant under global transformations of $SU_F(6)$ (or $U(6)$). However, it is worth pointing out that so far no such experiment exists (except possibly at LHC or at big bang).

To be a little bit more specific, let us define the matrices

$$
(J^\mu(x))^\alpha{}_\beta = \bar{q}_{\alpha,i}(x)\gamma^\mu q_{\beta,i}(x), \quad T^\alpha{}_\beta(x) = \bar{q}_\alpha(x)q_\beta(x), \tag{8.78}
$$

in the flavor space where $\alpha, \beta = 1, 2, \cdots, 6$ and $i = 1, 2, 3$ with repeated indices summed. From the dynamical equations for the fermions following from (8.72), we can obtain

$$
\partial_\mu (J^\mu(x))^\alpha{}_\beta = (m_\alpha - m_\beta)T^\alpha{}_\beta(x), \tag{8.79}
$$

for fixed α, β. It is clear from (8.79) that if $m_\alpha = m_\beta$, namely, if the quark masses are flavor independent, then the flavor current matrix defined in (8.78) will be conserved,

$$
\partial_\mu (J^\mu(x))^\alpha{}_\beta = 0. \tag{8.80}
$$

As a result, we can define a conserved flavor charge matrix

$$
X^\alpha{}_\beta = \int d^3x \, (J^0(x))^\alpha{}_\beta, \tag{8.81}
$$

which will be independent of x^0 (time). Furthermore, the equal-time $(x^0 = y^0)$ canonical anti-commutation relations for the quark fields give

$$
[q_{\alpha,i}(x), q_{\beta,j}(y)]_+ = 0 = [q^\dagger_{\alpha,i}(x), q^\dagger_{\beta,j}(y)]_+,
$$

$$
[q_{\alpha,i}(x), q^\dagger_{\beta,j}(y)]_+ = \delta_{\alpha\beta}\delta_{ij}\delta^3(x-y), \tag{8.82}
$$

where $\alpha, \beta = 1, 2, \cdots, 6$, $i, j = 1, 2, 3$ and the anti-commutator of two operators A, B is defined to be $[A, B]_+ = AB + BA$. Furthermore, we have suppressed the spinor indices (of the Dirac fields as well as the anti-commutation relations) in (8.82) only to bring out their behavior in the internal space. With the help of (8.82) it can readily be checked that the charges satisfy the $u(6)$ Lie algebra (see section 3.2.3),

$$[X^\alpha_{\ \beta}, X^\gamma_{\ \sigma}] = \delta^\alpha_{\ \sigma} X^\gamma_{\ \beta} - \delta^\gamma_{\ \beta} X^\alpha_{\ \sigma}, \tag{8.83}$$

where $\alpha, \beta, \gamma, \sigma = 1, 2, \cdots, 6$. We can also check that

$$[X^\alpha_{\ \beta}, T^\gamma_{\ \sigma}(x)] = \delta^\alpha_{\ \sigma} T^\gamma_{\ \beta}(x) - \delta^\gamma_{\ \sigma} T^\alpha_{\ \sigma}(x). \tag{8.84}$$

However, we will only be concerned with experiments in the intermediate energies where only the u, d and the s quarks are relevant, namely, we will restrict the theory in (8.72) only to these three quark flavors so that the flavor indices $\alpha, \beta = 1, 2, 3$. Since ($\alpha = 1, 2, 3$ and repeated indices are summed)

$$B = X^\alpha_{\ \alpha}, \tag{8.85}$$

corresponds to the quark or the baryon number (generator of the $U(1)$ subgroup), it is more convenient to define the $su_F(3)$ sub Lie algebra by defining the (traceless) generators (charges) (see (3.52) as well as (3.54))

$$Q^\alpha_{\ \beta} = X^\alpha_{\ \beta} - \frac{1}{3} \delta^\alpha_{\ \beta} B, \tag{8.86}$$

which satisfies

$$[Q^\alpha_{\ \beta}, Q^\gamma_{\ \sigma}] = \delta^\alpha_{\ \sigma} Q^\gamma_{\ \beta} - \delta^\gamma_{\ \beta} Q^\alpha_{\ \sigma}, \quad Q^\alpha_{\ \alpha} = 0, \tag{8.87}$$

for $\alpha, \beta, \gamma, \sigma = 1, 2, 3$. Similarly, we can define

$$T(x) = T^\alpha_{\ \alpha}, \quad S^\alpha_{\ \beta}(x) = T^\alpha_{\ \beta}(x) - \frac{1}{3} \delta^\alpha_{\ \beta} T(x), \tag{8.88}$$

so that $T(x)$ and $S^\alpha_{\ \beta}(x)$ would belong respectively to the singlet and the octet representations of $SU_F(3)$. With this notation, we can write the mass term in the Lagrangian density in (8.89) which violates the $SU_F(3)$ symmetry as

$$\mathcal{L}_m = -m_\alpha T^\alpha_{\ \alpha} = -\frac{1}{3}(m_u + m_d + m_s)T$$

$$-\frac{1}{2}(m_u - m_d)\left(S^1_{\ 1}(x) - S^2_{\ 2}(x)\right)$$

$$-\frac{1}{2}(2m_s - m_u - m_d)S^3_{\ 3}(x), \tag{8.89}$$

where we have identified $m_1 = m_u$, $m_2 = m_d$, $m_3 = m_s$. We note that the $SU_F(3)$ violating terms in $\mathcal{L}_m$ consist of only the octet terms involving mass differences (the singlet term is invariant under $SU_F(3)$). As we will see in the next chapter, this leads to the $SU_F(3)$ mass formula for the baryons. The presence of the singlet terms in the Lagrangian density will not affect this discussion as we will see in the next chapter.

8.4 References

P. Bechers, M. Böhm and H. Joos, *Gauge Theories of Strong and Electroweak Interactions*, John Wiley (1984).

A. Das, *Lectures on Quantum Field Theory*, World Scientific (2008).

CHAPTER 9

Quark model and $SU_F(3)$ symmetry

9.1 SU_F flavor symmetry

As we have seen in the preceding chapter, the QCD Lagrangian density (8.72), when restricted to the first three flavors of quarks u, d, s, is invariant under the flavor symmetry group $SU_F(3)$ if we can ignore the mass differences between m_u, m_d and m_s. In this case we can assign the u, d and s quarks to belong to the triplet (fundamental) representation of $SU_F(3)$ while all other quarks c, b and t as well as the gluons belong to the singlet representation of the group. Here we will discuss about the sector consisting only of the u, d and s quarks.

First we note that according to QCD, all mesons (spin integer hadrons) can be built as color singlet states out of these three quarks and their anti-particles. Hence such an element (object) can be written as $\bar{q}_\alpha(x)q_\beta(x), \alpha, \beta = 1, 2, 3$, where $q_\alpha(x)$ denotes the αth quark feld. For example, $q_1(x) = u(x), q_2(x) = d(x), q_3(x) = s(x)$. We have ignored the color index $i = 1, 2, 3$ of the quarks for simplicity (in a meson field, the constituent q_α and $\bar{q}_\alpha$ fields have to have the same i index or its sum for color neutrality). It follows now that the mesons bilinear in the quark anti-quark fields can either be singlets of $SU_F(3)$ of the form (repeated indices are summed)

$$\bar{q}_\alpha(x)q_\alpha(x), \quad \alpha = 1, 2, 3, \tag{9.1}$$

or belong to the octet representation of the form

$$S^\alpha{}_\beta(x) = \bar{q}_\alpha(x)q_\beta(x) - \frac{1}{3}\delta^\alpha{}_\beta\left(\bar{q}_\lambda q_\lambda\right), \quad \alpha, \beta = 1, 2, 3. \tag{9.2}$$

In contrast, we note that all baryons (spin half integer hadrons) consist of color singlet combinations of three quarks of the form $q_\alpha(x)q_\beta(x)q_\gamma(x), \alpha, \beta, \gamma = 1, 2, 3$. As we have noted earlier, this tensor is not irreducible (see (5.94)). If we identify $\square$ to represent the 3 dimensional (fundamental) representation of the $SU_F(3)$ group, then the tensor $T_{\alpha\beta\gamma} \sim q_\alpha q_\beta q_\gamma$ decomposes as (see (5.95))

$$\square \otimes \square \otimes \square = \boxed{\square\square\square} \oplus \begin{matrix}\square\\\square\\\square\end{matrix} \oplus \square \oplus \square \;, \qquad (9.3)$$

Algebraically we can write this also as

$$3 \otimes 3 \otimes 3 = 10 \oplus 1 \oplus 8 \oplus 8, \qquad (9.4)$$

so that the baryons must belong to either the trivial (singlet) one dimensional or 10 dimensional (decouplet) or 8 dimensional (octet) representations of the $SU_F(3)$ group.

Let us consider the octet representation denoted by $B_{\alpha\beta\gamma}$ which is specified by the Young tableaux symbol $(2,1,0)$ satisfying the constraints

$$B_{\alpha\beta\gamma} = -B_{\beta\alpha\gamma},$$

$$B_{\alpha\beta\gamma} + B_{\beta\gamma\alpha} + B_{\gamma\alpha\beta} = 0, \quad \alpha, \beta, \gamma = 1, 2, 3. \qquad (9.5)$$

Moreover we have already seen (see (5.81)) that any irreducible representation corresponding to the Young tableaux (f_1, f_2, f_3) with $f_1 \geq f_2 \geq f_3$ is equivalent to that with $(f_1 + \Delta, f_2 + \Delta, f_3 + \Delta)$ for any integer Δ. Hence choosing $\Delta = -1$, we see that the representation $(2, 1, 0)$ is equivalent to $(1, 0, -1)$ which is described by a traceless tensor $B^\alpha{}_\beta, \alpha, \beta = 1, 2, 3$ satisfying

$$B^\alpha{}_\alpha = 0, \qquad (9.6)$$

just as in the bosonic case. The connection between the two is given explicitly by

$$B^\alpha{}_\beta = \frac{1}{2} \epsilon^{\alpha\mu\nu} B_{\mu\nu\beta}, \quad \alpha, \beta, \mu, \nu = 1, 2, 3, \qquad (9.7)$$

as we have noted earlier also. Here $\epsilon^{\alpha\mu\nu}$ is the completely anti-symmetric Levi-Civita tensor. Since the use of $B^\alpha{}_\beta$ is more conventional for practical purposes, we use it hereafter to represent the octet of baryons, rather than the notation $B_{\alpha\beta\gamma}$ satisfying (9.5).

Let us consider the Lie algebra $su_F(3)$ of the $SU_F(3)$ group, given by the commutation relations between the generators (see (3.54) and (3.56))

$$[A^\mu{}_\nu, A^\alpha{}_\beta] = -\delta^\mu{}_\beta A^\alpha{}_\nu + \delta^\alpha{}_\nu A^\mu{}_\beta, \quad \alpha, \beta, \mu, \nu = 1, 2, 3, \qquad (9.8)$$

which also satisfy the traceless condition

$$A^\mu_{\ \mu} = 0, \tag{9.9}$$

Furthermore the unitarity condition for the representation requires that the generators be Hermitian, namely,

$$(A^\mu_{\ \nu})^\dagger = A^\nu_{\ \mu}. \tag{9.10}$$

If $q_\alpha(x) = (u(x), d(x), s(x))$ for $\alpha = 1, 2, 3$ represents the αth quark field, the creation operator, say for the u quark, corresponds to its conjugate field $(q_1(x))^\dagger = q^1(x)$. Hence, we represent the basis vector for the $q_\alpha(x)$ field as $|q^\alpha\rangle$, on which the generators $A^\mu_{\ \nu}$ operate as

$$A^\mu_{\ \nu}|q^\alpha\rangle = \delta^\alpha_{\ \nu}|q^\mu\rangle - \frac{1}{3}\delta^\mu_{\ \nu}|q^\alpha\rangle. \tag{9.11}$$

This simply describes how the vectors in the fundamental representation transform under an infinitesimal $SU_F(3)$ transformation (see, for example, (3.64)). We can now readily verify that this operation is consistent with the $su_F(3)$ Lie algebra (9.8)

$$[A^\mu_{\ \nu}, A^\lambda_{\ \rho}]|q^\alpha\rangle = A^\mu_{\ \nu}(A^\lambda_{\ \rho}|q^\alpha\rangle) - A^\lambda_{\ \rho}(A^\mu_{\ \nu}|q^\alpha\rangle)$$

$$= A^\mu_{\ \nu}\left(\delta^\alpha_{\ \rho}|q^\lambda\rangle - \frac{1}{3}\delta^\lambda_{\ \rho}|q^\alpha\rangle\right) - A^\lambda_{\ \rho}\left(\delta^\alpha_{\ \nu}|q^\mu\rangle - \frac{1}{3}\delta^\mu_{\ \nu}|q^\alpha\rangle\right)$$

$$= \delta^\alpha_{\ \rho}\left(\delta^\lambda_{\ \nu}|q^\mu\rangle - \frac{1}{3}\delta^\mu_{\ \nu}|q^\lambda\rangle\right) - \frac{1}{3}\delta^\lambda_{\ \rho}\left(\delta^\alpha_{\ \nu}|q^\mu\rangle - \frac{1}{3}\delta^\mu_{\ \nu}|q^\alpha\rangle\right)$$

$$- \delta^\alpha_{\ \nu}\left(\delta^\mu_{\ \rho}|q^\lambda\rangle - \frac{1}{3}\delta^\lambda_{\ \rho}|q^\alpha\rangle\right) + \frac{1}{3}\delta^\mu_{\ \nu}\left(\delta^\alpha_{\ \rho}|q^\lambda\rangle - \frac{1}{3}\delta^\lambda_{\ \rho}|q^\alpha\rangle\right)$$

$$= \delta^\alpha_{\ \rho}\delta^\lambda_{\ \nu}|q^\mu\rangle - \delta^\alpha_{\ \nu}\delta^\mu_{\ \rho}|q^\lambda\rangle$$

$$= -\delta^\mu_{\ \rho}\left(\delta^\alpha_{\ \nu}|q^\lambda\rangle - \frac{1}{3}\delta^\lambda_{\ \nu}|q^\alpha\rangle\right) + \delta^\lambda_{\ \nu}\left(\delta^\alpha_{\ \rho}|q^\mu\rangle - \frac{1}{3}\delta^\mu_{\ \rho}|q^\alpha\rangle\right)$$

$$= \left(-\delta^\mu_{\ \rho}A^\lambda_{\ \nu} + \delta^\lambda_{\ \nu}A^\mu_{\ \rho}\right)|q^\alpha\rangle. \tag{9.12}$$

Let us next consider the subgroup $SU(2) \otimes U(1)$ of $SU_F(3)$ corresponding to the standard isotopic spin and the hypercharge subgroups so that we can identify

$$I_+ = A^2_{\ 1}, \quad I_- = A^1_{\ 2}, \quad I_3 = \frac{1}{2}\left(A^1_{\ 1} - A^2_{\ 2}\right),$$

$$Y = -A^3_{\ 3} = A^1_{\ 1} + A^2_{\ 2}, \tag{9.13}$$

where (the three) Is denote the generators of the isotopic spin group $SU(2)$ while Y corresponds to the generator of the hypercharge subgroup $U(1)$. Using (9.8) we can now verify that these generators satisfy

$$[I_+, I_-] = 2I_3, \quad [I_\pm, I_3] = \pm I_\pm, \quad [Y, I_\pm] = [Y, I_3] = 0. \tag{9.14}$$

Moreover, from the relation (9.11) it is clear that

$$I_3|q^1\rangle = \frac{1}{2}|q^1\rangle, \quad I_3|q^2\rangle = -\frac{1}{2}|q^2\rangle, \quad I_3|q^3\rangle = 0, \tag{9.15}$$

while

$$Y|q^1\rangle = \frac{1}{3}|q^1\rangle, \quad Y|q^2\rangle = \frac{1}{3}|q^2\rangle, \quad Y|q^3\rangle = -\frac{2}{3}|q^3\rangle. \tag{9.16}$$

Recalling that $q_1 = u, q_2 = d$ and $q_3 = s$, this implies that the (u, d) quarks belong to an isotopic spin doublet ($I = \frac{1}{2}$) with $Y = \frac{1}{3}$ while the s quark is an isotopic spin singlet ($I = 0$) with $Y = -\frac{2}{3}$. Moreover according to the Gell-Mann-Nakano-Nishijima formula, the electric charge operator can be identified with (see (9.13))

$$Q = I_3 + \frac{Y}{2} = A^1{}_1, \tag{9.17}$$

and leads to

$$Q|u\rangle = Q|q^1\rangle = \frac{2}{3}|u\rangle, \quad Q|d\rangle = Q|q^2\rangle = -\frac{1}{3}|d\rangle,$$

$$Q|s\rangle = Q|q^3\rangle = -\frac{1}{3}|s\rangle. \tag{9.18}$$

Therefore, we conclude that the u, d and the s quarks carry the fractional electric charges $\frac{2}{3}, -\frac{1}{3}$ and $-\frac{1}{3}$ respectively (in units of the magnitude of the electron charge).

Similarly, for the eight baryonic state vectors $|B^\alpha{}_\beta\rangle$, corresponding to the octet of (traceless) baryon fields $B^\alpha{}_\beta$ with $J^P = \frac{1}{2}^+$, we have (see, for example, (3.92))

$$A^\mu{}_\nu|B^\alpha{}_\beta\rangle = \delta^\alpha{}_\nu|B^\mu{}_\beta\rangle - \delta^\mu{}_\beta|B^\alpha{}_\nu\rangle. \tag{9.19}$$

Computing the actions of I_3 and Y on such states we can identify the states as

$$N = (p, n) : \quad I = \frac{1}{2}, Y = 1, \quad |p\rangle = |B^1{}_3\rangle, \quad |n\rangle = |B^2{}_3\rangle,$$

$$\Xi = (\Xi^0, \Xi^-) : \quad I = \frac{1}{2}, Y = -1, \quad |\Xi^0\rangle = |B^3{}_2\rangle, \quad |\Xi^-\rangle = |B^3{}_1\rangle,$$

$$\Lambda = \Lambda^0 : \quad I = 0, Y = 0, \quad |\Lambda^0\rangle = -\sqrt{\frac{3}{2}}|B^3{}_3\rangle,$$

$$\Sigma = (\Sigma^+, \Sigma^0, \Sigma^-) : \quad I = 1, Y = 0,$$

$$|\Sigma^+\rangle = |B^1{}_2\rangle, |\Sigma^0\rangle = \frac{1}{\sqrt{2}}|B^1{}_1 - B^2{}_2\rangle, |\Sigma^-\rangle = |B^2{}_1\rangle. \tag{9.20}$$

The nontrivial normalization constants in the definition of states in (9.20) arise from the $SU_F(3)$ invariant normalization condition that

$$\langle B^\mu{}_\nu | B^\alpha{}_\beta\rangle = \delta^\alpha{}_\nu \delta^\mu{}_\beta - \frac{1}{3}\delta^\mu{}_\nu \delta^\alpha{}_\beta, \quad \alpha, \beta, \mu, \nu = 1, 2, 3, \tag{9.21}$$

which also satisfies the traceless condition $|B^\alpha{}_\alpha\rangle = 0$ (α is summed). With the normalization constants in (9.20) and using (9.21), we can readily verify that if we denote the octet of states collectively as $|B_i\rangle, i = 1, 2, \cdots, 8$, then they satisfy the orthonormality relation

$$\langle B_i | B_j\rangle = \delta_{ij}, \quad i, j = 1, 2, \cdots, 8. \tag{9.22}$$

The electric charges carried by these states follow from (9.17) to be (already denoted explicitly in the specification of the states)

$$Q_p = I_3 + \frac{Y}{2} = \frac{1}{2} + \frac{1}{2} = 1, \quad Q_n = -\frac{1}{2} + \frac{1}{2} = 0,$$

$$Q_{\Xi^0} = \frac{1}{2} - \frac{1}{2} = 0, \quad Q_{\Xi^-} = -\frac{1}{2} - \frac{1}{2} = -1,$$

$$Q_{\Lambda^0} = 0 + 0 = 0, \tag{9.23}$$

$$Q_{\Sigma^+} = 1 + 0 = 1, \quad Q_{\Sigma^0} = 0 + 0 = 0, \quad Q_{\Sigma^-} = -1 + 0 = -1.$$

A flavor transformation belonging to the group $SU_F(3)$ of the form

$$|B^\alpha{}_\beta\rangle \rightarrow |B^\alpha{}_\beta\rangle' = a^\alpha{}_\gamma a_\beta{}^\delta |B^\gamma{}_\delta\rangle, \tag{9.24}$$

induces a transformation on the states $|B_i\rangle$ of the form

$$|B_i\rangle \rightarrow |B_i\rangle' = U_{ij}|B_j\rangle, \tag{9.25}$$

where U_{ij} represent 8×8 matrices and the orthonormality of the transformed states determines that the matrices U are unitary, namely,

$$U^\dagger U = \mathbb{1}_8. \tag{9.26}$$

Let us now consider the branching rule for $SU_F(3) \to U(2) \simeq SU(2) \otimes U(1)$ in which the irreducible representation (f_1, f_2, f_3) decomposes as (see, for example, (5.123) and (5.124))

$$(f_1, f_2, f_3) = \oplus \sum_{f_1', f_2'} \theta(f_1', f_2'), \tag{9.27}$$

for all (f_1', f_2') satisfying $f_1 \geq f_1' \geq f_2 \geq f_2' \geq f_3$ as we have already noted earlier. In this case, the sub-quantum numbers specified by the isotopic spin and I and the hypercharge Y can be identified with

$$I = \frac{1}{2}\left(f_1' - f_2'\right), \quad Y = f_1' + f_2' - 2. \tag{9.28}$$

If we apply this to the original baryon field $B_{\alpha\beta\gamma}$ as described in (9.5) with $(f_1, f_2, f_3) = (2, 1, 0)$, then from the requirement $f_1 \geq f_1' \geq f_2 \geq f_2' \geq f_3 = 2 \geq f_1' \geq 1 \geq f_2' \geq 0$, we conclude that there are four combined possibilities for the values of f_1', f_2', namely,

$$f_1' = 2, 1, \quad f_2' = 1, 0. \tag{9.29}$$

Each of the four possibilities leads to the multiplets identified earlier, namely,

$$(f_1', f_2') = (2, 1): \quad I = \frac{1}{2}, Y = 1, \quad \text{as in } (p, n),$$

$$(f_1', f_2') = (2, 0): \quad I = 1, Y = 0, \quad \text{as in } (\Sigma^+, \Sigma^0, \Sigma^-),$$

$$(f_1', f_2') = (1, 1): \quad I = 0, Y = 0, \quad \text{as in } \Lambda^0,$$

$$(f_1', f_2') = (1, 0): \quad I = \frac{1}{2}, Y = 0, \quad \text{as in } (\Xi^0, \Xi^-), \tag{9.30}$$

corresponding to the quantum numbers already discussed in (9.20). It is consistent to depict this octet as a hexagon diagram in the (I_3, Y) plane as shown in Fig. 9.1.

Similarly, consider the 10-dimensional irreducible tensor, describing baryons with $J^P = \frac{3}{2}^+$, which decomposes as

$$(3, 0, 0) = \oplus \sum_{f_1', f_2'} (f_1', f_2'), \quad 3 \geq f_1' \geq 0 \geq f_2' \geq 0, \tag{9.31}$$

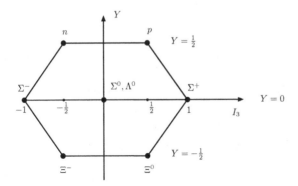

Figure 9.1: Six of the octet of particles can be represented at the six corners of the hexagon in the $I_3 - Y$ plane with the other two naturally fitting into the origin with the right quantum numbers.

which clearly determines $f'_2 = 0$ and $f'_1 = 3, 2, 1, 0$. The four solutions lead to the four multiplets (see (9.28))

$$(f'_1, f'_2) = (3, 0): \quad I = \frac{3}{2}, Y = 1, \text{ as in } \Delta = (\Delta^{++}, \Delta^+, \Delta^0, \Delta^-),$$

$$(f'_1, f'_2) = (2, 0): \quad I = 1, Y = 0, \quad \text{as in } Y^\star = (Y^{\star+}, Y^{\star 0}, Y^{\star-}),$$

$$(f'_1, f'_2) = (1, 0): \quad I = \frac{1}{2}, Y = -1, \quad \text{as in } \Xi^\star = (\Xi^{\star 0}, \Xi^{\star-}),$$

$$(f'_1, f'_2) = (0, 0): \quad I = 0, Y = -2, \quad \text{as in } \Omega^-. \tag{9.32}$$

We note from (9.28) that, since $f'_2 = 0$, in this case I and Y are related as

$$I = \frac{Y}{2} + 1. \tag{9.33}$$

The decouplet of baryons can be accomodated on a triangle in the (I_3, Y) plane as shown in Fig. 9.2. We normalize the state vector of the decouplet $|S^{\alpha\beta\gamma}\rangle$ for the totally symmetric tensor field $S_{\alpha\beta\gamma}$ as

$$\langle S^{\mu\nu\lambda}|S^{\alpha\beta\gamma}\rangle = \frac{1}{3!} \sum_{\text{permutations}} \delta^{\mu\alpha}\delta^{\nu\beta}\delta^{\lambda\gamma}, \tag{9.34}$$

where the summation is over the (3!) permutations of the indices (α, β, γ) so as to maintain the total symmetric nature of the tensor state $|S^{\alpha\beta\gamma}\rangle$. With this normalization, we can now identify the baryon

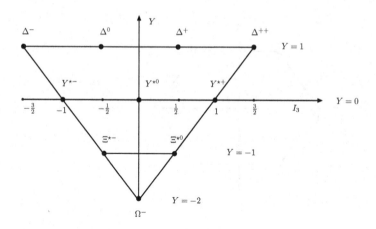

Figure 9.2: The decouplet of baryons can be arranged naturally on a triangle in the (I_3, Y) plane with the right quantum numbers.

states as

$$|\Delta^{++}\rangle = \frac{1}{\sqrt{6}}|S^{111}\rangle, \quad |\Delta^{+}\rangle = \frac{1}{\sqrt{2}}|S^{112}\rangle, \quad |\Delta^{0}\rangle = \frac{1}{\sqrt{2}}|S^{122}\rangle,$$

$$|\Delta^{-}\rangle = \frac{1}{\sqrt{6}}|S^{222}\rangle,$$

$$|Y^{\star+}\rangle = \frac{1}{\sqrt{2}}|S^{113}\rangle, \quad |Y^{\star0}\rangle = |S^{123}\rangle, \quad |Y^{\star-}\rangle = \frac{1}{\sqrt{2}}|S^{223}\rangle,$$

$$|\Xi^{\star0}\rangle = \frac{1}{\sqrt{2}}|S^{133}\rangle, \quad |\Xi^{\star-}\rangle = \frac{1}{\sqrt{2}}|S^{233}\rangle,$$

$$|\Omega^{-}\rangle = \frac{1}{\sqrt{6}}|S^{333}\rangle. \tag{9.35}$$

9.2 $SU_F(3)$ flavor symmetry breaking

As we have noted earlier, if we ignore the mass differences between the (u, d, s) quarks, then the QCD Lagrangian density is invariant under the global flavor $SU_F(3)$ group. In fact, all particles belonging to an irreducible representation of $SU_F(3)$ must have the same mass. Experimentally, however, the masses of the particles in the baryon octet, namely, $p, n, \Xi^0, \Xi^-, \Lambda^0, \Sigma^+, \Sigma^0, \Sigma^-$ are slightly different from each other. The differences can be attributed to $SU_F(3)$ violating mass terms in the Hamiltonian of the form

$$H_1 = \alpha \left(S^1{}_1 - S^2{}_2 \right) + \beta S^3{}_3, \tag{9.36}$$

where (see (8.89) as well as (9.2))

$$S^\alpha{}_\beta = \int d^3x \, S^\alpha{}_\beta(x)$$

$$= \int d^3x \left(\bar{q}_\alpha(x) q_\beta(x) - \frac{1}{3} \delta^\alpha{}_\beta \, \bar{q}_\lambda(x) q_\lambda(x) \right), \tag{9.37}$$

and the coefficients α, β in (9.36) are given by (see (8.89))

$$\alpha = \frac{1}{2} \left(m_u - m_d \right), \quad \beta = \frac{1}{2} \left(2m_s - m_u - m_d \right). \tag{9.38}$$

We note that $S^\alpha{}_\beta$ in (9.37) satisfies the commutation relations (with the generators of the group)

$$[A^\mu{}_\nu, S^\alpha{}_\beta] = \delta^\alpha{}_\nu S^\mu{}_\beta - \delta^\mu{}_\beta S^\alpha{}_\nu, \tag{9.39}$$

which has the same form as the relation satisfied by the generators of the group (see (9.8)). We call any such tensor $S^\alpha{}_\beta$ satisfying (9.39) as the adjoint tensor operator of the $SU_F(3)$ group.

This is a generalization of the vector operators in the case of $SU(2)$ group. For example, let us recall that the angular momentum Lie algebra for the $SU(2)$ group is given by

$$[J_i, J_j] = i\epsilon_{ijk} J_k, \quad i, j, k = 1, 2, 3, \tag{9.40}$$

with $J_+ = J_1 + iJ_2 = A^2{}_1$, $J_- = J_1 - iJ_2 = A^1{}_2$ and $J_3 = \frac{1}{2} \left(A^1{}_1 - A^2{}_2 \right)$ with $A^a{}_b$, $a, b = 1, 2$ denoting matrix generators for $SU(2)$. We know that any vector $\mathbf{V} = (V_1, V_2, V_3)$ satisfies the commutation relation

$$[J_i, V_j] = i\epsilon_{ijk} V_k, \tag{9.41}$$

which basically shows how any vector transforms under an infinitesimal rotation and coincides with the transformation of the generators in (9.40). In this case, the vector $\mathbf{V}$ is known as a vector operator. Labelling the irreducible representations for states as $|j, m\rangle$ with $m = j, j - 1, \cdots, -j + 1, -j$ for a fixed value of j (integer or half integer) we note that these states completely specify the vector space of the $su(2)$ algebra. In this case, it is well known that we can write

$$\langle j, m | \mathbf{V} | j, m' \rangle = a(j) \, \langle j, m | \mathbf{J} | j, m' \rangle, \tag{9.42}$$

where the value of the constant a depends on the vector $\mathbf{V}$ and j, but not on the azimuthal quantum number m. For simplicity, we write (9.42) symbolically as (even though it is not an expectation value)

$$\langle \mathbf{V} \rangle = a \langle \mathbf{J} \rangle. \tag{9.43}$$

An analogous theorem also exists for the $su(3)$ Lie algebra. In this case, the main quantum number analogous to j (in $su(2)$) is (f_1, f_2, f_3) while I, I_3, Y correspond to the sub-quantum number m (of $su(2)$). For a given irreducible representation specified by (f_1, f_2, f_3), the matrix element of the adjoint operator S^α_β can be written (analogous to (9.43)) as

$$\langle S^\alpha_\beta \rangle = a_1 \langle A^\alpha_\beta \rangle + a_2 \langle A^\alpha_\gamma A^\gamma_\beta - \frac{1}{3} \delta^\alpha_\beta A^\mu_\nu A^\nu_\mu \rangle, \tag{9.44}$$

which we will prove later. Here a_1 and a_2 are constants which depend on the main quantum numbers (f_1, f_2, f_3), but not on the sub-quantum numbers (I, I_3, Y). Moreover, for any triangular irreducible representation which satisfies $f_1 = f_2$ or $f_2 = f_3$ we have

$$\langle A^\alpha_\lambda A^\lambda_\beta - \frac{1}{3} \delta^\alpha_\beta A^\mu_\nu A^\nu_\mu \rangle = b \langle A^\alpha_\beta \rangle, \tag{9.45}$$

where b is a constant depending on the particular irreducible representation (f_1, f_2, f_3). Therefore, using (9.44) and (9.45) we can write in a triangular irreducible representation,

$$\langle S^\alpha_\beta \rangle = a_0 \langle A^\alpha_\beta \rangle, \tag{9.46}$$

where a_0 is a constant.

We are now in a position to explain the mass differences among the particles (states) in the octet and the decouplet representations. For this, we first ignore the small mass difference $(m_u - m_d)$. This would imply the exact isotropic spin invariance so that masses of all the particles within a given isotopic spin I will be identical irrespective of their I_3 quantum number (projection). We can consider $\overline{H_1} = \beta S^3_3$ in (9.36) as a small perturbation so that the mass of the particle with a given (I, Y) can now be computed for the general case as (the first term in (9.36) is proportional to I_3, see (9.13), and, therefore, does not contribute to the mass difference within a multiplet)

$$M(I, Y) = M_0 + \beta \langle S^3_3 \rangle + O(\beta^2)$$

$$= M_0 + \beta \left(a_1 \langle A^3_3 \rangle + a_2 \langle A^3_\lambda A^\lambda_3 - \frac{1}{3} A^\mu_\nu A^\nu_\mu \rangle \right) + O(\beta^2). \tag{9.47}$$

In particular, for the triangular representation we can use (9.45) to write (ignoring higher order terms in β)

$$M(I, Y) = M_0 + \beta \bar{a} \langle A^3{}_3 \rangle = M_0 - \beta \bar{a} \langle Y \rangle, \tag{9.48}$$

where $\bar{a}$ is a constant (whose value is not important for our discussion) and we have used the identification in (9.13). This leads to

$$M(\Delta) = M_0 - \beta\bar{a}, \quad M(Y^\star) = M_0, \quad M(\Xi^\star) = M_0 + \beta\bar{a},$$
$$M(\Omega^-) = M_0 + 2\beta\bar{a}, \tag{9.49}$$

which yields the correct mass (difference) formula for the decouplet, nnamely,

$$M(\Omega^-) - M(\Xi^\star) = M(\Xi^\star) - M(Y^\star) = M(Y^\star) - M(\Delta). \tag{9.50}$$

This mass (difference) formula for the decouplet agrees quite well with the experimental measurements of the masses of the baryons which leads to

$$1672 - 1530 \simeq 1530 - 1382 \simeq 1382 - 1232, \tag{9.51}$$

where the masses of the baryons are expressed in terms of MeV (we are assuming $c = 1$).

Let us now return to the general case and define

$$J_2 = A^\alpha{}_\beta A^\beta{}_\alpha, \tag{9.52}$$

which corresponds to the second order Casimir invariant of the $su(3)$ Lie algebra. (We do not denote it by I_2 as we have done earlier in order to avoid possible confusion with the isotopic spin generators.) Namely, it satisfies

$$[A^\alpha{}_\beta, J_2] = 0, \tag{9.53}$$

so that, in a given irreducible representation denoted by (f_1, f_2, f_3), we will have a constant value for J_2 of the form

$$J_2 = (\ell_1^2 + \ell_2^2 + \ell_3^2) - 2(\ell_1 + \ell_2 + \ell_3) + 1, \tag{9.54}$$

where

$$\ell_1 = f_1 + 2, \quad \ell_2 = f_2 + 1, \quad \ell_3 = f_3, \tag{9.55}$$

which we will show later.

We note that we can write the third term in (9.47) as

$$A^3{}_\lambda A^\lambda{}_3 - \frac{1}{3} A^\mu{}_\nu A^\nu{}_\mu$$

$$= A^3{}_\lambda A^\lambda{}_3 - \frac{1}{2} A^\mu{}_\nu A^\nu{}_\mu + \frac{1}{6} A^\mu{}_\nu A^\nu{}_\mu$$

$$= A^3{}_\lambda A^\lambda{}_3 - \frac{1}{2} \left(A^3{}_\nu A^\nu{}_3 + A^i{}_\nu A^\nu{}_i \right) + \frac{1}{6} J_2 \qquad (i,j = 1,2)$$

$$= \frac{1}{2} A^3{}_\lambda A^\lambda{}_3 - \frac{1}{2} \left(A^i{}_3 A^3{}_i + A^i{}_j A^j{}_i \right) + \frac{1}{6} J_2$$

$$= \frac{1}{2} \left(A^3{}_3 \right)^2 + \frac{1}{2} \left[A^3{}_i, A^i{}_3 \right] - \frac{1}{2} A^i{}_j A^j{}_i + \frac{1}{6} J_2$$

$$= \frac{1}{2} \left(A^3{}_3 \right)^2 + \frac{3}{2} A^3{}_3 - \frac{1}{2} A^i{}_j A^j{}_i + \frac{1}{6} J_2$$

$$= \frac{1}{2} Y^2 - \frac{3}{2} Y - \frac{1}{2} A^i{}_j A^j{}_i + \frac{1}{6} J_2, \qquad (9.56)$$

where we have used the commutation relations in (9.8), the traceless-ness of the generators $A^\mu{}_\mu = 0$ (see (9.9)) as well as the identifications in (9.13). Let us further simplify (9.56) by noting that

$$-\frac{1}{2} A^i{}_j A^j{}_i = -\frac{1}{2} \left((A^1{}_1)^2 + (A^2{}_2)^2 + A^1{}_2 A^2{}_1 + A^2{}_1 A^1{}_2 \right)$$

$$= -\frac{1}{2} \left(\frac{1}{2}(A^1{}_1 + A^2{}_2)^2 + \frac{1}{2}(A^1{}_1 - A^2{}_2)^2 + I_+ I_- + I_- I_+ \right)$$

$$= -\frac{1}{4} Y^2 - (I_1^2 + I_2^2 + I_3^2), \qquad (9.57)$$

where we have used the identifications in (9.13). Thus, substituting this into (9.56), we obtain

$$A^3{}_\lambda A^\lambda{}_3 - \frac{1}{3} A^\mu{}_\nu A^\nu{}_\mu$$

$$= \frac{1}{4} Y^2 - \mathbf{I}^2 - \frac{3}{2} Y + \frac{1}{6} J_2. \qquad (9.58)$$

Noting that $A^3{}_3 = -Y$ (see (9.13)), we can write the formula in (9.47) as (taking expectation values in a state with a given (I, Y) quantum numbers)

$$M(I,Y) = C_0 + C_1 Y + C_2 \left(\frac{1}{4} Y^2 - I(I+1) \right), \qquad (9.59)$$

where we can identify

$$C_0 = M_0 + \frac{1}{6}\beta a_2 J_2,$$

$$C_1 = -\beta\left(a_1 + \frac{3}{2}a_2\right),$$

$$C_2 = \beta a_2. \tag{9.60}$$

For the triangular irreducible representation with $f_1 = f_2$ or $f_2 = f_3$ (see, for example, (9.33)), we have $I = \frac{Y}{2} + 1$ (and $\frac{Y^2}{4} - I(I+1) = -\frac{3Y}{2} - 2$) so that the general mass formula (9.59) for $M(I, Y)$ reduces to the previous case (9.48), namely,

$$M(I, Y) = C_0' + C_1'Y. \tag{9.61}$$

For the baryon octet, the formula (9.59) gives (see also (9.20))

$$M(N) = C_0 + C_1 + C_2\left(\frac{1}{4} - \frac{1}{2}\left(\frac{1}{2} + 1\right)\right) = C_0 + C_1 - \frac{1}{2}C_2,$$

$$M(\Xi) = C_0 - C_1 + C_2\left(\frac{1}{4} - \frac{1}{2}\left(\frac{1}{2} + 1\right)\right) = C_0 - C_1 - \frac{1}{2}C_2,$$

$$M(\Lambda) = C_0,$$

$$M(\Sigma) = C_0 + C_2\left(0 - 1(1+1)\right) = C_0 - 2C_2. \tag{9.62}$$

The constants C_0, C_1 and C_2 can be determined from the last two relations as well as a combination (difference) of the first two to yield

$$C_0 = M(\Lambda), \quad C_1 = \frac{1}{2}\left(M(N) - M(\Xi)\right),$$

$$C_2 = \frac{1}{2}\left(M(\Lambda) - M(\Sigma)\right). \tag{9.63}$$

Furthermore, substituting these values for the constants into the orthogonal combination (sum) of the first two relations in (9.62) leads to the mass formula

$$M(N) + M(\Xi) = \frac{1}{2}\left(3M(\Lambda) + M(\Sigma)\right). \tag{9.64}$$

For the charge neutral members of the multiplets, relation (9.64) implies that

$$M(n) + M(\Xi^0) = \frac{1}{2}\left(3M(\Lambda^0) + M(\Sigma^0)\right), \tag{9.65}$$

and this is quite well satisfied by the exerimental values for the masses (in MeV), namely,

$$940 + 1315 \simeq \frac{1}{2}\left(3 \times 1116 + 1193\right). \tag{9.66}$$

Let us next consider the electromagnetic current operator in this sector (note the charge assignments for the quarks in (9.18))

$$
\begin{aligned}
J^\mu_{em}(x) &= \frac{e}{3}\left(2\bar{q}_1(x)\gamma^\mu q_1(x) - \bar{q}_2(x)\gamma^\mu q_2(x) - \bar{q}_3(x)\gamma^\mu q_3(x)\right) \\
&= e\left(\bar{q}_1(x)\gamma^\mu q_1(x) - \frac{1}{3}\bar{q}_\alpha(x)\gamma^\mu q_\alpha(x)\right) \\
&= e\left(V^\mu(x)\right)^1{}_1,
\end{aligned}
\tag{9.67}
$$

where we have defined the traceless current matrix

$$(V(x))^\alpha{}_\beta = \bar{q}_\alpha(x)\gamma^\mu q_\beta(x) - \frac{1}{3}\delta^\alpha{}_\beta\, \bar{q}_\gamma(x)\gamma^\mu q_\gamma(x), \tag{9.68}$$

with $\alpha, \beta, \gamma = 1, 2, 3$. This is an adjoint operator of the $SU_F(3)$ group and we emphasize here that we have ignored the contributions to the current from the c, b and the t quarks whose effects are expected to be very small. If we do add their contributions, the current will contain a $SU_F(3)$ singlet part in addition to the octet term, but we will continue working in the sector of the u, d and s quarks.

Given the electromagnetic current operator, the magnetic moment operator μ is defined by (the z component of the first moment)

$$\mu = \frac{1}{2}\int d^3x \left(\mathbf{x} \times \mathbf{J}_{em}\right)_z, \tag{9.69}$$

so that we can write it as

$$\mu = \mu^1{}_1, \tag{9.70}$$

of the traceless adjoint operator $\mu^\alpha{}_\beta$ satisfying $\mu^\alpha{}_\alpha = 0$. So, we can calculate its eigenvalues (expectation values in specific eigenstates) much the same way as before (see (9.44)) and write

$$\langle\mu\rangle = \langle\mu^1{}_1\rangle = a_1\langle A^1{}_1\rangle + a_2\langle A^1{}_\lambda A^\lambda{}_1 - \frac{1}{3}A^\mu{}_\nu A^\nu{}_\mu\rangle, \tag{9.71}$$

where a_1, a_2 are two constants (different from those in (9.44)). For any triangular irreducible representation, following (9.47) and (9.48), we can again show that

$$\langle\mu\rangle = a_1'\langle A^1{}_1\rangle = a_1'Q, \tag{9.72}$$

where Q denotes the electric charge (of the state) and we have made use of the identification in (9.17) in the last step. This immediately leads to the relation

$$\mu(\Delta^{++}) = 2\mu(\Delta^+) = -2\mu(\Delta^-) = -2\mu(\Omega^-), \tag{9.73}$$

among the members of the baryon decouplet. Unfortunately, only Ω^- is semi-stable and only the value of $\mu(\Omega^-)$ is known experimentally so that the relation (9.73) cannot yet be tested.

For the $J^P = \frac{1}{2}^+$ baryon octet, we can proceed as in the case of the mass formula (discussed earlier) and evaluate $\langle\mu\rangle$. However, here we will use a simpler method to evaluate this. Let $\left(\overline{B}\right)^\mu_{\ \nu} = \overline{B_\nu^{\ \mu}}$ be the anti-baryon tensor. Then, in a given octet representation we have

$$\langle\mu\rangle = C_1\langle\overline{B}^1_{\ \lambda}B^\lambda_{\ 1} - \frac{1}{3}\overline{B}^\mu_{\ \nu}B^\nu_{\ \mu}\rangle$$
$$+ C_2\langle\overline{B}^\lambda_{\ 1}B^1_{\ \lambda} - \frac{1}{3}\overline{B}^\mu_{\ \nu}B^\nu_{\ \mu}\rangle, \tag{9.74}$$

for constants C_1, C_2. We note that (see normalization in (9.20))

$$B^1_{\ 1} = \frac{1}{2}\left(B^1_{\ 1} + B^2_{\ 2}\right) + \frac{1}{2}\left(B^1_{\ 1} - B^2_{\ 2}\right) = -\frac{1}{2}B^3_{\ 3} + \frac{1}{2}\left(B^1_{\ 1} - B^2_{\ 2}\right)$$
$$= \frac{1}{\sqrt{6}}\Lambda^0 + \frac{1}{\sqrt{2}}\Sigma^0, \tag{9.75}$$

so that we obtain from (9.74)

$$\mu = \langle\mu^1_{\ 1}\rangle = (C_1 + C_2)\langle\overline{B}^1_{\ 1}B^1_{\ 1} - \frac{1}{3}\overline{B}^\alpha_{\ \beta}B^\beta_{\ \alpha}\rangle$$
$$+ C_1\langle\overline{B}^1_{\ 2}B^2_{\ 1} + \overline{B}^1_{\ 3}B^3_{\ 1}\rangle + C_2\langle\overline{B}^2_{\ 1}B^1_{\ 2} + \overline{B}^3_{\ 1}B^1_{\ 3}\rangle$$
$$= (C_1 + C_2)\langle\left(\left(\frac{1}{\sqrt{6}}\overline{\Lambda^0} + \frac{1}{\sqrt{2}}\overline{\Sigma^0}\right)\left(\frac{1}{\sqrt{6}}\Lambda^0 + \frac{1}{\sqrt{2}}\Sigma^0\right)\right.$$
$$- \frac{1}{3}\left(\overline{p}p + \overline{n}n + \overline{\Lambda^0}\Lambda^0 + \overline{\Sigma^+}\Sigma^+ + \overline{\Sigma^0}\Sigma^0\right.$$
$$\left.+ \overline{\Sigma^-}\Sigma^- + \overline{\Xi^0}\Xi^0 + \overline{\Xi^-}\Xi^-\right)\right)$$
$$+ C_1\left(\overline{\Sigma^-}\Sigma^- + \overline{\Xi^-}\Xi^-\right) + C_2\left(\overline{\Sigma^+}\Sigma^+ + \overline{p}p\right)\rangle. \tag{9.76}$$

Calculating this expectation value in definite baryon states we obtain

$$\mu(p) = \mu(\Sigma^+) = C_2 - \frac{1}{3}(C_1 + C_2),$$

$$\mu(\Sigma^-) = \mu(\Xi^-) = C_1 - \frac{1}{3}(C_1 + C_2),$$

$$\mu(n) = \mu(\Xi^0) = -\frac{1}{3}(C_1 + C_2),$$

$$\mu(\Lambda^0) = -\mu(\Sigma^0) = -\frac{1}{6}(C_1 + C_2). \tag{9.77}$$

Similarly calculating the matrix element in the states $\langle \Sigma^0 |$ and $|\Lambda^0 \rangle$ states we obtain the transition magnetic moment in the decay $\Sigma^0 \to \Lambda^0 + \gamma$ to be

$$\mu_T(\Lambda^0 \to \Sigma^0) = \frac{1}{2\sqrt{3}}(C_1 + C_2). \tag{9.78}$$

The Coleman-Glashow relations for the magnetic moments follow from (9.77) and (9.78), namely,

$$\mu(p) = \mu(\Sigma^+), \quad \mu(\Sigma^-) = \mu(\Xi^-),$$

$$\mu(n) = \mu(\Xi^0) = 2\mu(\Lambda^0) = -2\mu(\Sigma^0) = -\frac{1}{\sqrt{3}}\mu_T(\Lambda^0 \to \Sigma^0),$$

$$\mu(\Sigma^+) + \mu(\Sigma^-) = 2\mu(\Sigma^0). \tag{9.79}$$

Experimentally, the values of the magnetic moments are known to be

$$\mu(p) = 2.793\mu_N, \quad \mu(n) = -1.913\mu_N,$$

$$\mu(\Sigma^+) = 2.458 \pm 0.010\mu_N, \quad \mu(\Sigma^-) = -1.160 \pm 0.025\mu_N,$$

$$\mu(\Xi^0) = -1.250 \pm 0.014\mu_N, \quad \mu(\Xi^-) = -0.6507 \pm 0.0025\mu_N,$$

$$\mu(\Lambda^0) = -0.0613 \pm 0.004\mu_N,$$

$$|\mu_T(\Lambda^0 \to \Sigma^0)| = 1.61 \pm 0.08\mu_N, \tag{9.80}$$

where $\mu_N \simeq 2.89 \times 10^{14}$ MeV/T (MeV/tesla) denotes the nuclear magneton. Equation (9.79) now determines

$$\mu(\Sigma^0) = \frac{1}{2}\left(\mu(\Sigma^+) + \mu(\Sigma^-)\right) = 0.75 \pm 0.01\mu_N. \tag{9.81}$$

The experimental values in (9.80) can now be checked to fit the relations in (9.79) fairly well except perhaps $\mu(\Sigma^-) = \mu(\Xi^-)$ and

$\mu(n) = \mu(\Xi^0)$. However, we may note the following two points in this regard. First, the relation $\mu(n) = 2\mu(\Lambda^0)$ holds only when the electromagnetic current is purely an octet of $SU_F(3)$. But, in reality the current contains a singlet term as well (as described earlier) of the form

$$\frac{e}{3}\left(2\bar{c}(x)\gamma^\mu c(x) - \bar{b}(x)\gamma^\mu b(x) + 2\bar{t}(x)\gamma^\mu t(x)\right). \tag{9.82}$$

Since the bare quark mass for the bottom quark is $m_b \simeq 1\text{GeV}$, which is quite comparable to $m_s \simeq 0.15\text{GeV}$, we expect that the contribution of the singlet current to $\mu(\Lambda^0)$ can be as much as 15%. Indeed if we assume that the electromagnetic current includes the $SU_F(3)$ singlet term, then we note (without going into details) that the relation $\mu(\Lambda^0) = \frac{1}{2}\mu(n)$ will modify to

$$\mu(\Lambda^0) = \frac{1}{6}\left((\mu(\Sigma^+ + \Sigma^-) + 4\mu(n))\right), \tag{9.83}$$

which improves the prediction over $\mu(\Lambda^0) = \frac{1}{2}\mu(n)$. However, relations such as $\mu(\Sigma^-) = \mu(\Xi^-), \mu(n) = \mu(\Xi^0)$ continue to hold shedding no light on larger violations. This suggests that we have to take into account the $SU_F(3)$ violating contributions for these calculations. In such a case, the effective electromagnetic current will have a $SU_F(3)$ tensor structure of the form T^{13}_{13} in addition to T^1_1 which leads to the relations

$$-\sqrt{3}\mu_T(\Lambda^0 \to \Sigma^0) = \frac{1}{2}(3\mu(\Lambda^0) + \mu(\Sigma^0) - (\mu(n) + \mu(\Xi^0))),$$

$$\mu(\Sigma^0) = \frac{1}{2}(\mu(\Sigma^+) + \mu(\Sigma^-)). \tag{9.84}$$

The first of these two relations in (9.84) leads to $\mu_T(\Lambda^0 \to \Sigma^0) = -1.51\mu(n)$ improving the result compared with the earlier prediction of $\mu_T = -1.06\mu(n)$ in the sense that it is now closer to the relation $\mu_T = -\sqrt{3}\mu(n)$ in (9.79).

Although we cannot explain the large experimental violations of relations such as $\mu(n) = \mu(\Xi^0)$ and $\mu(\Sigma^-) = \mu(\Xi^-)$ (see (9.79)-(9.80)), we may argue as follows. Note that $\mu_N = \frac{e\hbar}{2M_N c}$, the nuclear magneton, is the Bohr magneton of the nucleon. Since the magnetic moment is not dimensionless, we argue that the exact $SU_F(3)$ group theoretical relations should be applicable not directly to $\mu(B)$, where B denotes a baryon, but to a dimensionless quantity $\mu_0(B)$ defined as

$$\mu(B) = \frac{e\hbar}{2M_B c}\mu_0(B), \tag{9.85}$$

so that we have, for example,

$$\mu_0(n) = \mu_0(\Xi^0) = 2\mu(\Lambda^0), \quad \mu_0(\Sigma^-) = \mu_0(\Xi^-). \tag{9.86}$$

This would lead to

$$\mu(\Xi^0) = \frac{M(n)}{M(\Xi^0)}\mu(n), \quad \Rightarrow \quad -1.25 = -1.37,$$

$$\mu(\Lambda^0) = \frac{M(n)}{2M(\Lambda^0)}\mu(n), \quad \Rightarrow \quad -0.61 = -0.80,$$

$$\mu(\Xi^-) = \frac{M(\Sigma^-)}{M(\Xi^-)}\mu(\Sigma^-), \quad \Rightarrow \quad -0.65 = -1.05, \tag{9.87}$$

which improves the results significantly over the predictions in (9.79) except for the last relation in (9.87). There is no clear idea regarding this discrepancy at the present time.

The predictions of $SU_F(3)$ symmetry are better for the weak leptonic decay modes of Λ^0, Σs and Ξs. However, since we have not discussed about the electro-weak theory, we point out only the electromagnetic self-energy effect in order to account for the mass difference between $M(p)$ and $M(n)$ as well as between $M(\Sigma^+)$ and $M(\Sigma^-)$ etc. Note that the electromagnetic self-energy contains photon exchange diagrams, so that its $SU_F(3)$ tensor behaves like a complicated tensor V_{11}^{11}. Also since $(m_u - m_d)S_1^1$ term (see, for example, (9.36)) can be written as $(m_u - m_d)S_1^1 S_1^1$, both terms can be combined into a single tensor T_{11}^{11}. Therefore, the effective mass term has the form (see, for example, (9.47))

$$M = M_0 + \beta\langle S_3^3\rangle + \langle T_{11}^{11}\rangle. \tag{9.88}$$

For the baryon octet, the same argument then shows that we may parameterize them as

$$M = a_1\overline{B}^\alpha{}_\beta B^\beta{}_\alpha + a_2\overline{B}^3{}_\alpha B^\alpha{}_3 + a_3\overline{B}^\alpha{}_3 B^3{}_\alpha$$

$$+ a_4 S_1^1\overline{B}^1{}_\alpha B^\alpha{}_1 + a_5 S_1^1\overline{B}^\alpha{}_1 B^1{}_\alpha + a_6\overline{B}^1{}_1 B^1{}_1, \tag{9.89}$$

in terms of six constants $a_1, a_2, \cdots, a_6$. Therefore, for the baryon octet, we can now have two mass relations of which one involves physically unobservable transition mass $M_T(\Lambda^0 \to \Sigma^0)$, so that effectively there is one mass relation of interest. With some computation, it leads to the Coleman-Glashow mass relation given by

$$M(\Xi^-) - M(\Xi^0) = M(\Sigma^-) - M(\Sigma^0) + M(p) - M(n), \tag{9.90}$$

which compares quite well with experimental values (masses in MeV since $c = 1$)

$$6.48\text{MeV} \simeq 8.08\text{MeV} - 1.29\text{MeV} \, (= 6.73\text{MeV}). \tag{9.91}$$

For completeness we remark here that dynamical calculations of masses and magnetic moments have also been satisfactorily performed within the framework of lattice gauge theory.

In ending this section, let us briefly mention 3×3 Gell-Mann matrices which are the $SU(3)$ analogs of the 2×2 Pauli matrices,

$$\lambda_1 = \begin{pmatrix} 0 & 1 & 0 \\ 1 & 0 & 0 \\ 0 & 0 & 0 \end{pmatrix}, \quad \lambda_2 = \begin{pmatrix} 0 & -i & 0 \\ i & 0 & 0 \\ 0 & 0 & 0 \end{pmatrix}, \quad \lambda_3 = \begin{pmatrix} 1 & 0 & 0 \\ 0 & -1 & 0 \\ 0 & 0 & 0 \end{pmatrix},$$

$$\lambda_4 = \begin{pmatrix} 0 & 0 & 1 \\ 0 & 0 & 0 \\ 1 & 0 & 0 \end{pmatrix}, \quad \lambda_5 = \begin{pmatrix} 0 & 0 & -i \\ 0 & 0 & 0 \\ i & 0 & 0 \end{pmatrix}, \quad \lambda_6 = \begin{pmatrix} 0 & 0 & 0 \\ 0 & 0 & 1 \\ 0 & 1 & 0 \end{pmatrix},$$

$$\lambda_7 = \begin{pmatrix} 0 & 0 & 0 \\ 0 & 0 & -i \\ 0 & i & 0 \end{pmatrix}, \quad \lambda_8 = \begin{pmatrix} \frac{1}{\sqrt{3}} & 0 & 0 \\ 0 & \frac{1}{\sqrt{3}} & 0 \\ 0 & 0 & -\frac{2}{\sqrt{3}} \end{pmatrix}, \tag{9.92}$$

which satisfy

$$\lambda_i^\dagger = \lambda_i, \quad \text{Tr}\,(\lambda_i \lambda_j) = 2\delta_{ij}, \quad i, j = 1, 2, 3. \tag{9.93}$$

Introducing the real structure constants

$$f_{ijk} = -\frac{i}{4}\,\text{Tr}\,(\lambda_k(\lambda_i\lambda_j - \lambda_j\lambda_i)),$$

$$d_{ijk} = \frac{1}{4}\,\text{Tr}\,(\lambda_k(\lambda_i\lambda_j + \lambda_j\lambda_i)), \tag{9.94}$$

it is straightforward to verify that these structure constants are totally anti-symmetric and symmetric respectively under the interchange of any pair of indices $i, j, k = 1, 2, \cdots, 8$. Moreover, it can be checked easily that the matrices in (9.92) satisfy respectively the commutation and the anti-commutation relations given by

$$[\lambda_i, \lambda_j] = (\lambda_i\lambda_j - \lambda_j\lambda_i) = 2if_{ijk}\,\lambda_k,$$

$$[\lambda_i, \lambda_j]_+ = (\lambda_i\lambda_j + \lambda_j\lambda_i) = \frac{4}{3}\,\delta_{ij}\,\mathbb{1} + 2d_{ijk}\lambda_k. \tag{9.95}$$

In particular, we note from the first relation in (9.95) that $\frac{1}{2}\lambda_i, i = 1, 2, \cdots, 8$ denote the 3×3 matrix representation of the generators of the $SU(3)$ group. The explicit numerical values for the structure constants can be found in the book by M. Gell-Mann and Y. Neeman. Furthermore, if we define

$$X_i = \frac{1}{2}(\lambda_i)_{ab}A^a{}_b, \quad a, b = 1, 2, 3, \quad i = 1, 2, \cdots, 8, \tag{9.96}$$

then it follows, using (9.95), that

$$[X_i, X_j] = \frac{1}{4}[(\lambda_i)_{ab}A^a{}_b, (\lambda_j)_{cd}A^c{}_d]$$

$$= \frac{1}{4}(\lambda_i)_{ab}(\lambda_j)_{cd}[A^a{}_b, A^c{}_d]$$

$$= \frac{1}{4}(\lambda_i)_{ab}(\lambda_j)_{cd}\left(-\delta^a{}_d A^c{}_b + \delta^c{}_b A^a{}_d\right)$$

$$= -\frac{1}{4}(\lambda_j\lambda_i)_{cb}A^c{}_b + \frac{1}{4}(\lambda_i\lambda_j)_{ad}A^a{}_d$$

$$= \frac{1}{4}[\lambda_i, \lambda_j]_{ab} A^a{}_b = \frac{1}{4}2if_{ijk}(\lambda_k)_{ab}A^a{}_b$$

$$= if_{ijk}X_k, \tag{9.97}$$

for a general matrix representation of X_i .

9.3 Some applications in nuclear physics

The structure of a nucleus is generally modelled after the Fermi gas model or the nuclear shell model. As possible applications of group theory to nuclear physics, let us first consider Wigner's $U(4)$ multiplet theory. Suppose that the dominant nuclear force between nucleons is independent of the ordinary spin as well as the isotopic spin of the nucleons. Then the Hamiltonian of the system of interacting nucleons may be written in the form

$$H = H_0 + H_1, \tag{9.98}$$

where H_0 represents the part of the Hamiltonian of the system which is invariant under the $U(4)$ group of transformations,

$$\psi' = U\psi, \quad \psi = \begin{pmatrix} p_\uparrow \\ p_\downarrow \\ n_\uparrow \\ n_\downarrow \end{pmatrix}, \tag{9.99}$$

where U denotes an arbitrary space-time independent matrix belonging to $U(4)$ and $p_\uparrow, p_\downarrow, n_\uparrow, n_\downarrow$ refer to the spin up and spin down states of the proton and the neutron respectively. On the other hand, the Hamiltonian H_1 may violate the $U(4)$ symmetry and its contributions are assumed to be small compared to those of H_0. In such a case, we can classify the nuclear states by the irreducible representations of the $U(4)$ group and the $U(4)$ violating part of the Hamiltonian H_1 can be treated as a perturbation as we have discussed in the case of $SU_F(3)$ symmetry of the quark model. Unfortunately, such a $U(4)$ model does not turn out to be realistic (does not agree with observations) and, therefore, we will not go into the details of this model.

As another nuclear model, let us consider the shell model in which we have n singlet particle levels (or shells) with the energy of the level $\epsilon_\mu, \mu = 1, 2, \cdots, n$. Let us assume that the μth shell is further labelled by another micro quantum number p such as the angular momentum or spin although we will not specify here what p represents. Let $a_{\mu,p}$ with $\mu = 1, 2, \cdots, n, p = 1, 2, \cdots, \ell$ denote the annihilation operator for a nucleon in the state specified by the quantum number (μ, p) and that the Hamiltonian for the system is given by (repeated indices are summed)

$$H = \epsilon_\mu a^\dagger_{\mu,p} a_{\mu,p} + g^{\mu\nu}_{\alpha\beta} a^\dagger_{\alpha,p} a_{\mu,p} a^\dagger_{\beta,q} a_{\nu,q}, \tag{9.100}$$

where $\alpha, \beta, \mu, \nu = 1, 2, \cdots, n$ while $p, q = 1, 2, \cdots, \ell$. (Here $g^{\mu\nu}_{\alpha\beta}$ is a constant tensor.) If we now identify

$$X^\alpha_\beta = a^\dagger_{\alpha,p} a_{\beta,p}, \tag{9.101}$$

then the Hamiltonian in (9.100) can be rewritten as

$$H = \epsilon_\mu X^\mu_\mu + g^{\mu\nu}_{\alpha\beta} X^\alpha_\mu X^\beta_\nu. \tag{9.102}$$

Moreover, since $a_{\alpha,p}$ and $a^\dagger_{\beta,q}$ satisfy the anti-commutation relations

$$a_{\alpha,p} a_{\beta,q} + a_{\beta,q} a_{\alpha,p} = 0 = a^\dagger_{\alpha,p} a^\dagger_{\beta,q} + a^\dagger_{\beta,q} a^\dagger_{\alpha,p},$$

$$a_{\alpha,p} a^\dagger_{\beta,q} + a^\dagger_{\beta,q} a_{\alpha,p} = \delta_{\alpha\beta} \delta_{pq}, \tag{9.103}$$

it follows that X^β_α satisfy the commutation relations of the $u(n)$ Lie algebra (see (3.50))

$$[X^\mu_\nu, X^\alpha_\beta] = -\delta^\mu_\beta X^\alpha_\nu + \delta^\alpha_\nu X^\mu_\beta, \tag{9.104}$$

as well as the unitarity condition (see (9.101))

$$(X^\mu_\nu)^\dagger = \left(a^\dagger_{\mu,p}a_{\nu,p}\right)^\dagger = a^\dagger_{\nu,p}a_{\mu,p} = X^\nu_\mu. \tag{9.105}$$

Here $\mu, \nu, \alpha, \beta = 1, 2, \cdots, n$.

Let us note that the Hamiltonian H does not commute with X^μ_ν in general so that it is not invariant under the $U(n)$ group of transformations. However, it does commute with all the Casimir operators $J_\mu, \mu = 1, 2, \cdots, n$ of the $u(n)$ Lie algebra given by

$$J_1 = X^\mu_\mu, \quad J_2 = X^\mu_\nu X^\nu_\mu, \quad J_3 = X^\mu_\nu X^\nu_\lambda X^\lambda_\mu, \tag{9.106}$$

and so on. In fact, if we think of X^μ_ν as matrices, then the Casimir operators can be written as

$$J_\alpha = \mathrm{Tr}\, X^\alpha, \tag{9.107}$$

and it follows from the commutation relations (9.104) that

$$[J_\alpha, X^\mu_\nu] = [\mathrm{Tr}\, X^\alpha, X^\mu_\nu]$$

$$= \sum_{m=1}^{\alpha} \left(X^{\alpha-m}\right)^\lambda_\rho [X^\rho_\delta, X^\mu_\nu] \left(X^{m-1}\right)^\delta_\lambda$$

$$= \sum_{m=1}^{\alpha} \left(X^{\alpha-m}\right)^\lambda_\rho \left(-\delta^\rho_\nu X^\mu_\delta + \delta^\mu_\delta X^\rho_\nu\right) \left(X^{m-1}\right)^\delta_\lambda$$

$$= \sum_{m=1}^{\alpha} \left[-\left(X^{\alpha-m}\right)^\lambda_\nu (X^m)^\mu_\lambda + \left(X^{\alpha-m+1}\right)^\lambda_\nu \left(X^{m-1}\right)^\mu_\lambda\right]$$

$$= \sum_{m=1}^{\alpha} \left[-(X^\alpha)^\mu_\nu + (X^\alpha)^\mu_\nu\right] = 0, \tag{9.108}$$

which shows that these are indeed the Casimir operators for the Lie algebra of $u(n)$. Furthermore, using (9.108) it follows immediately that

$$[J_\alpha, H] = 0, \tag{9.109}$$

and (J_α, H) define a set of commuting operators which can be diagonalized simultaneously.

In other words, we can classify the eigenstates of H by the irreducible representations of the $u(n)$ Lie algebra. Since any $d \times d$

irreducible matrix representation of the $u(n)$ Lie algebra can be readily constructed, we can express H as a $d \times d$ matrix which can then
be diagonalized to determine its energy eigenvalues. (Note that any
irreducible $d \times d$ matrix representation of $u(n)$ can be specified by n
eigenvalues corresponding to the n Casimir invariants.) As a simple
illustration, let us make the (rather drastic) assumption that all the
eigenvalues are degenrate,

$$\epsilon_1 = \epsilon_2 = \epsilon_3 = \cdots = \epsilon_n = \epsilon,$$

$$g^{\mu\nu}_{\alpha\beta} = g_1 \delta^\mu_\alpha \delta^\nu_\beta + g_2 \delta^\mu_\beta \delta^\nu_\alpha + g_3 \delta^{\mu\nu} \delta_{\alpha\beta}, \tag{9.110}$$

where g_1, g_2, g_3 are constants. With these, the Hamiltonian (9.102)
takes the form

$$H = \epsilon X^\mu_{\ \mu} + (g_1 \delta^\mu_\alpha \delta^\nu_\beta + g_2 \delta^\mu_\beta \delta^\nu_\alpha + g_3 \delta^{\mu\nu} \delta_{\alpha\beta}) X^\alpha_{\ \mu} X^\beta_{\ \nu}$$

$$= \epsilon J_1 + g_1 (J_1)^2 + g_2 J_2 + g_3 X^\mu_{\ \nu} X^\mu_{\ \nu}$$

$$= E_0 + g_3 X^\mu_{\ \nu} X^\mu_{\ \nu}, \tag{9.111}$$

where we have identified, for simplicity,

$$E_0 = \epsilon J_1 + g_1 (J_1)^2 + g_2 J_2. \tag{9.112}$$

Furthermore, let us note that if we define

$$J_{\mu\nu} = X^\mu_{\ \nu} - X^\nu_{\ \mu} = -J_{\nu\mu}, \tag{9.113}$$

then using (9.104) it is straightforward to verify that it satisfies the
Lie algebra of $so(n)$ and, therefore, corresponds to the generators of
the infinitesimal rotations in n-dimensional space. It follows from
(9.113) that

$$J_{\mu\nu} J_{\mu\nu} = \left(X^\mu_{\ \nu} - X^\nu_{\ \mu} \right) \left(X^\mu_{\ \nu} - X^\nu_{\ \mu} \right)$$

$$= 2 \left(X^\mu_{\ \nu} X^\mu_{\ \nu} - (X^2)^\mu_{\ \mu} \right),$$

or, $\quad X^\mu_{\ \nu} X^\mu_{\ \nu} = J_2 + \dfrac{1}{2} J_{\mu\nu} J_{\mu\nu}. \tag{9.114}$

As a result, we can write the Hamiltonian (9.111) also as

$$H = E'_0 + \frac{g_3}{2} J_{\mu\nu} J_{\mu\nu}, \tag{9.115}$$

where we have identified

$$E'_0 = E_0 + g_3 J_2. \tag{9.116}$$

We note that

$$C_2 = \frac{1}{2} J_{\mu\nu} J_{\mu\nu}, \tag{9.117}$$

corresponds to the quadratic Casimir invariant of the $so(n)$ sub-Lie algebra of $u(n)$ and commutes with both J_1 and J_2. As a result, the energy eigenvalues of H are now completely determined by the values of J_1, J_2 and C_2. However, these models are not realistic enough. Perhaps the most successful nuclear model is based on the interacting boson model where the nucleus consists of nucleon pairs tightly bound to each other behaving effectively like bosons. As a result, the nucleus (with an even number of nucleons) may be regarded as a boson gas (see Iachello et al.).

Another interesting topic in nuclear physics is known as the nuclear-boson expansion theory in which we can express products of two fermion operators such as $a_{\mu,p} a_{\nu,q}$ as a polynomial of some boson operators to justify replacing fermionic systems by bosonic ones. One of the related ideas in this regard is the Holstein-Primakoff representation of the $su(2)$ Lie algebra. Let us write $f = 2j = 0, 1, 2, 3, \cdots$ where j denotes the angular momentum quantum number characterizing a state. Let a and $a^\dagger$ denote two bosonic annihilation and creation operators satisfying the commutation relation

$$[a, a^\dagger] = 1. \tag{9.118}$$

If we now define

$$J_+ = J_1 + iJ_2 = a^\dagger (f - a^\dagger a)^{\frac{1}{2}},$$

$$J_- = J_1 - iJ_2 = (f - a^\dagger a)^{\frac{1}{2}} a,$$

$$J_3 = a^\dagger a - \frac{f}{2}, \tag{9.119}$$

then it can be directly checked that they satisfy the $su(2)$ Lie algebra with

$$\mathbf{J}^2 = J_1^2 + J_2^2 + J_3^2 = \frac{1}{2} (J_+ J_- + J_- J_+) + J_3^2$$

$$= \frac{f}{2} \left(\frac{f}{2} + 1 \right). \tag{9.120}$$

This clearly works if we restrict ourselves to the sector $f \geq a^\dagger a \geq 0$.

9.4 References

M. Gell-Mann and Y. Neeman, *Eightfold way*, W. A. Benjamin, NY (1964).

F. Iachello and A. Arima, *Interacting boson model*, Cambridge University Press (1987).

CHAPTER 10

Casimir invariants and adjoint operators

To start with let us consider, as an example, the $u(N)$ Lie algebra given by

$$[X^\mu{}_\nu, X^\alpha{}_\beta] = -\delta^\mu{}_\beta X^\alpha{}_\nu + \delta^\alpha{}_\nu X^\mu{}_\beta, \tag{10.1}$$

for $\mu, \nu, \alpha, \beta = 1, 2, \cdots, N$. In the universal enveloping algebra of $u(N)$ or in some matrix representation of $u(N)$, then the $N \times N$ traceless matrix $S^\alpha{}_\beta$ defined by (see section 3.2.3)

$$S^\alpha{}_\beta = T^\alpha{}_\beta - \frac{1}{N} \delta^\alpha{}_\beta T^\gamma{}_\gamma, \tag{10.2}$$

will also satisfy the commutation relation

$$[X^\mu{}_\nu, S^\alpha{}_\beta] = -\delta^\mu{}_\beta S^\alpha{}_\nu + \delta^\alpha{}_\nu S^\mu{}_\beta, \tag{10.3}$$

where

$$[X^\mu{}_\nu, T^\alpha{}_\beta] = -\delta^\mu{}_\beta T^\alpha{}_\nu + \delta^\alpha{}_\nu T^\mu{}_\beta. \tag{10.4}$$

We have called such an operator an adjoint operator (sometimes also known as a vector operator) in the earlier chapter. This notion can be generalized to any Lie algebra L satisfying (we are assuming that the generators X_α are Hermitian)

$$[X_\alpha, X_\beta] = i C^\gamma_{\alpha\beta} X_\gamma, \tag{10.5}$$

where $\alpha, \beta, \gamma = 1, 2, \cdots, M$ and $M =$ Dimension of the Lie algebra L. For example, for $u(N)$, we know that $M = N^2$. Furthermore, we recall that

$$C^\gamma_{\alpha\beta} = -C^\gamma_{\beta\alpha}, \tag{10.6}$$

are known as the structure constants of the algebra. In this case, any operator S_α satisfying

$$[X_\alpha, S_\beta] = i C^\gamma_{\alpha\beta} S_\gamma, \tag{10.7}$$

is known as a covariant adjoint operator of the Lie algebra L. In a similar manner, we can define a contravariant adjoint operator through the relation

$$[X_\alpha, T^\beta] = -iC^\beta_{\alpha\gamma} T^\gamma. \tag{10.8}$$

It now follows that any quantity of the form

$$I = S_\alpha T^\alpha, \quad \text{or,} \quad I = T^\alpha S_\alpha, \tag{10.9}$$

will satisfy (say, for the first form in (10.9))

$$\begin{aligned}
[X_\alpha, I] &= [X_\alpha, S_\beta T^\beta] = [X_\alpha, S_\beta]T^\beta + S_\alpha[X_\alpha, T^\beta] \\
&= iC^\gamma_{\alpha\beta} S_\gamma T^\beta - iC^\beta_{\alpha\gamma} S_\beta T^\gamma = iC^\gamma_{\alpha\beta}(S_\gamma T^\beta - S_\gamma T^\beta) \\
&= 0.
\end{aligned} \tag{10.10}$$

Here we have renamed the indices in the second term in the intermediate step. This shows that the generators of the Lie algebra will commute with quantities of the forms in (10.9) and, consequently, such quantities define Casimir invariants of the Lie algebra L.

Let us assume from now on that there exists a symmetric (second rank) constant tensor (metric) $g_{\alpha\beta} = g_{\beta\alpha}$ (see (3.111) and (3.115)) which satisfies

$$g_{\alpha\delta} C^\delta_{\beta\gamma} = f_{\alpha\beta\gamma}, \tag{10.11}$$

with $f_{\alpha\beta\gamma}$ denoting the completely anti-symmetric structure constants of the Lie algebra. Furthermore, we assume that the metric is nonsingular so that the inverse $g^{\alpha\beta}$ exists and satisfies

$$g_{\alpha\gamma} g^{\gamma\beta} = \delta^\beta_\alpha. \tag{10.12}$$

As we have noted earlier, these conditions hold for any reductive Lie algebra, namely, the direct sum of semi-simple and Abelian Lie algebras. For such a case, given a covariant adjoint operator S_α, we can always obtain a contravariant adjoint operator S^α as

$$S^\alpha = g^{\alpha\beta} S_\beta. \tag{10.13}$$

As a result, the Casimir invariant (10.9) of the Lie algebra L can now be written as

$$I = S_\alpha T^\alpha = g^{\alpha\beta} S_\alpha T_\beta. \tag{10.14}$$

In particular, if $T^\alpha = S^\alpha$, this takes the form

$$I = I_2 = g^{\alpha\beta} S_\alpha S_\beta = S_\alpha S^\alpha = S^\alpha S_\alpha, \tag{10.15}$$

and would correspond to a quadratic Casimir operator for $S_\alpha = X_\alpha$. In general, if we can write ($g^{\alpha\alpha_1\alpha_2\cdots\alpha_{p-1}}$ is a constant tensor which is a generalization of the metric tensor)

$$T^\alpha = g^{\alpha\alpha_1\alpha_2\cdots\alpha_{p-1}} X_{\alpha_1} X_{\alpha_2} \cdots X_{\alpha_{p-1}}, \tag{10.16}$$

then the Casimir invariant in (10.9) with $S_\alpha = X_\alpha$ can be written as

$$I_p = g^{\alpha_1\alpha_2\cdots\alpha_p} X_{\alpha_1} X_{\alpha_2} X_{\alpha_3} \cdots X_{\alpha_p}, \tag{10.17}$$

would define a pth order Casimir invariant of the Lie algebra.

To be specific, let us assume that we are considering an irreducible matrix representation space ρ of the Lie algebra L. Then, it will be shown below that since the Casimir invariants commute with all the generators of the group, they must be constant multiples of the unit (identity) matrix $\mathbb{1}$, namely,

$$I = I(\rho)\mathbb{1}, \tag{10.18}$$

where the constant $I(\rho)$ corresponds to the eigenvalue of I in the (representation) vector space $V = \rho$. To see this more directly, let us look at the eigenvalue equation

$$Iv = \lambda v, \tag{10.19}$$

which has a nontrivial solution if the characteristic equation

$$\det(I - \lambda\mathbb{1}) = 0, \tag{10.20}$$

is satisfied. Furthermore, since I commutes with all the generators (see (10.10)), it follows that

$$[X_\alpha, I]v = (X_\alpha I - IX_\alpha)v = 0,$$

$$\text{or,} \quad I(X_\alpha v) = \lambda(X_\alpha v). \tag{10.21}$$

More generally, if $P(X_\alpha)$ denotes an arbitrary function of the generators X_α, it would follow from (10.21) that

$$I(P(X_\alpha)v) = \lambda(P(X_\alpha)v). \tag{10.22}$$

This would imply that if v is an eigenstate of I with eigenvalue λ, then $(P(X_\alpha)v)$ is also an eigenstate of I with the same eigenvalue λ

for any arbitrary function $P(X_\alpha)$. Let V_0 denote the subspace of the vector space V spanned by vectors of the form $P(X_\alpha)v$. This would imply, in particular, that $X_\alpha P(X_\alpha)v \subseteq V_0$ (since we can think of $P'(X_\alpha) = X_\alpha P(X_\alpha)$). Namely, V_0 would correspond to an invariant subspace of V. On the other hand, by assumption V is an irreducible vector space and if V_0 contains $v \neq 0$, we must have $V_0 = V$ so that any vector $v \in V$ is an eigenvector of I with the same eigenvalue which implies that

$$I = \lambda \mathbb{1}. \tag{10.23}$$

Denoting $\lambda = I(\rho)$ we obtain (10.18).

10.1 Computation of the Casimir invariant $I(\rho)$

Let $\{\rho\}$ denote the generic irreducible representation of the Lie algebra L while $\{\rho_0\}$ represent a fixed irreducible representation which we call a reference representation. We will choose $\{\rho_0\}$ to be the simplest, namely, the lowest dimensional irreducible representation for a reason to be explained shortly. Let X_α and x_α denote the matrix representations of the generators in $\{\rho\}$ and $\{\rho_0\}$ respectively. Let us consider the tensor product $\{\rho\} \otimes \{\rho_0\}$ and define

$$Y_\alpha = X_\alpha \otimes \mathbb{1}_0 + \mathbb{1} \otimes x_\alpha, \tag{10.24}$$

where $\mathbb{1}$ and $\mathbb{1}_0$ represent the identity matrices in $\{\rho\}$ and $\{\rho_0\}$ respectively. We note that

$$[Y_\alpha, Y_\beta] = [X_\alpha \otimes \mathbb{1}_0 + \mathbb{1} \otimes x_\alpha, X_\beta \otimes \mathbb{1}_0 + \mathbb{1} \otimes x_\beta]$$
$$= [X_\alpha \otimes \mathbb{1}_0, X_\beta \otimes \mathbb{1}_0] + [X_\alpha \otimes \mathbb{1}_0, \mathbb{1} \otimes x_\beta]$$
$$+ [\mathbb{1} \otimes x_\alpha, X_\beta \otimes \mathbb{1}_0] + [\mathbb{1} \otimes x_\alpha, \mathbb{1} \otimes x_\beta]. \tag{10.25}$$

The individual commutators are easy to work out.

$$[X_\alpha \otimes \mathbb{1}_0, X_\beta \otimes \mathbb{1}_0] = (X_\alpha X_\beta) \otimes (\mathbb{1}_0 \mathbb{1}_0) - (X_\beta X_\alpha) \otimes (\mathbb{1}_0 \mathbb{1}_0)$$
$$= ([X_\alpha, X_\beta]) \otimes \mathbb{1}_0 = iC_{\alpha\beta}^\gamma (X_\gamma \otimes \mathbb{1}_0),$$
$$[X_\alpha \otimes \mathbb{1}_0, \mathbb{1} \otimes x_\beta] = (X_\alpha \mathbb{1}) \otimes (\mathbb{1}_0 x_\beta) - (\mathbb{1} X_\alpha) \otimes (x_\beta \mathbb{1}_0)$$
$$= X_\alpha \otimes x_\beta - X_\alpha \otimes x_\beta = 0,$$
$$[\mathbb{1} \otimes x_\alpha, X_\beta \otimes \mathbb{1}_0] = (\mathbb{1} X_\beta) \otimes (x_\alpha \mathbb{1}_0) - (X_\beta \mathbb{1}) \otimes (x_\alpha \mathbb{1}_0)$$
$$= X_\beta \otimes x_\alpha - X_\beta \otimes x_\alpha = 0,$$

$$[\mathbb{1} \otimes x_\alpha, \mathbb{1} \otimes x_\beta] = (\mathbb{1} \times \mathbb{1}) \otimes (x_\alpha x_\beta) - (\mathbb{1} \times \mathbb{1}) \otimes (x_\beta x_\alpha)$$
$$= \mathbb{1} \otimes [x_\alpha, x_\beta] = iC^\gamma_{\alpha\beta}(\mathbb{1} \otimes x_\gamma). \tag{10.26}$$

Here we have used the commutation relations in (10.6) and using these in (10.25) we obtain

$$[Y_\alpha, Y_\beta] = iC^\gamma_{\alpha\beta}(X_\gamma \otimes \mathbb{1}_0 + \mathbb{1} \otimes x_\gamma) = iC^\gamma_{\alpha\beta}Y_\gamma. \tag{10.27}$$

Therefore, Y_α in (10.24) also satisfies the same algebra as the generators in (10.6).

As we know we can raise the indices by applying $g^{\alpha\beta}$ so that we can write

$$X^\alpha = g^{\alpha\beta}X_\beta, \quad x^\alpha = g^{\alpha\beta}x_\beta, \tag{10.28}$$

and so on. This allows us to define the operator

$$Q = X_\alpha \otimes x^\alpha = X^\alpha \otimes x_\alpha = g^{\alpha\beta}X_\alpha \otimes x_\beta, \tag{10.29}$$

which has no indices. We can now calculate

$$[Y_\alpha, Q] = [X_\alpha \otimes \mathbb{1}_0 + \mathbb{1} \otimes x_\alpha, X_\beta \otimes x^\beta]$$
$$= [X_\alpha, X_\beta] \otimes x^\beta + X_\beta \otimes [x_\alpha, x^\beta]$$
$$= (iC^\gamma_{\alpha\beta}X_\gamma) \otimes x^\beta + X_\beta \otimes (-iC^\beta_{\alpha\gamma}x^\gamma)$$
$$= iC^\gamma_{\alpha\beta}\left(X_\gamma \otimes x^\beta - X_\gamma \otimes x^\beta\right) = 0, \tag{10.30}$$

where we have used (10.9). In particular, if

$$Q^p = \underbrace{QQ\cdots Q}_{p \text{ times}}, \tag{10.31}$$

denotes the pth matrix product of Qs, we have from (10.30)

$$[Y_\alpha, Q^p] = 0, \quad p = 1, 2, \cdots . \tag{10.32}$$

This can be rewritten as

$$[X_\alpha \otimes \mathbb{1}_0 + \mathbb{1} \otimes x_\alpha, X_{\beta_1} X_{\beta_2} \cdots X_{\beta_p} \otimes x^{\beta_1} x^{\beta_2} \cdots x^{\beta_p}] = 0,$$

or, $\quad [X_\alpha, X_{\beta_1} X_{\beta_2} \cdots X_{\beta_p}] \otimes x^{\beta_1} x^{\beta_2} \cdots x^{\beta_p}$
$$+ X_{\beta_1} X_{\beta_2} \cdots X_{\beta_p} \otimes [x_\alpha, x^{\beta_1} x^{\beta_2} \cdots x^{\beta_p}] = 0. \tag{10.33}$$

We can now take a partial trace Tr_{ρ_0} of both sides of the equation only over the reference representation ρ_0 (and not over the representation space ρ). Then the second term in (10.33) will vanish (the trace of a commutator vanishes) and defining

$$\text{Tr}_{\rho_0}\left(x^{\beta_1}x^{\beta_2}\cdots x^{\beta_p}\right) = g^{\beta_1\beta_2\cdots\beta_p},\tag{10.34}$$

the partial trace yields

$$[X_\alpha, g^{\beta_1\beta_2\cdots\beta_p}X_{\beta_1}X_{\beta_2}\cdots X_{\beta_p}] = 0,\tag{10.35}$$

for any $X_\alpha \in L$. In other words, we can identify the pth order Casimir invariant of the Lie algebra L with

$$I_p = g^{\alpha_1\alpha_2\cdots\alpha_p}X_{\alpha_1}X_{\alpha_2}\cdots X_{\alpha_p},\tag{10.36}$$

so that we can write

$$I_p = I_p(\rho)\,\mathbb{1},\tag{10.37}$$

for all possible eigenvalues $I_p(\rho)$ of the Casimir invariant.

Furthermore, since by definition

$$Q^p = X_{\alpha_1}X_{\alpha_1}\cdots X_{\alpha_p}\otimes x^{\alpha_1}x^{\alpha_2}\cdots x^{\alpha_p},\tag{10.38}$$

the complete trace (trace over both the spaces) leads to

$$\begin{aligned}
\text{Tr}\,Q^p &= \text{Tr}_\rho\left(X_{\alpha_1}X_{\alpha_2}\cdots X_{\alpha_p}\right)\text{Tr}_{\rho_0}\left(x^{\alpha_1}x^{\alpha_2}\cdots x^{\alpha_p}\right)\\
&= g^{\alpha_1\alpha_2\cdots\alpha_p}\text{Tr}_\rho\left(X_{\alpha_1}X_{\alpha_2}\cdots X_{\alpha_p}\right)\\
&= \text{Tr}_\rho\left(g^{\alpha_1\alpha_2\cdots\alpha_p}X_{\alpha_1}X_{\alpha_2}\cdots X_{\alpha_p}\right)\\
&= \text{Tr}_\rho I_p = I_p(\rho)\,\text{Tr}_\rho\,\mathbb{1} = d(\rho)I_p(\rho),
\end{aligned}\tag{10.39}$$

where $d(\rho) = \text{Tr}_\rho\,\mathbb{1}$ denotes the dimension of the irreducible representation space $\{\rho\}$. Here we have used the definitions (10.34) as well as (10.36) in the intermediate steps. It follows now that we can write

$$I_p(\rho) = \frac{1}{d(\rho)}\,\text{Tr}\,Q^p,\tag{10.40}$$

and, therefore, to determine the Casimir invariant, we only need to calculate $\text{Tr}\,Q^p$.

This can be done as follows. Let us assume that L is a reductive Lie algebra and hence $\{\rho\}\otimes\{\rho_0\}$ can be decomposed into a direct sum of n irreducible representations $\{\rho_j\}, j = 1, 2, \cdots, n$ as (see section 2.4)

$$\{\rho\} \otimes \{\rho_0\} = \sum_{j=1}^{n} \oplus \{\rho_j\}. \tag{10.41}$$

In this case, we can write

$$Y_\alpha = X_\alpha \otimes 1_0 + 1 \otimes x_\alpha$$
$$= \sum_{j=1}^{n} X_\alpha^{(j)} P_j, \tag{10.42}$$

where $X_\alpha^{(j)}$ is a function of X_α while P_j (which contains the x_α dependence) denotes the projection operator onto the space $\{\rho_j\}$ so that

$$\sum_{j=1}^{n} \oplus P_j = 1 \otimes 1_0. \tag{10.43}$$

In other words, P_j can be written in the block diagonal form ($E_j \equiv 1_j$)

$$P_j = \begin{pmatrix} \boxed{0} & & & & & & \\ & \boxed{0} & & & & & \\ & & \ddots & & & & \\ & & & \boxed{E_j} & & & \\ & & & & \ddots & & \\ & & & & & \boxed{0} & \end{pmatrix}. \tag{10.44}$$

For example, denoting the second order Casimir invariant by I_2, we have

$$g^{\alpha\beta} X_\alpha^{(j)} X_\beta^{(j)} = I_2(\rho_j) P_j, \quad \text{for fixed } j = 1, 2, \cdots, n,$$
$$g^{\alpha\beta} X_\alpha X_\beta = I_2(\rho) 1,$$
$$g^{\alpha\beta} x_\alpha x_\beta = I_2(\rho_0) 1_0, \tag{10.45}$$

so that we can rewrite (see (10.29) as well as (10.42))

$$Q = g^{\alpha\beta} X_\alpha \otimes x_\beta$$
$$= \frac{1}{2} \left(g^{\alpha\beta} Y_\alpha Y_\beta - X_\alpha X_\beta \otimes \mathbb{1}_0 - \mathbb{1} \otimes x_\alpha x_\beta \right)$$
$$= \frac{1}{2} \sum_{j=1}^{n} I_2(\rho_j) P_j - \frac{1}{2} \left(I_2(\rho) + I_2(\rho_0) \right) \mathbb{1} \otimes \mathbb{1}_0$$
$$= \frac{1}{2} \sum_{j=1}^{n} \left(I_2(\rho_j) - I_2(\rho) - I_2(\rho_0) \right) P_j, \tag{10.46}$$

where we have used (10.43). For simplicity, let us denote

$$\xi_j = \frac{1}{2} \left(I_2(\rho_j) - I_2(\rho) - I_2(\rho_0) \right), \tag{10.47}$$

so that we can write

$$Q = \sum_{j=1}^{n} \xi_j P_j. \tag{10.48}$$

It follows now that (remember that P_j is a projection operator)

$$Q^p = \sum_{j=1}^{n} (\xi_j)^p P_j, \tag{10.49}$$

and the trace leads to (see (10.40))

$$\mathrm{Tr}\, Q^p = \sum_{j=1}^{n} (\xi_j)^p \, \mathrm{Tr}\, P_j = \sum_{j=1}^{n} (\xi_j)^p d(\rho_j),$$

$$\text{or,} \quad I_p(\rho) = \frac{1}{d(\rho)} \mathrm{Tr}\, Q^p = \frac{1}{d(\rho)} \sum_{j=1}^{n} (\xi_j)^p d(\rho_j). \tag{10.50}$$

Therefore, if we know the Clebsch-Gordan decomposition law for $\{\rho\} \otimes \{\rho_0\} = \sum_{j=1}^{n} \oplus \{\rho_j\}$, we can calculate the pth order Casimir invariant $I_p(\rho)$ for any generic irreducible representation $\{\rho\}$ since the eigenvalues $I_2(\rho_j)$ of the second order Casimir invariant can be easily computed.

As an illustration, let us compute $I_p(\rho)$ for the Lie algebra $u(N)$. In this case the dimensionality of the group is given by $M = N^2$. Let the generic irreducible representation $\{\rho\}$ be labelled by the Young tableaux symbol

$$\{\rho\} = (f_1, f_2, \cdots, f_N), \quad f_1 \geq f_2 \geq \cdots \geq f_N, \tag{10.51}$$

and we choose the reference representation to correspond to (the fundamenta representation)

$$\{\rho_0\} = \boxed{} = (1, 0, 0, \cdots, 0). \tag{10.52}$$

In this case, we can decompose the product of representations $\{\rho\} \otimes \{\rho_0\}$ as (see section 5.4)

$$\{\rho\} \otimes \{\rho_0\} = (f_1, f_2, \cdots, f_N) \otimes (1, 0, 0, \cdots, 0)$$
$$= (f_1 + 1, f_2, \ldots, f_N) \oplus (f_1, f_2 + 1, f_3, \cdots, f_N)$$
$$\oplus \cdots \oplus (f_1, f_2, \cdots, f_N + 1). \tag{10.53}$$

If $f_j < f_{j+1} + 1$, we can drop the representation

$$(f_1, f_2, \cdots, f_j, f_{j+1} + 1, \cdots, f_N), \tag{10.54}$$

since it violates the condition required for a Young tableaux (see, for example, (10.51)).

Furthermore, Weyl's dimensional formula (see, for example, the book, "Group theory and quantum mechanics", by H. Weyl) gives

$$d(\rho) = \frac{1}{(1!)(2!) \cdots ((N-1)!)} \prod_{j<k}^{N} (\ell_j - \ell_k). \tag{10.55}$$

where $\ell_j = f_j + N - j, j = 1, 2, \cdots, N$. Note that by construction $\ell_1 > \ell_2 > \cdots > \ell_N$. Moreover, we can calculate $I_2(\rho)$ as

$$I_2(\rho) = \sum_{j=1}^{N} \ell_j (\ell_j - N + 1) + \frac{1}{6} N(N-1)(N-2), \tag{10.56}$$

which we will show later. Then, ξ_j can be computed using (10.47) and hence $I_p(\rho)$ from (10.50). We can, therefore, determine the pth order Casimir invariant to be

$$I_p(\rho) = \sum_{j=1}^{N} (\ell_j)^p \prod_{k=1}^{N}{}' \left(1 + \frac{1}{\ell_k - \ell_j} \right)$$
$$= \sum_{j=1}^{N} (\ell_j)^p \prod_{k=1}^{N}{}' \frac{1 + \ell_k - \ell_j}{\ell_k - \ell_j}, \tag{10.57}$$

where the prime implies excluding the term $j = k$ in the product.

In spite of its fractional representation in (10.57), $I_p(\rho)$ is really a pth order symmetric polynomial in $\ell_1, \ell_2, \cdots \ell_N$. In order to show this, let us assume that z is a complex variable and consider an analytic function

$$f(z) = \prod_{k=1}^{N} \left(1 + \frac{1}{\ell_k - z} \right) = \prod_{k=1}^{N} \frac{1 + \ell_k - z}{\ell_k - z}, \tag{10.58}$$

which has simple poles at $z = \ell_j, j = 1, 2, \cdots, N$ with residues

$$\text{Res } f(z = \ell_j) = -\prod_{k=1}^{N}{}' \frac{1 + \ell_k - \ell_j}{\ell_k - \ell_j}, \tag{10.59}$$

and $f(z) \to 1$ for $z \to \infty$. Therefore, we can rewrite $f(z)$ also as

$$f(z) = 1 - \sum_{j=1}^{N} \frac{b_j}{z - \ell_j}, \tag{10.60}$$

with the coefficients determined from the residues to be (see (10.59))

$$b_j = \prod_{k=1}^{N}{}' \frac{1 + \ell_k - \ell_j}{\ell_k - \ell_j}. \tag{10.61}$$

Therefore, near $z \to \infty$. we can expand this function as

$$f(z) = 1 - \sum_{j=1}^{N} \frac{b_j}{z - \ell_j} = 1 - \sum_{p=0}^{\infty} \sum_{j=1}^{N} \frac{(\ell_j)^p b_j}{z^{p+1}}$$

$$= 1 - \sum_{p=0}^{\infty} \frac{J_p(\rho)}{z^{p+1}}, \tag{10.62}$$

where we have used the relations in (10.57) and (10.61). On the other hand, rewriting the function $f(z)$ in (10.58) near $z \to \infty$ as

$$f(z) = \prod_{k=1}^{N} \left(1 - \frac{1}{z - \ell_k} \right) = \prod_{k=1}^{N} \left(1 - \sum_{p=0}^{\infty} \frac{(\ell_k)^p}{z^{p+1}} \right), \tag{10.63}$$

and comparing with (10.62) we can now compute $I_p(\rho)$ for lower orders as

$$I_0(\rho) = N = \binom{N}{1},$$

$$I_1(\rho) = \sum_{k=1}^{N} \ell_k - \binom{N}{2} = f_1 + f_2 + \cdots + f_N,$$

$$I_2(\rho) = \sum_{k=1}^{N} \ell_k(\ell_k - N + 1) + \binom{N}{3},$$

$$I_3(\rho) = \sum_{k=1}^{N}(\ell_k)^3 - \frac{2N-3}{2}\sum_{k=1}^{N}(\ell_k)^2 - \frac{1}{2}\left(\sum_{k=1}^{N}\ell_k\right)^2$$

$$+ \frac{1}{2}N(N-1)\sum_{k=1}^{N}\ell_k - \binom{N}{4}, \tag{10.64}$$

and so on, where we have identified

$$\binom{N}{k} = \frac{N!}{(N-k)!k!}, \tag{10.65}$$

with the binomial coefficients.

We can rewrite the Casimir invariants I_p in terms of the generators of the group X^μ_ν as follows. Let us write $x^\alpha, \alpha = 1, 2, \cdots, M = N^2$ as x^a_b where $a, b = 1, 2, \cdots, N$. Thought of as $N \times N$ matrices, its elements are given by

$$(x^\alpha)_{cd} = (x^a_b)_{cd} = \delta^a_c \delta_{bd}, \quad a, b, c, d = 1, 2, \cdots, N, \tag{10.66}$$

which leads to

$$(x^a_b x^c_d)_{pq} = (x^a_b)_{pr}(x^c_d)_{rq} = \delta^a_p \delta^c_b \delta_{dq} = \delta^c_b (x^a_d)_{pq}. \tag{10.67}$$

It follows now that these indeed satisfy the commutation relations (10.1) of the Lie algebra, namely,

$$[x^a_b, x^c_d] = -\delta^a_d x^c_b + \delta^c_b x^a_d. \tag{10.68}$$

Furthermore, using (10.67) we can calculate the metric

$$g^{\alpha\beta} = \mathrm{Tr}\,(x^\alpha x^\beta) = \mathrm{Tr}\,(x^a_b x^c_d) = \delta^a_d \delta^c_b, \tag{10.69}$$

as well as

$$g^{\alpha\beta\gamma} = \mathrm{Tr}\left(x^{a_1}_{b_1} x^{a_2}_{b_2} x^{a_3}_{b_3}\right) = \delta^{a_1}_{b_3} \delta^{a_2}_{b_1} \delta^{a_2}_{b_1}. \tag{10.70}$$

As a result, we have (see discussions around (10.29)-(10.37))

$$I_1 = \text{Tr}\,(x^a{}_b)\,X^b{}_a = \delta^a{}_b X^b{}_a = X^a{}_a,$$

$$I_2 = g^{\alpha\beta} X_\alpha X_\beta = \delta^a{}_d \delta^c{}_b X^b{}_a X^d{}_c = X^a{}_b X^b{}_a,$$

$$I_3 = g^{\alpha\beta\gamma} X_\alpha X_\beta X_\gamma = \delta^{a_1}{}_{b_3} \delta^{a_2}{}_{b_1} \delta^{a_3}{}_{b_2} X^{b_1}{}_{a_1} X^{b_2}{}_{a_2} X^{b_3}{}_{a_3}$$

$$= X^a{}_b X^b{}_c X^c{}_a, \tag{10.71}$$

and so on. We can similarly compute eigenvalues of higher order Casimir operators for the Lie algebras $so(N), sp(2N), G_2$ (for details, see S. Okubo, J. Math. Phys. 18, 2382 (1977)). In the next section, we will enumerate all adjoint operators in a given irreducible representation which has been solved in the above mentioned paper and has been utilized in the derivation of the $SU_F(3)$ mass formula discussed in the previous chapter.

10.2 Symmetric Casimir invariants

One problem with the Casimir invariants I_p of the $u(N)$ Lie algebra discussed in the last section is that they are not unique. For example, let us consider the second order Casimir invariant I_2. It is clear that we can always define a family of second order (in the generators) invariants of the form

$$I_2' = I_2 + c\,(I_1)^2\,, \tag{10.72}$$

where c is an arbitrary constant. The nonuniqueness in the defintion becomes much more serious for higher order Casimir invariants. For I_3, for example, we can have a two parameter family of invariants of the form

$$I_3' = I_3 + c_1\,(I_1)^3 + c_2 I_1 I_2, \tag{10.73}$$

for arbitrary constants c_1 and c_2.

This question of nonuniqueness can be easily fixed if we consider the $su(N)$ sub-Lie algebra of $u(N)$. More generally, let L denote a simple (or, in general, semi-simple) Lie algebra. Then we can show that it does not have any first order Casimir invariant in the following manner. Suppose that we have a first order Casimir invariant

$$I_1 = \xi^\alpha X_\alpha, \tag{10.74}$$

for constants $\xi^\alpha, \alpha = 1, 2, \cdots, M$ where $M = \dim L$ denotes the dimension of the Lie algebra. Being an invariant, I_1 commutes with all the generators (see (3.103) for the definition of the adjoint action)

$$[I_1, X_\alpha] = 0, \quad \alpha = 1, 2, \cdots, M,$$

or, $\quad ad\,(I_1)X_\alpha = 0.$ \hfill (10.75)

Using (10.6) and (10.75), it follows that for any $\alpha, \beta = 1, 2, \cdots, M$

$$[I_1, [X_\alpha, X_\beta]] = 0,$$

or, $\quad ad\,(I_1)ad\,(X_\alpha)X_\beta = 0,$

or, $\quad ad\,(I_1)ad\,(X_\alpha) = 0,$

or, $\quad \mathrm{Tr}\,ad\,(X_\alpha)ad\,(I_1) = 0.$ \hfill (10.76)

On the other hand, by definition (10.74),

$$ad\,(I_1) = \xi^\beta ad\,(X_\beta),$$ \hfill (10.77)

so that (10.76) leads to

$$\xi^\beta \mathrm{Tr}\,ad\,(X_\alpha)ad\,(X_\beta) = 0.$$ \hfill (10.78)

Since X_α belongs to an arbitrary representation, we can choose it in the adjoint representation to obtain

$$g_{\alpha\beta}\xi^\beta = 0,$$ \hfill (10.79)

with $g_{\alpha\beta} = \mathrm{Tr}\,(ad\,(X_\alpha)ad\,(X_\beta))$ (see (3.111)), which implies that $\xi^\alpha = 0, \alpha = 1, 2, \cdots, M$ since for any semi-simple Lie algebra the metric is nonsingular (has an inverse). This proves that $I_1 = 0$ for any semi-simple Lie algebra. We will show shortly (see also (3.111)) that for any irreducible representation $\{\rho\}$ of V we have

$$\mathrm{Tr}\,X_\alpha X_\beta = \frac{d(\rho)}{d(\rho_{adj})}\,I_2(\rho)\,g_{\alpha\beta},$$ \hfill (10.80)

where $d(\rho)$ denotes the dimensionality of the irreducible representation $\{\rho\}$ while $d(\rho_{adj}) = M$ corresponds to the dimensionality of the adjoint representation of L. As a result, the types of nonuniqueness problems associated with the $u(N)$ Lie algebra described above do not arise in any semi-simple Lie algebra simply because $I_1 = 0$. However, we note here that even for the $SU(N)$ group, if we consider the fourth order Casimir invariant $I_4(\rho)$, we can always add to it lower order invariants of the form

$$I_4'(\rho) = I_4(\rho) + c \left(I_2(\rho)\right)^2 , \tag{10.81}$$

for an arbitrary constant c. $I_4'(\rho)$ will also denote a fourth order Casimir invariant and we will discuss below how such non uniqueness can be removed from the definition of higher order Casimir invariants in a systematic manner.

There may be other types of ambiguities present in the definition of higher order Casimir invariants. Let us consider the third order Casimir invariant given by

$$I_3 = g^{\alpha\beta\gamma} X_\alpha X_\beta X_\gamma, \quad g^{\alpha\beta\gamma} = \mathrm{Tr}\, (x^\alpha x^\beta x^\gamma), \tag{10.82}$$

where by cyclicity of the trace we have

$$g^{\alpha\beta\gamma} = g^{\beta\gamma\alpha} = g^{\gamma\alpha\beta}. \tag{10.83}$$

Let us decompose $g^{\alpha\beta\gamma}$ into totally symmetric and totally anti-symmetric parts as

$$g^{\alpha\beta\gamma} = h^{\alpha\beta\gamma} + f^{\alpha\beta\gamma}, \tag{10.84}$$

where (using (10.82) it is easy to see that they are respectively totally symmetric and anti-symmetric in all the indices)

$$h^{\alpha\beta\gamma} = \frac{1}{2} \left(g^{\alpha\beta\gamma} + g^{\beta\alpha\gamma}\right) = \frac{1}{2}\,\mathrm{Tr}\,\left((x^\alpha x^\beta + x^\beta x^\alpha)x^\gamma\right),$$

$$f^{\alpha\beta\gamma} = \frac{1}{2} \left(g^{\alpha\beta\gamma} - g^{\beta\alpha\gamma}\right) = \frac{1}{2}\,\mathrm{Tr}\,\left((x^\alpha x^\beta - x^\beta x^\alpha)x^\gamma\right). \tag{10.85}$$

Therefore, the third order invariant in (10.82) can be decomposed as

$$I_3 = I_3^{(S)} + I_3^{(A)}, \tag{10.86}$$

where

$$I_3^{(S)} = h^{\alpha\beta\gamma} X_\alpha X_\beta X_\gamma,$$

$$I_3^{(A)} = f^{\alpha\beta\gamma} X_\alpha X_\beta X_\gamma. \tag{10.87}$$

Moreover, using the anti-symmetry property of $f^{\alpha\beta\gamma}$, we can reduce $J_3^{(A)}$ into a lower order Casimir invariant as follows

$$I_3^{(A)} = f^{\alpha\beta\gamma} X_\alpha X_\beta X_\gamma = \frac{1}{2} f^{\alpha\beta\gamma} [X_\alpha, X_\beta] X_\gamma$$

$$= \frac{1}{2}\, f^{\alpha\beta\gamma} \left(i C_{\alpha\beta}^\delta X_\delta\right) X_\gamma$$

$$= \frac{i}{2}\, g^{\delta\tau} \left(f^{\alpha\beta\gamma} f_{\alpha\beta\tau}\right) X_\delta X_\gamma. \tag{10.88}$$

Furthermore, using the symmetry property as well as the definition of the structure constants we note that

$$f_{\alpha\beta\tau} = f_{\tau\alpha\beta} = C^{\sigma}_{\tau\alpha} g_{\sigma\beta} = (ad\,(x_\tau))^{\sigma}_{\alpha}\, g_{\sigma\beta} = (ad\,(x_\tau))_{\alpha\beta}, \quad (10.89)$$

and, similarly,

$$f^{\alpha\beta\gamma} = -f^{\beta\alpha\gamma} = -(ad\,(x^\gamma))^{\beta\alpha}, \tag{10.90}$$

which leads to

$$f^{\alpha\beta\gamma}f_{\alpha\beta\tau} = -(ad\,(x^\gamma))^{\beta\alpha}(ad\,(x_\tau))_{\alpha\beta} = -\mathrm{Tr}\,(ad\,(x^\gamma)ad\,(x_\tau))$$

$$= -I_2(\rho_{adj})\,\delta^{\gamma}_{\tau}. \tag{10.91}$$

With this we can now write (10.88) as

$$I_3^{(A)} = -\frac{i}{2}\,I_2(\rho_{adj})\,g^{\delta\tau}\delta^{\gamma}_{\tau}X_\delta X_\gamma$$

$$= -\frac{i}{2}\,I_2(\rho_{adj})\,g^{\delta\tau}X_\delta X_\tau = -\frac{i}{2}\,I_2(\rho_{adj})I_2, \tag{10.92}$$

so that we can write

$$I_3 = I_3^{(S)} - \frac{i}{2}\,I_2(\rho_{adj})I_2. \tag{10.93}$$

Clearly the lower order invariants in (10.93) can be eliminated if we define the Casimir invariant in terms of the symmetrized structure $h^{\alpha\beta\gamma}$. However, even when we do use a symmetrized structure $g^{\alpha_1\alpha_2\cdots}$ in the definition of the Casimir invariants, there may still be nonuniqueness in the definition at higher orders. For example, for the fourth order Casimir invariant we have

$$I_4 = g^{\alpha\beta\gamma\delta}X_\alpha X_\beta X_\gamma X_\delta, \quad g^{\alpha\beta\gamma\delta} = \mathrm{Tr}\,(x^\alpha x^\beta x^\gamma x^\delta). \tag{10.94}$$

As before, it can be shown that this invariant contains lower order invariants as in (10.93). Introducing the completely symmetric structure

$$h^{\alpha\beta\gamma\delta} = \frac{1}{4!}\sum_P g^{\alpha\beta\gamma\delta}, \tag{10.95}$$

where "P" denotes summing over all possible (4!) possible permutations of the indices $\alpha, \beta, \gamma, \delta$ and defining as before

$$I_4^{(S)} = h^{\alpha\beta\gamma\delta}X_\alpha X_\beta X_\gamma X_\delta, \tag{10.96}$$

it is clear that the lower order terms coming from the anti-symmetric parts of the structure $g^{\alpha\beta\gamma\delta}$ will not be present in (10.96). However, we note that

$$h^{\alpha\beta\gamma\delta} \to \overline{h}^{\alpha\beta\gamma\delta} = h^{\alpha\beta\gamma\delta} + c\left(g^{\alpha\beta}g^{\gamma\delta} + g^{\alpha\gamma}g^{\beta\delta} + g^{\alpha\delta}g^{\beta\gamma}\right), \quad (10.97)$$

is also a completely symmetric structure for any arbitrary constant c (namely, $h^{\alpha\beta\gamma\delta}$ is not unique) and would generate lower order Casimir invariants in the definition of $I_4^{(S)}$. This ambiguity can be avoided by requiring that the symmetric structure satisfies

$$\overline{h}^{\alpha\beta\gamma\delta}g_{\alpha\beta}g_{\gamma\delta} = 0, \tag{10.98}$$

so that the lower order invariants will not be present in the definition of I_4. Therefore, we conclude that we can define I_4 with a symmetric structure

$$I_4 = I_4^{(S)} = \overline{h}^{\alpha\beta\gamma\delta}X_\alpha X_\beta X_\gamma X_\delta, \tag{10.99}$$

where $\overline{h}^{\alpha\beta\gamma\delta}$ is completely symmetric in all the indices and satisfies (10.98). As we will indicate shortly, this form of I_4 leads to the unique (up to a multiplicative constant) fourth order Casimir invariant of any simple Lie algebra except for the case of $so(8)$ Lie algebra where there are two linearly independent fourth order invariants.

For higher order Casimir invariants we can do something similar. For example, for the sixth order Casimir invariant of the form

$$I_6 = g^{\alpha\beta\gamma\delta\sigma\tau}X_\alpha X_\beta X_\gamma X_\delta X_\sigma X_\tau, \tag{10.100}$$

if we require

(i) $g^{\alpha\beta\gamma\delta\sigma\tau}$ to be totally symmetric in all the indices,

(ii) $g^{\alpha\beta\gamma\delta\sigma\tau}g_{\alpha\beta}g_{\gamma\delta}g_{\sigma\tau} = 0,$

        $g^{\alpha\beta\gamma\delta\sigma\tau}g_{\alpha\beta\gamma\delta}g_{\sigma\tau} = 0,$

        $g^{\alpha\beta\gamma\delta\sigma\tau}g_{\alpha\beta\gamma}g_{\delta\sigma\tau} = 0, \tag{10.101}$

then it follows that I_6, if not trivial, would be independent of lower order invariants I_2, I_3 and I_4. Here $g_{\alpha\beta\gamma}$, $g_{\alpha\beta\gamma\delta}$ denote respectively the completely symmetric third and fourth order tensor structures. We point out here that the three conditions in (ii) in (10.101) imply (in some sense) the absence of respectively terms of the forms $(I_2)^3$, $I_2 I_4$

and $(I_3)^2$ in the definition (10.100) of I_6. This construction can be generalized to obtain any higher order symmetric Casimir invariant.

Without going into technical details, we simply note here that the classification of algebraically independent Casimir invariants is known in mathematical literature to be as follows:

1. For Lie algebra $A_n = su(n+1), n \geq 1$:

   $I_2, I_3, I_4, I_5, \cdots, I_{n+1}$.

2. For Lie algebra $B_n = so(2n+1), n \geq 2$:

   $I_2, I_4, I_6, \cdots, I_{2n}$.

3. For Lie algebra $C_n = sp(2n), n \geq 3$:

   $I_2, I_4, I_6, \cdots, I_{2n}$.

4. For Lie algebra $D_n = so(2n), n \geq 4$:

   $I_2, I_4, I_6, \cdots, I_{2(n-1)}$ and $\widetilde{I}_n$. (We will explain $\widetilde{I}_n$ shortly.)

5. For (exceptional) Lie algebra G_2:

   I_2, I_6.

6. For (exceptional) Lie algebra F_4:

   I_2, I_6, I_8, I_{12}

7. For (exceptional) Lie algebra E_6:

   $I_2, I_5, I_6, I_8, I_9, I_{12}$.

8. For (exceptional) Lie algebra E_7:

   $I_2, I_6, I_8, I_{10}, I_{12}, I_{14}, I_{18}$.

9. For (exceptional) Lie algebra E_8:

   $I_2, I_8, I_{12}, I_{14}, I_{18}, I_{20}, I_{24}, I_{30}$.

Note that the total number of Casimir invariants for a simple Lie algebra coincides with the rank of the Lie algera. We did not consider here the Lie algebras $so(4), so(6)$ and $sp(4)$ because they are equivalent to $so(3) \oplus so(3), su(6)$ and $so(5)$ respectively. Also we note that, for a given Lie algebra as described above, I_p is unique except for the case of $D_{2m} = so(4m)$ where it has two independent Casimir invariants at the $(2m)$th order which we denote as I_{2m} and $\widetilde{I}_{2m}$. However, there exists a well defined individualization of these invariants (see, for example, S. Okubo and J. Patera, J. Math. Phys. 25, 219 (1984)).

10.3 Casimir invariants of $so(N)$

Let us consider the Casimir invariants of the Lie algebra $so(N)$ described by the generators J_{ab} satisfying (see, for example, (3.26) and (3.27))

$$J_{ab} = -J_{ba},$$

$$[J_{ab}, J_{cd}] = \delta_{ac}J_{bd} + \delta_{bd}J_{ac} - \delta_{ad}J_{bc} - \delta_{bc}J_{ad}, \tag{10.102}$$

where $a, b, c, d = 1, 2, \cdots, N$. This algebra possesses a p-th order unsymmetrized Casimir invariant given by

$$I_p = \frac{1}{p!} \mathrm{Tr}\left(J_{a_1a_2}J_{a_2a_3} \cdots J_{a_pa_1}\right), \tag{10.103}$$

for $p \geq 2$. However, if $N = 2n$ is even, then, in addition to the invariant I_n given by (10.103), there exists an additional n-th order Casimir invariant of the form

$$\tilde{J}_n = \frac{1}{(2n)!} \epsilon^{a_1b_1a_2b_2\cdots a_nb_n} \mathrm{Tr}\left(J_{a_1b_1}J_{a_2b_2} \cdots J_{a_nb_n}\right), \tag{10.104}$$

where $\epsilon^{a_1b_1\cdots a_nb_n}$ denotes the completely anti-symmetric Levi-Civita tensor in the $2n$ dimensional space. This marks the difference between the two cases $B_n = so(2n+1)$ and $D_n = so(2n)$. Moreover if p is an odd integer, then I_p can be reduced to a sum of lower order Casimir invariants (of even order). For example, let us consider the case of $p = 3$ with

$$I_3 = \frac{1}{3!} \mathrm{Tr}\left(J_{a_1a_2}J_{a_2a_3}J_{a_3a_1}\right)$$

$$= \frac{(-1)^3}{3!} \mathrm{Tr}\left(J_{a_2a_1}J_{a_3a_2}J_{a_1a_3}\right), \tag{10.105}$$

where we have used $J_{ab} = -J_{ba}$. Furthermore, using the commutation relation between the generators as well as the anti-symmetry of the generators given in (10.102), we can write

$$J_{a_2a_1}J_{a_3a_2} = J_{a_3a_2}J_{a_2a_1} + [J_{a_2a_1}, J_{a_3a_2}]$$

$$= J_{a_3a_2}J_{a_2a_1} + \delta_{a_2a_3}J_{a_1a_2} + \delta_{a_1a_2}J_{a_2a_3} - \delta_{a_2a_2}J_{a_1a_3}$$

$$= J_{a_3a_2}J_{a_2a_1} + J_{a_1a_3} + J_{a_1a_3} - NJ_{a_1a_3}$$

$$= J_{a_3a_2}J_{a_2a_1} - (N-2)J_{a_1a_3}. \tag{10.106}$$

Substituting this into (10.105) we obtain

$$I_3 = -\frac{1}{3!} \text{Tr} \left((J_{a_3 a_2} J_{a_2 a_1} - (N-2) J_{a_1 a_3}) J_{a_1 a_3} \right)$$

$$= -\frac{1}{3!} \text{Tr} \left(J_{a_3 a_2} J_{a_2 a_1} J_{a_1 a_3} \right) - \frac{(N-2)}{3!} \text{Tr} \left(J_{a_1 a_3} J_{a_1 a_3} \right)$$

$$= -I_3 - \frac{(N-2)}{3!} \text{Tr} \left(J_{a_1 a_3} (-J_{a_3 a_1}) \right)$$

$$= -I_3 + \frac{(N-2)}{3!} (2! I_2), \tag{10.107}$$

so that we can write

$$I_3 = \frac{(N-2)}{3!} I_2 = \frac{(N-2)}{6} I_2. \tag{10.108}$$

However, if p is even with $p \geq 4$, the invariants I_p are not ortho-symmetrized. Setting $\alpha = (a, b)$ with $a > b$ and with $J_{ab} = X_\alpha$, the even order Casimir invariant I_p can always be written in the form

$$I_p = h^{\alpha_1 \alpha_2 \cdots \alpha_p} \text{Tr} \left(X_{\alpha_1} X_{\alpha_2} \cdots X_{\alpha_p} \right), \tag{10.109}$$

with some coefficients $h^{\alpha_1 \alpha_2 \cdots \alpha_p}$ which are not symmetric in the indices $\alpha_1, \alpha_2, \cdots \alpha_p$, in general. On the other hand, we can always reduce the coefficients into a sum of ortho-symmetrized tensors of the forms $g^{\alpha_1 \alpha_2 \cdots \alpha_p}$, $g^{\alpha_1 \alpha_2} g^{\alpha_3 \alpha_4 \cdots \alpha_p}$ and so on, so that I_p can be reduced to the standard ortho-symmetrized Casimir invariants I_p.

Next, let us return to the second order Casimir invariant I_2 for the general simple Lie algebra. We note that

$$\text{Tr} \left(X_\alpha X_\beta \right) = c g_{\alpha\beta}, \tag{10.110}$$

where c is a constant. Multiplying (10.110) by $g^{\alpha\beta}$ and summing over repeated indices we obtain

$$g^{\alpha\beta} \text{Tr} \left(X_\alpha X_\beta \right) = c g^{\alpha\beta} g_{\alpha\beta} = c\, d(\rho_{adj}),$$

$$\text{or,} \quad d(\rho) I_2(\rho) = c d(\rho_{adj}),$$

$$\text{or,} \quad c = \frac{d(\rho)}{d(\rho_{adj})} I_2(\rho). \tag{10.111}$$

Therefore, from (10.110) we recover the relation

$$\text{Tr} \left(X_\alpha X_\beta \right) = \frac{d(\rho)}{d(\rho_{adj})} I_2(\rho)\, g_{\alpha\beta}$$

$$= \frac{d(\rho)}{d(\rho_{adj})} I_2(\rho)\, \text{Tr} \left(x_\alpha x_\beta \right), \tag{10.112}$$

which we have already noted (and used) earlier in (10.80). Here x_α denotes the generators in the reference representation $\{\rho_0\}$.

As an application of this formula, let us consider the simple case of $L = so(3) \simeq su(2)$ where, as we have already seen, the angular momentum states are labelled by $j = 0, \frac{1}{2}, 1, \frac{3}{2}, \cdots$ and $m = -j, -j+ 1, \cdots, j-1, j$. The dimensionality of any representation labelled by a given j is, therefore, obtained from the number of values m can take and leads to

$$d(\rho) = d(j) = 2j + 1, \tag{10.113}$$

while for the adjoint representation we have

$$d(\rho_{adj}) = 2^2 - 1 = 3, \quad (\text{equivalently, } 2 \times 1 + 1 = 3), \tag{10.114}$$

corresponding to the dimensionality of the algebra or the number of generators (for $su(n)$ it is $n^2 - 1$). Let the reference representation $\{\rho_0\}$ correspond to the fundamental 2-dimensional representation for which $j = \frac{1}{2}$ so that $d(\rho_0) = 2$.

Let us note that the second order Casimir in the reference representation is given by

$$I_2(\rho_0) = \frac{d(\rho_{adj})}{d(\rho_0)}. \tag{10.115}$$

This follows from (10.112) if we set $\rho = \rho_0$ or it can be derived directly from the definition in (10.69) as follows

$$g^{\alpha\beta} \operatorname{Tr}(x_\alpha x_\beta) = g^{\alpha\beta} g_{\alpha\beta} = \delta^\alpha_\alpha = d(\rho_{adj}). \tag{10.116}$$

On the other hand, from the definition of the second order Casimir invariant (see, for example, (10.71)), we have

$$g^{\alpha\beta} \operatorname{Tr}(x_\alpha x_\beta) = \operatorname{Tr}(I_2(\rho_0)\mathbb{1}_0) = I_2(\rho_0)d(\rho_0), \tag{10.117}$$

so that (10.115) follows from (10.116) and (10.117). We note that, in the present example (of $so(3) \simeq su(2)$), we have

$$I_2(\rho_0) = \frac{d(\rho_{adj})}{d(\rho_0)} = \frac{3}{2}. \tag{10.118}$$

For the present example of $so(3) \simeq su(2)$, we also know (see, for example, (3.122)) that the second order Casimir invariant is given up to a normalization constant by

$$I_2(\rho) = I_2(j) = c'j(j + 1), \quad j = 0, \frac{1}{2}, 1, \cdots, \tag{10.119}$$

where c' is a normalization constant. Recalling that for the reference representation ρ_0, we have $j = \frac{1}{2}$ and $I_2(\rho_0) = \frac{3}{2}$, we determine the normalization constant (from (10.119) by identifying $\rho = \rho_0$) to be

$$c' = 2, \tag{10.120}$$

which leads to

$$I_2(\rho) = I_2(j) = 2j(j+1). \tag{10.121}$$

The multiplicative factor 2 in (10.116) (compared to (3.122) or the conventional definition) can be absorbed into the normalization of the Casimir invariant, for example, by defining the second order invariant by $I_2 \to \frac{1}{2} I_2$. If we now consider (10.112) for $X_\alpha = X_\beta = X_3 = J_3$, then, for any representation, we have

$$\mathrm{Tr}\,(X_3 X_3) = \mathrm{Tr}\,(J_3)^2 = \sum_{-j}^{j} m^2 = \frac{j(j+1)(2j+1)}{3}$$

$$= \frac{2j+1}{3}(2j(j+1))\,\frac{\frac{1}{2}(\frac{1}{2}+1)(1+1)}{3}$$

$$= \frac{d(\rho)}{d(\rho_{adj})}\,J_2(\rho)\,\mathrm{Tr}\,(x_3 x_3), \tag{10.122}$$

which agrees with (10.112).

Let us next comment on the third order Casimir invariant I_3. Among simple Lie algebras, $L = su(N), N \geq 3$ and $so(6) \simeq su(4)$ alone have unique third order Casimir invariants. We note here without going into details that this implies the absence of the triangular gauge anomaly for all simple Lie algebras except for these special algebras. Absence of gauge anomaly ensures the renormalizability (and consistency) of the underlying grand unified field (gauge) theory based on such a L.

As we have already noted, the Lie algebras $L = su(2), su(3), G_2,$ F_4, E_6, E_7 and E_8 do not possess the fourth order Casimir invariant. Let X_αs denote the generators of such a Lie algebra in the generic irreducible representation $\{\rho\}$ and let us define

$$g_{\alpha\beta\gamma\delta} = \frac{1}{4!} \sum_P \left(\mathrm{Tr}\,(X_\alpha X_\beta X_\gamma X_\delta) - c\,\mathrm{Tr}\,(X_\alpha X_\beta)\,\mathrm{Tr}\,(X_\gamma X_\delta) \right), \tag{10.123}$$

where c is a constant and the summation is over the (4!) possible permutations of the indices $\alpha, \beta, \gamma, \delta$. This, therefore, defines a totally symmetric structure and we require (see (10.98))

$$g^{\alpha\beta}g^{\gamma\delta}g_{\alpha\beta\gamma\delta} = 0, \tag{10.124}$$

which after some calculation determines

$$c = K_4(\rho) = \frac{d(\rho_{adj})}{2\left(2 + d(\rho_{adj})\right)d(\rho)}\left(6 - \frac{I_2(\rho_{adj})}{I_2(\rho)}\right). \tag{10.125}$$

Since the Lie algebras under consideration do not possess a genuine fourth order Casimir invariant, we must have $g_{\alpha\beta\gamma\delta}$ in (10.123) vanish identically, namely,

$$\frac{1}{4!}\sum_P \left(\mathrm{Tr}\left(X_\alpha X_\beta X_\gamma X_\delta\right) - K_4(\rho)\mathrm{Tr}\left(X_\alpha X_\beta\right)\mathrm{Tr}\left(X_\gamma X_\delta\right)\right) = 0. \tag{10.126}$$

We can rewrite this relation more compactly as follows. Let $\xi^\alpha, \alpha = 1, 2, \cdots, M$ be arbitrary constants and let us define

$$X = \xi^\alpha X_\alpha, \tag{10.127}$$

which would then correspond to an arbitrary element of the Lie algebra L in the irreducible representation $\{\rho\}$. Multiplying (10.126) with $\xi^\alpha\xi^\beta\xi^\gamma\xi^\delta$ and summing over all the indices we obtain the trace identity

$$\mathrm{Tr}\, X^4 = K_4(\rho)\left(\mathrm{Tr}\, X^2\right)^2, \tag{10.128}$$

which holds for these Lie algebras.

As an application, let us continue with the example of $L = su(2)$ with $X = J_3$ in the irreducible $(2j+1)$ dimensional representation space with $j = 0, \frac{1}{2}, 1, \frac{3}{2}, \cdots$, as we have already discussed. Then using (10.113)-(10.122) we have

$$K_4(\rho) = \frac{3}{2(2+3)(2j+1)}\left(6 - \frac{4}{2j(j+1)}\right)$$

$$= \frac{3}{5(2j+1)}\left(3 - \frac{1}{j(j+1)}\right),$$

$$\mathrm{Tr}\left(X^2\right) = \frac{j(j+1)(2j+1)}{3},$$

so that

$$K_4(\rho)(\operatorname{Tr} X^2)^2 = \frac{j(j+1)(2j+1)}{15}(3j(j+1)-1). \qquad (10.129)$$

On the other hand, we can evaluate directly (using the table of integrals)

$$\operatorname{Tr}(X^4) = \operatorname{Tr}(J_3)^4 = \sum_{m=-j}^{j} m^4$$

$$= \frac{j(j+1)(2j+1)}{15}(3j(j+1)-1), \qquad (10.130)$$

which coincides with (10.129) proving (10.128).

Finally, we note from the classification given earlier that both $su(2)$ and E_8 do not have any sixth order invariant (in addition to the third and the fifth order invariants). In these cases, proceeding as before, we can obtain the trace identity

$$\operatorname{Tr}(X^6) = K_6(\rho)\left(\operatorname{Tr} X^2\right)^3, \qquad (10.131)$$

where

$$K_6(\rho) = \frac{15}{(d(\rho_{adj})+2)(d(\rho_{adj})+4)}\left(\frac{d(\rho_{adj})}{d(\rho)}\right)^2$$

$$\times\left(1 - \frac{1}{2}\frac{I_2(\rho_{adj})}{I_2(\rho)} + \frac{1}{12}\left(\frac{I_2(\rho_{adj})}{I_2(\rho)}\right)^2\right). \qquad (10.132)$$

We note that $d(\rho_{adj}) = 248$ for the exceptional Lie algebra E_8 and relation (10.131) is useful in showing the absence of mixed gauge-gravitational anomaly for the heterotic string theory based on $E_8 \times E_8$. For the case of $su(2)$ we have (see (10.113)-(10.122) as well as (10.132))

$$\left(\operatorname{Tr} X^2\right)^3 = \sum_{m=-j}^{j} m^3 = \left(\frac{1}{3}j(j+1)(2j+1)\right)^3,$$

$$K_6(\rho) = \frac{15}{(3+2)(3+4)}\left(\frac{3}{2j+1}\right)^2$$

$$\times\left(1 - \frac{1}{2}\frac{1(1+1)}{j(j+1)} + \frac{1}{12}\left(\frac{1(1+1)}{j(j+1)}\right)^2\right)$$

$$= \frac{3^3}{21(j(j+1)(2j+1))^2}$$
$$\times \left(3(j(j+1))^2 - 3j(j+1) + 1\right), \qquad (10.133)$$

so that we have

$$K_6(\rho) \left(\operatorname{Tr} X^2\right)^3 = \frac{1}{21} j(j+1)(2j+1)$$
$$\times \left(3(j(j+1))^2 - 3j(j+1) + 1\right). \qquad (10.134)$$

On the other hand, a direct evaluation leads to

$$\operatorname{Tr} X^6 = \operatorname{Tr} (J_3)^6 = \sum_{m=-j}^{j} m^6$$

$$= \frac{1}{21} j(j+1)(2j+1)$$
$$\times \left(3(j(j+1))^2 - 3j(j+1) + 1\right), \qquad (10.135)$$

which coincides with (10.134) proving that (10.131) holds.

Since the Lie algebra satisfies (see (10.5))

$$[X_\alpha, X_\beta] = iC_{\alpha\beta}^{\gamma} X_\gamma, \qquad (10.136)$$

it follows from this that $\widetilde{X}_\alpha = -X_\alpha^T$ (where X_α^T denotes the matrix transpose of X_α), also satisfies

$$[\widetilde{X}_\alpha, \widetilde{X}_\beta] = iC_{\alpha\beta}^{\gamma} \widetilde{X}_\gamma. \qquad (10.137)$$

$\widetilde{X}_\alpha$ is known as the conjugate or the dual representation $\{\widetilde{\rho}\}$ of $\{\rho\}$. Then, under the transformation $X_\alpha \to \widetilde{X}_\alpha$, we have

$$I_p(\rho) = \frac{1}{d(\rho)} g^{\alpha_1 \alpha_2 \cdots \alpha_p} \operatorname{Tr} \left(X_{\alpha_1} X_{\alpha_2} \cdots X_{\alpha_p}\right)$$

$$\to I_p(\widetilde{\rho}) = \frac{1}{d(\widetilde{\rho})} g^{\alpha_1 \alpha_2 \cdots \alpha_p} \operatorname{Tr} \left(\widetilde{X}_{\alpha_1} \widetilde{X}_{\alpha_2} \cdots \widetilde{X}_{\alpha_p}\right)$$

$$= (-1)^p I_p(\rho), \qquad (10.138)$$

since $d(\widetilde{\rho}) = d(\rho)$. (Note that the structures $g^{\alpha_1 \alpha_2 \cdots \alpha_p}$ involve the fixed reference representation and, therefore, are not affected by this transformation.) Therefore, for even $p = 2m$ we have $I_{2m}(\widetilde{\rho}) = I_{2m}(\rho)$. Since for simple Lie algebras $su(2)$, $so(4m)$, $so(2m+1)$, $sp(2m)$,

G_2, F_4, E_7 and E_8 do not have any odd order Casimir invariant, we see that all the Casimir invariants of these algebras are invariant under $X_\alpha \to \widetilde{X}_\alpha$. This suggests that, for these algebras, the two irreducible representations $\{\rho\}$ and $\{\widetilde{\rho}\}$ are equivalent, namely, there exists a nonsingular matrix S (similarity transformation) which relates the generators in the two representations as

$$\widetilde{X}_\alpha = -X_\alpha^T = S^{-1}X_\alpha S. \tag{10.139}$$

Taking the transpose of both sides of (10.139) we obtain

$$- X_\alpha = S^T X_\alpha^T (S^{-1})^T = -S^T S^{-1} X_\alpha S(S^{-1})^T,$$

$$\text{or,} \quad [X_\alpha, S^T S^{-1}] = 0,$$

$$\text{or,} \quad S^T S^{-1} = \lambda \mathbb{1}, \quad \text{or,} \quad S^T = \lambda S, \tag{10.140}$$

for some constant λ. Taking the transpose of the last relation leads to

$$S = \lambda S^T = \lambda^2 S, \quad \text{or,} \quad \lambda = \pm 1, \tag{10.141}$$

so that the similarity transformation satisfies

$$S^T = \pm S. \tag{10.142}$$

Representations satisfying (10.142) are known respectively as real and pseudo-real representations depending on the two respective signs. It is known in mathematical literature that any irreducible representation of these Lie algebras without odd order Casimir invariants (mentioned above) is indeed self-conjugate, namely, $\{\widetilde{\rho}\}$ and $\{\rho\}$ are equivalent. As a result, all of their irreducible representations are either real or pseudo-real. On the other hand, for simple Lie algebras $su(n), n \geq 3$, $so(4n+2)$ and E_6 which possess at least one odd order Casimir invariant, some of their representations are complex, namely, the representations of these algebras are neither real nor pseudo-real. For the case of $su(2)$, it is known that any tensor representation is real, but spinor representations are pseudo-real.

The adjoint representation of a simple Lie algebra is always self-conjugate and real, namely, it satisfies

$$-(ad\, X_\alpha)^T = g^{-1}(ad\, X_\alpha)g, \quad g^T = g, \tag{10.143}$$

where g and g^{-1} are $M \times M$ matrices with $M = d(\rho_{adj})$ and are defined by the metric as

$$(g)_{\alpha\beta} = g_{\alpha\beta}, \quad (g^{-1})_{\alpha\beta} = g^{\alpha\beta}. \tag{10.144}$$

This can be seen as follows. Let us note that (see (3.112)-(3.115))

$$((ad\,X_\alpha)g)_{\beta\gamma} = (ad\,X_\alpha)_{\beta\delta}g_{\delta\gamma} = C^\delta_{\alpha\beta}g_{\delta\gamma} = f_{\alpha\beta\gamma}, \qquad (10.145)$$

which is completely anti-symmetric in all the indices. We can now take the transpose of (10.145) and using the anti-symmetry of the structure constant as well as the symmetry of the metric, we obtain

$$((ad\,X_\alpha)g)^T = -(ad\,X_\alpha)g,$$

$$\text{or,} \quad g(ad\,X_\alpha)^T = -(ad\,X_\alpha)g,$$

$$\text{or,} \quad -(ad\,X_\alpha)^T = g^{-1}(ad\,X_\alpha)g, \qquad (10.146)$$

which proves (10.143).

In ending this discussion we will give some examples of $I_p(\rho)$ for the $su(n)$ Lie algebra. Let X^μ_ν denote the generators in the Lie algebra $u(N)$ and define

$$A^\mu_\nu = X^\mu_\nu - \frac{1}{N}\delta^\mu_\nu X^\lambda_\lambda. \qquad (10.147)$$

Clearly, these are traceless generators and belong to $su(n)$. Using these we can obtain the unsymmetrized the p th order Casimir invariant of $su(N)$ from those of $u(N)$, symmetrize and orthonormalize them suitably as follows. If $\rho = (f_1, f_2, f_3, \cdots, f_N)$ with $f_1 \geq f_2 \geq \cdots \geq f_N$ represent the Young tableaux corresponding to the irreducible representation of $su(n)$, it is more convenient fo our purposes to introduce

$$\sigma_j = f_j + \frac{1}{2}(N+1) - j - \frac{1}{N}\sum_{k=1}^N f_k = \ell_j - \frac{1}{N}\sum_{k=1}^N \ell_k, \qquad (10.148)$$

for $j = 1, 2, \cdots, N$ and they satisfy $\sigma_1 > \sigma_2 > \cdots > \sigma_N$ as well as

$$\sum_{j=1}^N \sigma_j = 0. \qquad (10.149)$$

We note that σ_j defined in (10.148) is invariant under the shift $f_j \to$

$f_j + \Delta$ for any arbitrary Δ. We can now easily calculate

$$I_2(\rho) = N \sum_{j=1}^{N} (\sigma_j)^2 - \frac{N^2}{12}(N^2 - 1),$$

$$I_3(\rho) = N^2 \sum_{j=1}^{N} (\sigma_j)^3,$$

$$I_4(\rho) = N(N^2 + 1) \sum_{j=1}^{N} (\sigma_j)^4 - (2N^2 - 3)\left(\sum_{j=1}^{N}(\sigma_j)^2\right)^2$$

$$+ \frac{1}{720} N^2(N^2 - 1)(N^2 - 4)(N^2 - 9),$$

$$I_5(\rho) = N^2(N^2 + 5) \sum_{j=1}^{N} (\sigma_j)^5$$

$$- 5N(N^2 - 2)\left(\sum_{j=1}^{N}(\sigma_j)^2\right)\left(\sum_{k=1}^{N}(\sigma_k)^3\right),$$

$$I_6(\rho) = N(N^4 + 15N^2 + 8) \sum_{j=1}^{N} (\sigma_j)^6$$

$$- 6(N^2 - 4)(N^2 + 5)\left(\sum_{j=1}^{N}(\sigma_j)^2\right)\left(\sum_{k=1}^{N}(\sigma_k)^4\right)$$

$$+ 7N(N^2 - 7)\left(\sum_{j=1}^{N}(\sigma_j)^3\right)^2$$

$$- (3N^4 - 11N^2 + 80)\left(\sum_{j=1}^{N}(\sigma_j)^3\right)^2$$

$$- \frac{N^2(N^2 - 1)(N^2 - 4)(N^2 - 9)(N^2 - 16)(N^2 - 25)}{30240},$$

$$\text{(10.150)}$$

and so on. Here we have changed the normalization of $I_p(\rho)$ by multiplying suitable constants to make it simpler. Also the formula for the dimension is now given by (see (10.55))

$$d(\rho) = \frac{1}{(1!)(2!)\cdots((N-1)!)} \prod_{j<k}^{N} (\sigma_j - \sigma_k). \tag{10.151}$$

We note here that since the $su(3)$ Lie algebra, or example, does not possess the 4th, 5th and the 6th order Casimir invariants, we must have

$$I_4(\rho) = I_5(\rho) = I_6(\rho) = 0, \tag{10.152}$$

identically for $N = 3$. We can verify these directly as follows. For example, $J_4 = 0$ for $N = 3$ we obtain from (10.150)

$$30 \sum_{j=1}^{3} (\sigma_j)^4 - 15 \left(\sum_{j=1}^{3} (\sigma_j)^2 \right)^2 = 0,$$

$$\text{or,} \quad a^4 + b^4 + c^4 = \frac{1}{2}(a^2 + b^2 + c^2)^2, \tag{10.153}$$

for $a + b + c = 0$. Here we have identified $a = \sigma_1, b = \sigma_2, c = \sigma_3$. To verify the other cases, let us, for example, choose a totally symmetric tensor representation $(f_1, f_2, \cdots, f_N) = (f, 0, 0, \cdots, 0)$ where we have identified $f_1 = f$. In this case, it follows from (10.148) that

$$\sigma_j = \frac{(N\delta_{j1} - 1)}{N} f + \frac{N + 1 - 2j}{2}. \tag{10.154}$$

Let us denote this particular representation as Λ_f and note from (10.150) that it leads to

$$I_2(\Lambda_f) = (N-1)f(N+f),$$

$$I_3(\Lambda_f) = \frac{1}{2}(N-1)(N-2)f(N+f)(N+2f),$$

$$I_4(\Lambda_f) = \frac{1}{6}(N-1)(N-2)(N-3)f(N+f)$$
$$\times (N(N-1) + 6f(N+f)),$$

$$I_5(\Lambda_f) = \frac{1}{24}(N-1)(N-2)(N-3)(N-4)f(N+f)(N+2f)$$
$$\times (N(N-5) + 12f(N+f)),$$

$$I_6(\Lambda_f) = (N-1)(N-2)(N-3)(N-4)(N-5)f(N+f)$$

$$\times \left(f^2(N+f)^2 + \frac{N(N-3)f(N+f)}{4} \right.$$

$$\left. + \frac{N(N-1)(N^2 - 15N - 4)}{120} \right), \tag{10.155}$$

and so on. It is clear from (10.155) that $I_4(\Lambda_f) = 0$ for $N = 2, 3$; $I_5(\Lambda_f) = 0$ for $N = 2, 3, 4$; $I_6(\Lambda_f) = 0$ for $N = 2, 3, 4, 5$ in accordance with the classification given earlier.

For the totally anti-symmetric representation 1^f with the Young tableaux

$$1^f = (\underbrace{1, 1, 1, \cdots, 1}_{f}, 0, 0, \cdots, 0),$$

the corresponding p th order Casimir invariants are formally obtained by replacing $N \to -N$ in (10.155) for the case of $I_p(\Lambda_f)$ so that, for example, we find

$$I_3(1^f) = \frac{1}{2}(N+1)(N+2)f(N-f)(N-2f). \tag{10.156}$$

Remarkably, the same replacement $(N \to -N)$ also holds for the formula of the dimension (10.151) except for the additional multiplicative factor $(-1)^f$. For example, we note from (10.151) that for the symmetric representation with $f = 2$, we have $d(\Lambda_2) = \frac{1}{2}N(N+1)$ while (with the replacement) for the anti-symmetric representation we have $d(1^2) = \frac{(-1)^2}{2}N(N-1) = \frac{1}{2}N(N-1)$. (See P. Cvitanovic and A. D. Kennedy, Phys. Scripta 26, 5 (1982).)

10.4 Generalized Dynkin indices

Let $\{\rho\}$ be a representation of a simple Lie algebra which is not necessarily irreducible. For the symmetric orthonormalized p th order Casimir invariant J_p (in this section only we denote the Casimir invariants by J simply because the example of the standard model involves isospin generators which are labelled conventionally by $I_i, i = 1, 2, 3$) defined as in (10.101) (and discussion around there), we define the p th order Dynkin index for the representation $\{\rho\}$ by

$$D_p(\rho) = \text{Tr } J_p. \tag{10.157}$$

If $\{\rho\}$ is irreducible, then as we have already seen $J_p = J_p(\rho)\mathbb{1}$ so that we have

$$D_p(\rho) = J_p(\rho)d(\rho), \tag{10.158}$$

for any irreducible representation $\{\rho\}$. On the other hand, for a general representation, we decompose $\{\rho\}$ as a direct sum of its irreducible components as

$$\{\rho\} = \sum_{i=1}^{n} \oplus \{\rho_i\}, \tag{10.159}$$

corresponding to the decomposition (S denotes a similarity transformation)

$$\widetilde{X}_\alpha = S^{-1} X_\alpha S = \begin{pmatrix} \boxed{X_\alpha^{(1)}} & & & \\ & \boxed{X_\alpha^{(2)}} & & \\ & & \ddots & \\ & & & \boxed{X_\alpha^{(n)}} \end{pmatrix}, \tag{10.160}$$

for irreducible matrices $X_\alpha^{(i)}, i = 1, 2, \cdots, n$. Taking the p th power of this relation, tracing over the matrix indices and multiplying with the structure $g^{\alpha_1\alpha_2\cdots\alpha_p}$ we obtain the Dynkin index for the general representation to be $(J_p(\rho_i) = g^{\alpha_1\alpha_2\cdots\alpha_p} X_{\alpha_1}^{(i)} X_{\alpha_2}^{(i)} \cdots X_{\alpha_p}^{(i)})$

$$D_p(\rho) = \mathrm{Tr} \begin{pmatrix} \boxed{J_p(\rho_1)\mathbb{1}_{(1)}} & & & \\ & \boxed{J_p(\rho_2)\mathbb{1}_{(2)}} & & \\ & & \ddots & \\ & & & \boxed{J_p(\rho_n)\mathbb{1}_{(n)}} \end{pmatrix}$$

$$= \sum_{i=1}^{n} J_p(\rho_i)d(\rho_i) = \sum_{i=1}^{n} D_p(\rho_i), \tag{10.161}$$

where we have used (10.158). Together with the standard relation

$$d(\rho) = \sum_{i=1}^{n} d(\rho_i), \tag{10.162}$$

these relations are often sufficient to determine the decomposition of a reducible representation $\{\rho\}$ into a sum of its irreducible components.

As we have noted in the previous section, any simple Lie algebra which does not possess the third order Casimir invariant has no triangle anomaly. For an irreducible representation $\{\rho\}$, therefore, this corresponds to $D_3(\rho) = 0$. However, if the third order Casimir invariant does not vanish for a simple Lie algebra, then if $\{\rho\}$ is reducible, then the condition for the absence of the triangle anomaly is given by

$$D_3(\rho) = \sum_{i=1}^{n} D_3(\rho_i) = 0. \tag{10.163}$$

The well known example is the grand unified theory based on $su(5)$ which does possess a third order Casimir invariant. However, it is known that for a representation $(\{\rho\})$ consisting of a sum of the 5 and 10 dimensional irreducible representations of $su(5)$, we have $D_3(\rho) = 0$ and there is no triangle anomaly in such a theory. As we will show in the next chapter, this is related to the branching rule for the decomposition of $so(10)$ into $su(5)$. We note that $so(10)$ does not have a third order Casimir invariant and, therefore, does not have the triangle anomaly ($D_3(\rho) = 0$). The 16 dimensional spinor representation of $so(10)$ decomposes into a direct sum of the 5 and 10 dimensional irreducible representations of $su(5)$ and the branching sum rule immediately gives $D_3(\{5\}) + D_3(\{10\}) = 0$. (We note that there is also a trivial one dimensional representation in the decomposition, but $D_3(\{1\}) = 0$.)

As a slight digression, let us note that so far we have only discussed about simple Lie algebras. However, sometimes it is of interest to consider non-simple algebras in physical theories. A familiar example of such a theory is the standard electro-weak theory based on the $SU(2) \otimes U(1)$ gauge group. In this case, we do have a third order Casimir invariant given by $J_3 = J_2 \otimes J_1$ where J_1 and J_2 are respectively the first and the second order Casimir invariants of $L = su(2) \oplus u(1)$. In a given irreducible representation $\{\rho_i\}$, therefore, this would imply $D_3(\rho_i) = J_3(\rho_i)d(\rho_i) = J_2(\rho_i)J_1(\rho_i)d(\rho_i)$. As a result, the absence of triangle anomaly in a general reducible representation in this theory is determined from

$$D_3(\rho) = \sum_{i=1}^{n} D_3(\rho_i) = \sum_{i=1}^{n} J_2(\rho_i)J_1(\rho_i)d(\rho_i)$$

$$= \sum_{i=1}^{n} I_i(I_i + 1)Y_i(2I_i + 1) = 0, \tag{10.164}$$

where I_i, Y_i denote the isospin and hypercharge quantum numbers of the states in $\{\rho_i\}$. (Here we have identified $J_1 = \frac{Y_i}{2}$.) It is well known that in the electro-weak theory such a condition holds because of cancellation of the contribution due to quarks and leptons.

With this small digression, let us return to our discussion of simple Lie algebras. One interesting example is the Clebsch-Gordan decomposition of a product representation $\{\rho_A\} \otimes \{\rho_B\}$ of two irreducible representations $\{\rho_A\}$ and $\{\rho_B\}$ of L into its irreducible components,

$$\{\rho_A\} \otimes \{\rho_B\} = \sum_{i=1}^{n} \oplus\{\rho_i\}. \tag{10.165}$$

We will show now that in this case,

$$D_p(\rho_A \otimes \rho_B) = d(\rho_B)D_p(\rho_A) + d(\rho_A)D_p(\rho_B), \tag{10.166}$$

so that Dynkin's index formula leads to the sum rule

$$D_p(\rho_A \otimes \rho_B) = \sum_{i=1}^{n} D(\rho_i),$$

$$\text{or,} \quad d(\rho_B)D_p(\rho_A) + d(\rho_A)D_p(\rho_B) = \sum_{i=1}^{n} d(\rho_i)J_p(\rho_i),$$

$$\text{or,} \quad d(\rho_A)d(\rho_B)\left(J_p(\rho_A) + J_p(\rho_B)\right) = \sum_{i=1}^{n} d(\rho_i)J_p(\rho_i), \tag{10.167}$$

since ρ_A, ρ_B, ρ_i are all irreducible by assumption.

To prove (10.166) we note that for a tensor product representation of two irreducible representation, we can write

$$X_\alpha = X_\alpha^{(A)} \otimes \mathbb{1}_B + \mathbb{1}_A \otimes X_\alpha^{(B)}, \tag{10.168}$$

so that it follows

$$D_p(\rho_A \otimes \rho_B) = \sum_{\alpha_1,\alpha_2,\cdots\alpha_p=1}^{M} g^{\alpha_1\alpha_2\cdots\alpha_p} \mathrm{Tr}\left(X_{\alpha_1}X_{\alpha_2}\cdots X_{\alpha_p}\right)$$

$$= \sum_{\alpha_1,\alpha_2,\cdots\alpha_p=1}^{M} g^{\alpha_1\alpha_2\cdots\alpha_p} \mathrm{Tr}\left(X_{\alpha_1}^{(A)}\cdots X_{\alpha_p}^{(A)}\right) \mathrm{Tr}\left(\mathbb{1}_B \cdots \mathbb{1}_B\right) + (A \leftrightarrow B)$$

$$+ R$$

$$= d(\rho_B)D_p(\rho_A) + d(\rho_A)D_p(\rho_B) + R, \tag{10.169}$$

where R contains all the mixed terms involving the products of X_αs in the two representations. To show that $R = 0$, let us recall that, for simple Lie algebras, $\operatorname{Tr} X_\alpha = 0$. However, R may contain traces of quadratic or higher order products of the generators. For example, suppose R contains a term of the form

$$g^{\alpha_1\alpha_2\cdots\alpha_p} \operatorname{Tr}(X^{(A)}_{\alpha_1} X^{(A)}_{\alpha_2}) \operatorname{Tr}(X^{(B)}_{\alpha_3} X^{(B)}_{\alpha_4} \cdots X^{(B)}_{\alpha_p}), \qquad (10.170)$$

then we can write (the derivation is a bit involved so that we only give the result here)

$$\operatorname{Tr}(X^{(A)}_{\alpha_1} X^{(A)}_{\alpha_2}) = C_A\, g_{\alpha_1\alpha_2},$$

$$\operatorname{Tr}(X^{(B)}_{\alpha_3} \cdots X^{(B)}_{\alpha_p}) = C_B\, g_{\alpha_3\cdots\alpha_p} + C'_B g_{\alpha_3\alpha_4} g_{\alpha_5\cdots\alpha_p} + \cdots, \quad (10.171)$$

where $C_A, C_B, C'_B, \cdots$ are (representation dependent) constants. On the other hand, orthogonality conditions such as (see, for example, (10.101))

$$g^{\alpha_1\alpha_2\cdots\alpha_p}\, g_{\alpha_1\alpha_2} g_{\alpha_3\alpha_4\cdots\alpha_p} = 0, \qquad (10.172)$$

etc makes such a term vanish identically. In a similar manner, it can be shown that all the other kinds of terms in R involving higher powers of X_α also vanish so that $R = 0$ and we have the sum rule (10.166). It is worth emphasizing here that the orthogonality relations are crucial for the validity of this sum rule.

Another use of the Dynkin indices is within the context of the branching rule. Let L_0 be a simple sub-Lie algebra of a simple Lie algebra L (for example, $L = E_6$ and $L_0 = D_5 = so(10)$). Then, an irreducible representation $\{\rho\}$ of L, in general, becomes a reducible representation of L_0 so that we may decompose $\{\rho\}$ as a direct sum

$$\{\rho\} = \sum_i \oplus\{\rho_i^{(0)}\}, \qquad (10.173)$$

of the irreducible representations $\{\rho_i^{(0)}\}$ of the subalgebra L_0. For this case we can write

$$\xi_p D_p(\rho) = \sum_i D_p(\rho_i^{(0)}), \qquad (10.174)$$

where ξ_p is a constant which does not depend on the representation $\{\rho\}$. To prove this relation we note that, for generators X_α of the Lie algebra L we can write

$$\frac{1}{p!} \sum_P \operatorname{Tr}(X_{\alpha_1} X_{\alpha_2} \cdots X_{\alpha_p})$$

$$= C_p D_p(\rho) g_{\alpha_1\cdots\alpha_p} + C'_p g_{\alpha_1\alpha_2} g_{\alpha_3\cdots\alpha_p} + \cdots, \qquad (10.175)$$

where the sum over P denotes sum over the $p!$ permutations of the indices $\alpha_1, \alpha_2, \cdots, \alpha_p$. We recall that $g^{\alpha_1 \alpha_2 \cdots \alpha_p} X_{\alpha_1} \cdots X_{\alpha_p} = J_p$ where J_p is the p-th order symmetrized (but not orthogonalized) Casimir invariant. As a result, (10.175) basically expresses the fact that the left hand side can always be decomposed into a sum of $g_{\alpha_1 \cdots}$s satisfying relations as in (10.101). Let X_j with $j = 1, 2, \cdots, M_0(< M)$ denote the generators of L_0 (here dim $L_0 = M_0$). Then restricting the generators to this subalgebra, we can take the trace, multiply with the pth order structure $g_0^{j_1 j_2 \cdots j_p}$ of L_0 and sum over all the indices to obtain

$$g_0^{j_1 j_2 \cdots j_p} \, \text{Tr} \, (X_{j_1} X_{j_2} \cdots X_{j_p})$$
$$= C_p D_p(\rho) g_0^{j_1 \cdots j_p} g_{j_1 \cdots j_p} + C'_p g_0^{j_1 \cdots j_p} g_{j_1 j_2} g_{j_3 \cdots j_p} + \cdots, \qquad (10.176)$$

where we have used (10.175). Furthermore, we can express

$$g_{j_1 j_2} = \beta_2 g_{j_1 j_2}^{(0)}, \quad g_{j_1 j_2 j_3} = \beta_3 g_{j_1 j_2 j_3}^{(0)}, \quad \text{and so on}, \qquad (10.177)$$

where $\beta_1, \beta_2, \cdots$ are constants. In view of the orthogonality conditions (see, for example, (10.172))

$$g_0^{j_1 j_2 \cdots j_p} g_{j_1 j_2}^{(0)} g_{j_3 j_4 \cdots j_p}^{(0)} = 0, \qquad (10.178)$$

and so on, we obtain (10.174)

$$\sum_i D_p(\rho_i^{(0)}) = \xi_p D_p(\rho), \qquad (10.179)$$

with

$$\xi_p = C_p g_0^{j_1 j_2 \cdots j_p} g_{j_1 j_2 \cdots j_p}. \qquad (10.180)$$

Note that ξ_p could be identically zero, since L_0 may not possess the genuine pth order Casimir invariant. The value of ξ_p can be determined in general as follows. For the simplest irreducible representation $\{\rho\}$ of L, the decomposition $\{\rho\} = \sum_i \oplus \{\rho_i^{(0)}\}$ is relatively easy to determine and one can calculate ξ_p in this case from the sum rule. However, since ξ_p is independent of the representation $\{\rho\}$, this value can then be used for any other irreducible representation of L in the formula for the branching rule (10.174) (or (10.179)). (For some other types of indices which satisfy analogous sum rules for such decompositions, see S. Okubo and J. Patera, J. Math. Phys. 24, 2922 (1983); J. Math. Phys. 25, 219 (1984). Other indices are also useful in the study of anomaly free gauge field theories in the higher dimensional space-time, see, for example, Y. Tosa and S. Okubo, Phys. Rev. D36, 2484 (1984); Phys. Rev. D37, 996 (1988); Phys. Rev. D40, 1925 (1989).)

We have already noted the validity of the 4th order trace identity (10.128) when the simple Lie algebra L does not possess the genuine 4th order Casimir invariant. We note, however, that for any simple Lie algebra we have a more general relation given by

$$\text{Tr}\, X^4 - K_4(\rho) \left(\text{Tr}\, X^2\right)^2 = f_4(X) D_4(\rho), \tag{10.181}$$

where X is defined in (10.127) and $f_4(X)$ denotes a scalar function of X obtained as follows. We recall that the pth order Casimir invariant is defined as

$$J_p = g^{\alpha_1 \alpha_2 \cdots \alpha_p} X_{\alpha_1} X_{\alpha_2} \cdots X_{\alpha_p}. \tag{10.182}$$

In terms of these we can write

$$f_p(\xi) = \frac{1}{D(\rho_0)}\, g^{\alpha_1 \alpha_2 \cdots \alpha_p}\, \xi_{\alpha_1} \xi_{\alpha_2} \cdots \xi_{\alpha_p} (\equiv f_p(X)), \tag{10.183}$$

where $\{\rho_0\}$ denotes the reference representation. Similarly, we have

$$\text{Tr}\left(X^3\right) = f_3(X) D_3(\rho). \tag{10.184}$$

Let us conclude by noting that if $\widetilde{X}$ denotes the irreducible matrix representation of the generator X in the representation $\{\widetilde{\rho}\}$. Then $f_p(\widetilde{X}) = f_p(X)$ in view of the same value of J_ps. This leads to the relation

$$\frac{\text{Tr}\, \widetilde{X}^4 - K_4(\widetilde{\rho}) \left(\text{Tr}\, \widetilde{X}^2\right)^2}{\text{Tr}\, X^4 - K_4(\rho) \left(\text{Tr}\, X^2\right)^2} = \frac{D_4(\widetilde{\rho})}{D_4(\rho)},$$

and

$$\frac{\text{Tr}\, \widetilde{X}^3}{\text{Tr}\, X^3} = \frac{D_3(\widetilde{\rho})}{D_3(\rho)}, \quad \frac{\text{Tr}\, \widetilde{X}^2}{\text{Tr}\, X^2} = \frac{D_2(\widetilde{\rho})}{D_2(\rho)}, \tag{10.185}$$

and so on. These relations are valid for any irreducible matrix representations of X and $\widetilde{X}$ for any simple Lie algebra.

10.5 References

H. Weyl, *The Theory of Groups and Quantum Mechanics*, (translated from German by H. P. Robertson), Dover Publication, NY (1950).

S. Okubo, J. Math. Phys. 18, 2382 (1977).

P. Cvitanovic and A. D. Kennedy, Phys. Scripta 26, 5 (1982).

S. Okubo and J. Patera, J. Math. Phys. 24, 2922 (1983); J. Math. Phys. 25, 219 (1984).

Y. Tosa and S. Okubo, Phys. Rev. D36, 2484 (1984); Phys. Rev. D37, 996 (1988); Phys. Rev. D40, 1925 (1989).

CHAPTER 11

Root system of Lie algebras

We have derived the irreducible space of the Lie algebra $su(N)$ from the unitary group $U(N)$ by the method of Weyl. However, it may be desirable to obtain the same directly from the Lie algebra structure itself without utilizing the global group structure. This is the goal of the present chapter. The method to be used, for this purpose, is essentially special cases of the more general Cartan-Dynkin theory which is discussed in great detail in the literature and which we review briefly in the next section. The present discussion will allow us to see its connection with the Young tableaux method discussed earlier.

11.1 Cartan-Dynkin theory

In this section, we will give a brief description of some of the important results of the Cartan-Dynkin theory without proof which is beyond the scope of the present book. For a comprehensive account of the results, we refer the reader to the references at the end of this chapter.

Let G denote a finite dimensional simple Lie group with L representing its Lie algebra defined over the complex field. Let $t_a, a = 1, 2, \cdots, \dim L$ correspond to the basis elements of L where $\dim L$ denotes the dimension of L. For example, the dimension of the Lie algebra $su(N)$ is given by

$$\dim su(N) = N^2 - 1. \tag{11.1}$$

As we have discussed the elements of the algebra can always be chosen such that they define an orthonormal basis so that we can always diagonalize the metric with suitable normalization of t_a (see, for example, (8.71)) in the form

$$g_{ab} = \operatorname{Tr} ad\, t_a\, ad\, t_b = C\delta_{ab}. \tag{11.2}$$

However, even though this is the natural basis for the Lie algebra, various algebraic questions such as the representation theory are much better studied in an alternate basis known as the Cartan-Weyl basis. Here the generators are divided into two distinct classes in the following way.

1. **Cartan subalgebra**: There exist ℓ maximally commuting elements of L denoted by $H_1, H_2, \cdots, H_\ell$ such that

$$[H_i, H_j] = 0, \quad i, j = 1, 2, \cdots, \ell. \tag{11.3}$$

The integer ℓ is known as the rank of the Lie algebra, namely, rank $L = \ell$. For example, for $L = su(N)$, we have

$$\text{rank } su(N) = N - 1. \tag{11.4}$$

It is clear that the generators of the Cartan subalgebra are not unique. In fact, given a set of generators $\{H_i\}$ satisfying (11.3), we can always define another set given by

$$H_i' = \sum_{j=1}^{\ell} a_{ij} H_j, \quad a_{ij} = \text{constants}, \tag{11.5}$$

which would also satisfy (11.3). Therefore, we can, in general, choose a set of generators which would satisfy

$$g_{ij} = C_0 \, \text{Tr} \, ad \, H_i \, ad \, H_j = C\delta_{ij}, \quad i, j = 1, 2, \cdots, \ell. \tag{11.6}$$

However, we will not use this special form for a while.

2. **The complement**: We will assume that the generators of the Cartan subalgebra are Hermitian, namely,

$$H_i^\dagger = H_i, \tag{11.7}$$

partly because it facilitates our discussion, but mainly because Hermiticity is quite important in physics. In fact, the Hermiticity follows essentially as a consequence of the Cartan-Dynkin theory and has been utilized by Weyl to prove the full reducibility of any representation of a simple Lie algebra L. (This is the celebrated unitarity trick of Weyl. The full reducibility can now also be proved using cohomology theory without using Weyl's unitarity trick. But these topics are beyond the scope of this book.) We note that in the adjoint representation, we can also write (11.3) and (11.7) as

$$[ad\,H_i, ad\,H_j] = ad\,[H_i, H_j] = 0, \quad (ad\,H_i)^\dagger = ad\,H_i,$$
(11.8)

Since there are ℓ such commuting Hermitian generators, we can diagonalize them simultaneously with real eigenvalues a_i, namely (in the adjoint representation),

$$(ad\,H_i)X = a_i X, \quad a_i^* = a_i, \quad i = 1, 2, \cdots, \ell,$$
(11.9)

for some $X \in L$. The zero eigenvalue solutions with $a_1 = a_2 = \cdots = a_\ell = 0$ for this equation always exist for $X = H_j, j = 1, 2, \cdots, \ell$ and they naturally define a ℓ-dimensional vector space V_0. Let $\mathbf{e}^1, \mathbf{e}^2 \cdots, \mathbf{e}^\ell$ denote a set of complete basis vectors in this space. Then, any nontrivial vector $\boldsymbol{\alpha} \neq 0$ in V_0 can be written in this basis as

$$\boldsymbol{\alpha} = \sum_{i=1}^{\ell} a_i \mathbf{e}^i = a_1 \mathbf{e}^1 + a_2 \mathbf{e}^2 + \cdots + a_\ell \mathbf{e}^\ell,$$
(11.10)

where $a_i \neq 0$ denote the components of the vector $\boldsymbol{\alpha}$ in this basis.

With the help of these vectors, one can now try to organize the rest of the generators of L by taking complex combinations of them, much like the raising and lowering operators (step operators), $J_\pm$, of angular monemtum (see, for example, (3.59)). In this way, we can find generators E_α for a vector $\boldsymbol{\alpha} \neq 0$ to satisfy

$$[H_i, E_\alpha] = a_i E_\alpha,$$
$$\text{or,} \quad (ad\,H_i)E_\alpha = a_i E_\alpha, \quad i = 1, 2, \cdots, \ell,$$
(11.11)

so that a_i's represent the eigenvalues of H_i in the adjoint representation. We note here for future use that (11.11) is invariant under $E_\alpha \to C_\alpha E_\alpha$ where C_αs are arbitrary constants.

3. **Root vector**: We can think of the components a_i which satisfy the commutation relations (11.11) to correspond to structure constants in the Cartan-Weyl basis. A specific nontrivial vector α with components given by these structure constants (α_i

satisfying (11.11)) is known as a root vector and since there are
(dim L − rank L) number of generators in the complement, there
will be as many nontrivial root vectors in this basis. (In this
language, then, the Cartan subalgebra can be thought of as the
algebra of the generators with vanishing roots.) For example,
we note that the number of root vectors in $su(N)$ is given by

$$\dim su(N) - \operatorname{rank} su(N) = (N^2 - 1) - (N - 1)$$
$$= N(N - 1). \tag{11.12}$$

Note that this is an even integer.

4. **Properties of roots**: It is known that we can classify all the
   nontrivial roots of a Lie algebra into two groups. If the components of a root given by

   $$\boldsymbol{\alpha} = \sum_{i=1}^{\ell} a_i \mathbf{e}^i, \tag{11.13}$$

   satisfy $a_i \geq 0, i = 1, 2, \cdots, \ell$, then it is called a positive root
   (the generator E_α in such a case will correspond to raising (step
   up) operators) while if all the components are negative, $a_i \leq
   0, i = 1, 2, \cdots, \ell$, it is called a negative root (the corresponding
   generator E_α would represent a lowering (step down) operator).
   It is also known that a_is are all integers.

   Since the generators H_i are Hermitian and the components a_i
   are real, by taking the Hermitian conjugate of (11.11), we obtain

   $$[H_i, E_\alpha^\dagger] = -a_i E_\alpha^\dagger. \tag{11.14}$$

   This shows that if $\boldsymbol{\alpha}$ is a root vector, then $(-\boldsymbol{\alpha})$ is also a root
   vector and that we can identify

   $$E_\alpha^\dagger = E_{-\alpha}, \quad [H_i, E_{-\alpha}] = -a_i E_{-\alpha}, \tag{11.15}$$

   and the even number of generators in the complement (see,
   for example, (11.12)) can be arranged into pairs of raising and
   lowering (step) operators. (In addition, one can show that for
   a simple Lie algebra, the roots are not degenerate so that we
   cannot have two generators in the complement, E_α, E'_α, corresponding to the same root.)

It is also worth pointing out here that the dependence of the generators on the roots is not linear so that the sum of two roots is not necessarily a root, namely,

$$E_\alpha + E_\beta \neq E_{\alpha+\beta}, \quad E_{k\alpha} \neq kE_\alpha, \tag{11.16}$$

where

$$\alpha = \sum_i a_i \mathbf{e}^i, \quad \beta = \sum_i b_i \mathbf{e}^i, \tag{11.17}$$

denote two roots and k is an integer.

5. **Commutation relations**: We note from the Jacobi identity that the generators of L must satisfy

$$[H_i, [E_\alpha, E_\beta]] = [[H_i, E_\alpha], E_\beta] + [E_\alpha, [H_i, E_\beta]]$$

$$= (a_i + b_i)[E_\alpha, E_\beta], \tag{11.18}$$

where α and β are two roots with components a_i and b_i. This implies that if $\beta = -\alpha$, we have

$$[H_i, [E_\alpha, E_{-\alpha}]] = 0. \tag{11.19}$$

On the other hand, since H_is denote the maximal number of commuting elements in L, (11.19) implies that we can write

$$[E_\alpha, E_{-\alpha}] = \sum_{i=1}^{\ell} a^i H_i = H_\alpha, \tag{11.20}$$

for some constants a^i.

To determine the constants a^i and a_i, let us introduce the metric tensor

$$g_{ij} = C_0 \operatorname{Tr}(ad\, H_i\, ad\, H_j), \tag{11.21}$$

where C_0 is a normalization constant usually chosen to be unity. We can now calculate

$$\operatorname{Tr}(ad\, H_i\, [ad\, E_\alpha, ad\, E_{-\alpha}]) = \operatorname{Tr}(ad H_i\, ad\, [E_\alpha, E_{-\alpha}])$$

$$= \operatorname{Tr}(ad\, H_i\, ad(\sum_{j=1}^{\ell} a^j H_j)) = \sum_{j=1}^{\ell} a^j \operatorname{Tr}(ad\, H_i\, ad\, H_j)$$

$$= \frac{1}{C_0} \sum_{j=1}^{\ell} a^j g_{ij}. \tag{11.22}$$

Here we have used (11.20) and (11.21) in the intermediate steps. On the other hand, using the cyclicity of the trace we also have

$$\text{Tr}\,(ad\,H_i\,[ad\,E_\alpha, ad\,E_{-\alpha}]) = \text{Tr}\,([ad\,H_i, ad\,E_\alpha]\,ad\,E_{-\alpha})$$

$$= \text{Tr}\,(ad\,[H_i, E_\alpha]\,ad\,E_{-\alpha}) = a_i\,\text{Tr}\,(ad\,E_\alpha\,ad\,E_{-\alpha}), \quad (11.23)$$

so that from (11.22) and (11.23), we determine

$$\sum_{j=1}^{\ell} g_{ij} a^j = C_0 a_i\,\text{Tr}\,(ad\,E_\alpha\,ad\,E_{-\alpha}). \tag{11.24}$$

We note that the normalization of the generators E_α is arbitrary since the defining relation (11.11) is invariant under $E_\alpha \to C_\alpha E_\alpha$. Therefore, we can choose the constants C_α such that

$$C_0\,\text{Tr}\,(ad\,E_\alpha\,ad\,E_{-\alpha}) = 1, \tag{11.25}$$

and, in this case, (11.24) leads to

$$a_i = \sum_{j=1}^{\ell} g_{ij} a^j, \quad \text{or,} \quad a^i = \sum_{j=1}^{\ell} g^{ij} a_j, \tag{11.26}$$

where g^{ij} is the inverse of the metric g_{ij}, namely,

$$\sum_{k=1}^{\ell} g^{ik} g_{kj} = \delta_j^i, \quad i, j = 1, 2, \cdots, \ell. \tag{11.27}$$

Returning to (11.18), we note that if $\alpha + \beta \neq 0$, then, in general we can write

$$[E_\alpha, E_\beta] = N(\alpha, \beta) E_{\alpha+\beta}, \tag{11.28}$$

for some constant $N(\alpha, \beta)$. However, we also recognize that $\alpha + \beta$ may not be a root in general. In this case, either $E_{\alpha+\beta} = 0$ or $N(\alpha, \beta) = 0$.

Therefore, collecting all these results (together with (11.3) and (11.11)), we can write the complete algebra in the Cartan-Weyl basis in the form

$$[H_i, H_j] = 0 = [H_\alpha, H_\beta],$$

$$[H_i, E_\alpha] = a_i E_\alpha,$$

$$[E_\alpha, E_\beta] = \begin{cases} \sum_i a^i H_i = H_\alpha, & \text{if } \alpha + \beta = 0, \\ N(\alpha, \beta) E_{\alpha+\beta}, & \text{if } \alpha + \beta \text{ is a root,} \\ 0 & \text{otherwise.} \end{cases} \quad (11.29)$$

The constant $N(\alpha, \beta)$ can be chosen to be real and can be calculated explicitly if we know all the roots of the algebra. However, even without the explicit form, it follows from the algebra that they satisfy various relations such as

$$N(\alpha, \beta) = -N(\beta, \alpha) = -N(-\alpha, -\beta) = N(-\alpha, \alpha + \beta)$$

$$= -N(-\beta, \alpha + \beta) = -N(-\alpha - \beta, \beta), \quad (11.30)$$

which are known as Weyl's basis relations.

Let α and β be two nonzero roots with

$$\alpha = \sum_{i=1}^{\ell} a_i e^i, \quad \beta = \sum_{i=1}^{\ell} b_i e^i,$$

and introduce an inner product by the relation

$$(\alpha, \beta) = \text{Tr}\,(ad\, H_\alpha\, ad\, H_\beta) = \sum_{i,j=1}^{\ell} g^{ij} a_i b_j. \quad (11.31)$$

We then note from (11.29) that we can write

$$[H_\alpha, E_\beta] = [\sum_{i=1}^{\ell} a^i H_i, E_\beta] = (\sum_{i=1}^{\ell} a^i b_i) E_\beta$$

$$= (\alpha, \beta)\, E_\beta. \quad (11.32)$$

If we now identify

$$J_3^{(\alpha)} = \frac{1}{(\alpha, \alpha)} H_\alpha, \quad J_\pm^{(\alpha)} = \sqrt{\frac{2}{(\alpha, \alpha)}} H_{\pm\alpha}, \quad (11.33)$$

then, for any given α, they satisfy the $su(2)$ algebra

$$[J_3^{(\alpha)}, J_\pm^{(\alpha)}] = \pm J_\pm^{(\alpha)}, \quad [J_+^{(\alpha)}, J_-^{(\alpha)}] = 2J_3, \quad (11.34)$$

and the eigenvalues of $J_3^{(\alpha)}$ must have the form $\frac{1}{2}n^{(\alpha)}$ for integer $n^{(\alpha)}$. Therefore, we conclude that the eigenvalues of H_α must have the form

$$\frac{1}{2}(\alpha, \alpha)\, n^{(\alpha)}, \tag{11.35}$$

with $n^{(\alpha)}$ an integer.

Let m_i be an eigenvalue of H_i (see (11.55)), namely,

$$H_i\phi = m_i\phi. \tag{11.36}$$

This leads to

$$H_\alpha\phi = (\sum_{i=1}^{\ell} a^i H_i)\phi = (\sum_{i=1}^{\ell} a^i m_i)\phi. \tag{11.37}$$

If we now define a vector (see also (11.58))

$$\mathbf{M} = \sum_{i=1}^{\ell} m_i \mathbf{e}^i, \tag{11.38}$$

then it follows from (11.31) that

$$(\mathbf{M}, \alpha) = \sum_{i,j=1}^{\ell} g^{ij} m_i a_j = \sum_{i=1}^{\ell} m^i a_i. \tag{11.39}$$

It follows now from (11.35) and (11.37) that

$$\frac{2(\mathbf{M}, \alpha)}{(\alpha, \alpha)} = \text{integer}. \tag{11.40}$$

Similarly, let us consider the adjoint representation which gives

$$(ad\, H_\alpha)E_\beta = [H_\alpha, E_\beta] = (\alpha, \beta)\, E_\beta, \tag{11.41}$$

so that $ad\, H_\alpha$ (namely, H_α in the adjoint representation) has an eigenvalue (α, β). Comparing with (11.35) we see that

$$\frac{2(\alpha, \beta)}{(\alpha, \alpha)} = \text{integer}, \tag{11.42}$$

for any two nontrivial roots α and β.

6. **Simple roots**: It is clear that the structure of the Lie algebra is completely contained in its structure constants and, as we have noted, these are nothing other than the root vectors of the Lie algebra in the Cartan-Weyl basis. Furthermore, we have seen that the negative root vectors are merely negatives of the positive root vectors. Therefore, the positive root vectors really encode the structure of the Lie algebra.

Let us define a simple root (also known as a fundamental root) as a positive root vector which cannot be written as a sum of two positive root vectors. With a little bit of work it can be shown that the number of simple roots equal rank L and they are all linearly independent. It is clear from this description that we can always identify the simple (fundamental) roots with

$$\alpha_i = \mathbf{e}^i, \quad i = 1, 2, \cdots, \text{rank } L. \tag{11.43}$$

Furthermore, as we have already discussed, any positive root can be written as a linear superposition of fundamental roots with positive coefficients. Namely, if $\alpha_i, i = 1, 2, \cdots, \text{rank } L$ denotes a simple root system, then any positive root vector can be expressed as

$$\alpha = \sum_{i=1} a_i \alpha_i, \tag{11.44}$$

where a_i denote nonnegative integers. Therefore, the structure of the Lie algebra now resides completely in the simple root system of an algebra.

7. **Cartan matrix**: Let us recall from (11.31) that, for any two nonzero roots α and β, we can write

$$(\alpha, \beta) = \text{Tr}\,(ad\,H_\alpha\,ad\,H_\beta) = \sum_{i,j=1}^{\ell} a^i b^j \text{Tr}\,(ad\,H_i\,ad\,H_j)$$

$$= \sum_{i,j=1}^{\ell} a^i b^j g_{ij}, \tag{11.45}$$

where we have used (11.6) (with $C_0 = 1$). On the other hand, using the expansion in (11.44) we can express the left hand side of (11.45) as

$$(\alpha, \beta) = \sum_{i,j=1}^{\ell} a^i b^j (\alpha_i, \alpha_j), \tag{11.46}$$

so that comparing (11.45) and (11.46) we determine

$$(\boldsymbol{\alpha}_i, \boldsymbol{\alpha}_j) = g_{ij}. \tag{11.47}$$

With this, we can now define the Cartan matrix as (i, j are fixed indices and j is not summed)

$$K_{ij} = \frac{2(\boldsymbol{\alpha}_i, \boldsymbol{\alpha}_j)}{(\boldsymbol{\alpha}_j, \boldsymbol{\alpha}_j)}, \quad i, j = 1, 2, \cdots, \ell, \tag{11.48}$$

which is an integer because of (11.42). It is clear from (11.48) that all the diagonal elements of this matrix are 2, namely, (i not summed)

$$K_{ii} = 2, \quad i = 1, 2, \cdots, \ell. \tag{11.49}$$

On the other hand, the off-diagonal elements are known to take only four possible values, namely, for $i \neq j$

$$K_{ij} = 0, -1, -2, -3. \tag{11.50}$$

Similarly, we can introduce an angle between two fundamental roots $\boldsymbol{\alpha}_i, \boldsymbol{\alpha}_j$ as

$$\cos \theta_{ij} = \frac{(\boldsymbol{\alpha}_i, \boldsymbol{\alpha}_j)}{\sqrt{(\boldsymbol{\alpha}_i, \boldsymbol{\alpha}_i)(\boldsymbol{\alpha}_j, \boldsymbol{\alpha}_j)}}, \tag{11.51}$$

and, for $i \neq j$, they assume only four values

$$\theta_{ij} = 90°, 120°, 135°, 150°. \tag{11.52}$$

Note that we have $K_{ij} \neq K_{ji}$, in general, with

$$K_{ij} K_{ji} = 4(\cos \theta_{ij})^2. \tag{11.53}$$

The two sets of values in (11.50) and (11.52) can be shown to be in one to one correspondence. Diagrammatically, this is represented (for $i \neq j$) as in Fig. 11.1 where each vertex represents a root vector and the number of bonds connecting two vertices corresponds to

$$n_{ij} = n_{ji} = |K_{ij} K_{ji}| = \left| \frac{2(\boldsymbol{\alpha}_i, \boldsymbol{\alpha}_j)}{\sqrt{(\boldsymbol{\alpha}_j, \boldsymbol{\alpha}_j)(\boldsymbol{\alpha}_i, \boldsymbol{\alpha}_i)}} \right|^2. \tag{11.54}$$

For example, we note from (11.50) and (11.53) that when $\theta_{ij} = 90°$, we have $K_{ij} = K_{ji} = 0$, while for $\theta_{ij} = 120°$, we have $K_{ij} = K_{ji} = -1$ Similarly, when $\theta_{ij} = 135°$, we can have $K_{ij} = 2K_{ji} = -2$ or $K_{ji} = 2K_{ij} = -2$, while for $\theta_{ij} = 150°$ we have $K_{ij} = 3K_{ji} = -3$ or $K_{ji} = 3K_{ij} = -3$.

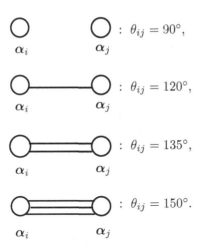

Figure 11.1: Diagrammatic representation of the off-diagonal elements of the Cartan matrix.

8. **Dynkin diagrams**: Dynkin diagrams are a diagrammatic way of representing the Cartan matrix and are built out of the elements known as nodes (basis vectors) and bonds representing the magnitude of the off-diagonal element of the Cartan matrix. We can always choose the normalization constant C in (11.6) such that the minimum of the norm among all of the ℓ vectors, namely, (α_i, α_i) for $i = 1, 2, \cdots, \ell$ is unity. In that case, it is known that the values that any of the norms can have are $1, 2, 3$. In some literature, these are represented diagrammatically as open circles with $1, 2$ or 3 on top denoting their norms. However, we will follow the slightly different convention (which is standard by now), namely, each node represented by an open circle denotes a vector α_i with norm unity while a filled (black) circle represents a vector α_i with norm 2 or 3 (with some fixed normalization). The number of nodes corresponds to the rank of the algebra (and, therefore, gives the dimensionality of the Cartan matrix) and the number of bonds between two nodes represent the magnitude of the off-diagonal elements in the matrix (remember that the nontrivial off-diagonal elements are all negative). We will explain this in more detail within context when we discuss different algebras later.

9. **Weight vector**: As we have mentioned before, since $[H_i, H_j] = 0$, we can simultaneously diagonalize these ℓ generators in any

irreducible representation V and we emphasize that $V \neq V_0$ except in the adjoint representation. Therefore, in any representation we can write

$$H_i \phi = m_i \phi, \quad i = 1, 2, \cdots, \ell, \tag{11.55}$$

where $\phi \in V$ is the simultaneous eiegenvector and the eigenvalues $m_1, m_2, \cdots, m_\ell$ are known as weights. If we wish, we may also represent the eigenstate in the Dirac notation as

$$\phi = |\rho, m_1, m_2, \cdots, m_\ell\rangle, \tag{11.56}$$

in analogy with the angular momentum states $|j, m\rangle$ of $su(2)$. Here ρ labels the irreducible representation space V (like j in angular momentum). Moreover, using the inverse of the metric g_{ij} (see (11.6)), to write

$$m^i = \sum_{j=1}^{\ell} g^{ij} m_j, \tag{11.57}$$

which allows us to introduce a vector in this space given by

$$\mathbf{M} = \sum_{i=1}^{\ell} m^i \boldsymbol{\alpha}_i = \sum_{i,j=1}^{\ell} g^{ij} m_i \boldsymbol{\alpha}_j. \tag{11.58}$$

This is known as a weight vector. Note that since the vectors $\boldsymbol{\alpha}_i = \mathbf{e}_i$ define a basis in V_0, the weight vector is a vector in the root space, but, in general, may not be a root vector (e.g. the weight vector M can be identically zero etc). This suggests that we may label the simultaneous eigenvector simply as ϕ_M or $|\mathbf{M}\rangle$ so that we can rewrite (11.55) as

$$H_i |\mathbf{M}\rangle = m_i |\mathbf{M}\rangle. \tag{11.59}$$

Together with (11.57), this also suggests that

$$H^i = \sum_{j=1}^{\ell} g^{ij} H_j, \tag{11.60}$$

satisfies the eigenvalue equation

$$H^i |\mathbf{M}\rangle = \sum_{j=1}^{\ell} g^{ij} H_j |\mathbf{M}\rangle = \sum_{j=1}^{\ell} g^{ij} m_j |\mathbf{M}\rangle$$

$$= m^i |\mathbf{M}\rangle. \tag{11.61}$$

Suppose now that $|\mathbf{N}\rangle$ denotes another weight vector with components $n^1, n^2, \cdots, n^\ell$, namely,

$$H^i|\mathbf{N}\rangle = n^i|\mathbf{N}\rangle. \qquad (11.62)$$

In this case, if at least one of the components of M is larger than the corresponding components of N while the other components being equal, we say that the weight vector M is higher than N and denote this by $M > N$. For example, if $m^1 > n^1$ while $m^i = n^i, i = 2, 3, \cdots, \ell$, we say $M > N$. Similarly, if $m^2 > n^2$ while $m^i = n^i, i \neq 2$, we still say $M > N$. This is connected with the fact that the choice of the Cartan generators is nonunique as we have pointed out in (11.5) and using this one can always reorder the generators. In a finite dimensional irreducible representation, there always exists a weight vector $\mathbf{\Lambda} = \sum_{i=1}^{\ell} \lambda^i \boldsymbol{\alpha}_i$ such that any other weight $\mathbf{M}$ is lower than $\mathbf{\Lambda}$, namely, $\mathbf{\Lambda} > \mathbf{M}$. In this case, one says that $\mathbf{\Lambda}$ defines the highest weight vector determined completely by the highest weight state $|\mathbf{\Lambda}\rangle = \phi_\Lambda$. This, of course, presupposes that the highest weight state is unique. For any finite dimensional irreducible representation, this can, in fact, be shown (as we will discuss in the next section).

The importance of the highest weight state $|\mathbf{\Lambda}\rangle = \phi_\Lambda$ comes from the fact that for any positive root $\boldsymbol{\alpha}$ we have

$$E_\alpha \phi_\Lambda = 0, \quad \boldsymbol{\alpha} > 0. \qquad (11.63)$$

This can be seen as follows. Using (11.29) and (11.55) we have

$$[H_i, E_{\pm\alpha}]\phi_M = (H_i E_{\pm\alpha} - E_{\pm\alpha})\phi_M = \pm a_i \phi_M,$$

$$\text{or,} \quad H_i(E_{\pm\alpha}\phi_M) = (m_i \pm a_i)(E_{\pm\alpha}\phi_M), \qquad (11.64)$$

so that $(E_{\pm\alpha}\phi_M)$ is also a simultaneous eigenstate with eigenvalues $(m_i \pm a_i)$. Correspondingly, we have

$$H^i(E_{\pm\alpha}\phi_M) = (m^i \pm a^i)(E_{\pm\alpha}\phi_M). \qquad (11.65)$$

On the other hand, for $\boldsymbol{\alpha} > 0$, we have $a^i \geq 0$, so that

$$m^i + a^i \geq m^i, \quad m^i - a^i \leq m^i, \qquad (11.66)$$

and we conclude that $(E_\alpha \phi_M)$ has a weight higher than ϕ_M while $(E_{-\alpha} \phi_M)$ has a weight lower than ϕ_M. Namely, E_α acts like a raising operator while $E_{-\alpha}$ acts like a lowering operator for the weights. If we identify $\mathbf{M} = \boldsymbol{\Lambda}$, this would say that $(E_\alpha \phi_\Lambda)$ will have a higher weight than ϕ_Λ which would contradict the assumption that $\boldsymbol{\Lambda}$ is the highest weight vector. Therefore, for the highest weight vector $\boldsymbol{\Lambda}$, we must have

$$E_\alpha \phi_\Lambda = 0.$$

Let us illustrate some of these ideas with the example of $L = su(2)$. This is a rank 1 algebra (namely, $\ell = 1$, see (11.4)) with the unique positive root vector $\boldsymbol{\alpha}$. From (11.29) and (11.45) we note that we can write the commutation relations as

$$[H_\alpha, E_{\pm\alpha}] = \pm(\boldsymbol{\alpha}, \boldsymbol{\alpha}) E_{\pm\alpha},$$
$$[E_\alpha, E_{-\alpha}] = H_\alpha, \tag{11.67}$$

so that if identify

$$J_3 = \frac{1}{(\boldsymbol{\alpha}, \boldsymbol{\alpha})} H_\alpha, \quad J_\pm = \sqrt{\frac{2}{(\boldsymbol{\alpha}, \boldsymbol{\alpha})}} E_{\pm\alpha}, \tag{11.68}$$

we recover the algebra of angular momentum operators, namely,

$$[J_3, J_\pm] = \pm J_\pm,$$
$$[J_+, J_-] = 2J_3. \tag{11.69}$$

We note from (11.29) and (11.67) that any given positive root in a simple Lie algebra satisfies a $su(2)$ algebra. In the case of $su(2)$, as we know, J_+ and J_- are respectively the raising and the lowering operators and the highest weight state in any representation has the form $|j, j\rangle$, namely the state with the maximal projection $m = j$.

For any weight vector $\mathbf{M}$, as we have shown in (11.40), we have

$$\frac{2(\mathbf{M}, \boldsymbol{\alpha})}{(\boldsymbol{\alpha}, \boldsymbol{\alpha})} = \text{integer}. \tag{11.70}$$

Moreover, we also know that there exist ℓ fundamental weight vectors $(\boldsymbol{\Lambda}_1, \boldsymbol{\Lambda}_2, \cdots, \boldsymbol{\Lambda}_\ell)$ satisfying

$$\frac{2(\boldsymbol{\Lambda}_i, \boldsymbol{\alpha}_j)}{(\boldsymbol{\alpha}_j, \boldsymbol{\alpha}_j)} = \delta_{ij}, \quad i, j = 1, 2, \cdots, \ell, \tag{11.71}$$

and the highest weight vector Λ of any finite dimensional irreducible representation can be expressed in terms of these fundamental weight vectors as

$$\Lambda = k_1\Lambda_1 + k_2\Lambda_2 + \cdots + k_\ell\Lambda_\ell, \tag{11.72}$$

where $k_1, k_2, \cdots, k_\ell$ are nonnegative integers. (Note that we did not use the symbol k^i here for conventional reason.) For explicit forms of Λ_j in terms of the fundamental root vectors α_i, the reader is advised to look at the book by Humphreys.

We remark here that given any weight vector $\mathbf{M}$ and any root vector α, we can define

$$\mathbf{M}_\alpha = \mathbf{M} - \frac{2(\mathbf{M}, \alpha)}{(\alpha, \alpha)}\alpha, \tag{11.73}$$

which also defines a weight vector. The transformation $\mathbf{M} \to \mathbf{M}_\alpha$ is known as the Weyl reflection of $\mathbf{M}$ by the root α (note that $(\mathbf{M}_\alpha, \alpha) = -(\mathbf{M}, \alpha)$). The Weyl reflection is very useful in deriving several results. As a simple example, let us note that if we choose (in the adjoint representation)

$$\mathbf{M} = \alpha_i, \quad \alpha = \alpha_j, \quad i \neq j, \tag{11.74}$$

then, (11.73) leads to

$$\mathbf{M}_\alpha = \alpha_{ij} = \alpha_i - \frac{2(\alpha_i, \alpha_j)}{(\alpha_j, \alpha_j)}\alpha_j, \tag{11.75}$$

which is expected to be a weight vector. However, if $(\alpha_i, \alpha_j) > 0$, the vector α_{ij} is neither positive nor negative. Therefore, this implies that

$$(\alpha_i, \alpha_j) \leq 0, \tag{11.76}$$

which, in turn, leads us to conclude that (see (11.48))

$$K_{ij} \leq 0, \quad i \neq j. \tag{11.77}$$

Without going into details, we simply note here that if we further use the Cauchy-Schwarz inequality, (11.50) results.

10. **Second order Casimir invariant of** L: Let us collectively denote $(H_i, E_\alpha, E_{-\alpha})$ as $X_\mu, \mu = 1, 2, \cdots, \dim L$ and set (see (8.71) or (11.2) as well as (3.117))

$$g_{\mu\nu} = \text{Tr}\,(ad\,X_\mu\,ad\,X_\nu),$$

$$I_2 = \sum_{\mu,\nu=1}^{\dim L} g^{\mu\nu} X_\mu X_\nu. \qquad (11.78)$$

Then, as we have discussed in section 3.3 (see (3.118) and (3.120)), I_2 defines the second order Casimir invariant of L, namely, we have

$$[I_2, X_\lambda] = 0, \qquad (11.79)$$

for any $X_\lambda \in L$. In terms of H_i and E_α we can write this explicitly as

$$I_2 = \sum_{i,j=1}^{\ell} g^{ij} H_i H_j + \sum_{\alpha>0}(E_\alpha E_{-\alpha} + E_{-\alpha}E_\alpha)$$

$$= \sum_{i,j=1}^{\ell} g^{ij} H_i H_j + 2\sum_{\alpha>0} E_{-\alpha}E_\alpha + \sum_{\alpha>0} H_\alpha, \qquad (11.80)$$

where we have used the fact (see (11.25)) that $g_{\alpha,-\alpha} = 1$ so that $g^{\alpha,-\alpha} = 1$ and $g^{\alpha\beta} = 0$ for $\alpha + \beta \neq 0$.

For the highest weight vector $\boldsymbol{\Lambda}$, we have (see (11.61) and the discussion in the following paragraph there)

$$H_i \phi_\Lambda = \lambda_i \phi_\Lambda, \quad E_\alpha \phi_\Lambda = 0, \quad \alpha > 0. \qquad (11.81)$$

This leads to

$$\sum_{i,j=1}^{\ell} g^{ij} H_i H_j \phi_\Lambda = (\sum_{i,j=1}^{\ell} g^{ij} \lambda_i \lambda_j)\phi_\Lambda = (\boldsymbol{\Lambda}, \boldsymbol{\Lambda})\phi_\Lambda, \qquad (11.82)$$

as well as

$$\sum_{\alpha>0} H_\alpha \phi_\Lambda = \sum_{\alpha>0}(\sum_{i=1}^{\ell} a^i H_i)\phi_\Lambda = \sum_{\alpha>0}(\sum_{i=1}^{\ell} a^i \lambda_i)\phi_\Lambda$$

$$= \sum_{\alpha>0}(\boldsymbol{\Lambda}, \boldsymbol{\alpha})\phi_\Lambda = 2(\boldsymbol{\Lambda}, \boldsymbol{\delta})\phi_\Lambda, \qquad (11.83)$$

where we have identified the vector $\boldsymbol{\delta}$ as

$$\boldsymbol{\delta} = \frac{1}{2} \sum_{\alpha > 0} \alpha. \tag{11.84}$$

Using (11.81)-(11.84) in the definition of I_2 in (11.80) we obtain

$$I_2 \phi_\Lambda = ((\boldsymbol{\Lambda}, \boldsymbol{\Lambda}) + 2(\boldsymbol{\Lambda}, \boldsymbol{\delta})) \, \phi_\lambda, \tag{11.85}$$

so that the eigenvalue of I_2, in the irreducible representation with the highest weight vector $\boldsymbol{\Lambda}$, is given by

$$\begin{aligned} I_2(\boldsymbol{\Lambda}) &= ((\boldsymbol{\Lambda}, \boldsymbol{\Lambda}) + 2(\boldsymbol{\Lambda}, \boldsymbol{\delta})) \\ &= (\boldsymbol{\Lambda} + \boldsymbol{\delta}, \boldsymbol{\Lambda} + \boldsymbol{\delta}) - (\boldsymbol{\delta}, \boldsymbol{\delta}). \end{aligned} \tag{11.86}$$

Weyl has shown that the dimension of the irreducible representation ϕ_Λ with the highest weight $\boldsymbol{\Lambda}$ is given by

$$\text{Dim}\,\phi_\Lambda = \frac{\prod\limits_{\alpha>0} (\alpha, \boldsymbol{\delta} + \boldsymbol{\Lambda})}{\prod\limits_{\alpha>0} (\alpha, \boldsymbol{\delta})}. \tag{11.87}$$

11.2 Lie algebra $A_\ell = su(\ell + 1)$:

We note right in the beginning that, for complex algebras which we are considering here, $su(\ell + 1) \simeq sl(\ell + 1)$ and, consequently, the identification $A_\ell = sl(\ell + 1)$ has been made in some literature. But, we will continue with $su(\ell+1)$ recognizing that the two are equivalent. It is convenient to study the general Lie algebra $su(\ell+1), \ell \geq 1$ which yields $su(N)$ for $N = \ell + 1$. The Lie algebra of the generators $A^\mu_{\ \nu}$ is given by (see (3.55) and (3.56))

$$\sum_{\mu=1}^{N} A^\mu_{\ \mu} = 0,$$

$$[A^\mu_{\ \nu}, A^\alpha_{\ \beta}] = \delta^\alpha_{\ \nu} A^\mu_{\ \beta} - \delta^\mu_{\ \beta} A^\alpha_{\ \nu}, \tag{11.88}$$

for $\mu, \nu, \alpha, \beta = 1, 2, \cdots, \ell+1$. We note from the commutation relation in (11.88) that $[A^\mu_{\ \mu}, A^\nu_{\ \nu}] = 0$ for any pair of fixed indices μ, ν. Therefore, we can diagonalize $A^1_{\ 1}, A^2_{\ 2}, \cdots, A^{\ell+1}_{\ \ell+1}$ simultaneously. (Actually, there are only ℓ such matrices which are independent because of the first condition in (11.88).)

Let us next consider the $su(2)$ sub-Lie algebra defined by the generators (see (3.57))

$$J_3 = \frac{1}{2}\left(A^1{}_1 - A^2{}_2\right), J_+ = A^1{}_2, \quad J_- = A^2{}_1, \tag{11.89}$$

which satisfy (see (3.58))

$$[J_3, J_\pm] = \pm J_\pm, \quad [J_+, J_-] = 2J_3. \tag{11.90}$$

We have already seen in chapter 3 that the eigenvalues of $A^1{}_1 - A^2{}_2 = 2J_3$ must always be integers. However, we note that by letting $1 \to 2 \to 3$, we can also look at the sub-algebra defined by

$$J_3' = \frac{1}{2}\left(A^2{}_2 - A^3{}_3\right), \quad J_+' = A^2{}_3, \quad J_-' = A^3{}_2, \tag{11.91}$$

and conclude that the eigenvalues of the generators $A^2{}_2 - A^3{}_3$ must also be integers. Since $A^\mu{}_\mu$ is diagonal for any fixed μ, it follows that $(A^\mu{}_\mu - A^{\mu+1}{}_{\mu+1})$ also commute for different values of μ and there must be a common eigenvector ϕ satisfying

$$(A^\mu{}_\mu - A^{\mu+1}{}_{\mu+1})\phi = m_\mu \phi, \quad \mu = 1, 2, \cdots, \ell, \tag{11.92}$$

for integral eigenvalues λ_μ. Let ϕ' denote another simultaneous eigenvector satisfying

$$(A^\mu{}_\mu - A^{\mu+1}{}_{\mu+1})\phi' = m_\mu' \phi'. \tag{11.93}$$

Let us assume that the two sets of eigenvalues $\mathbf{M} = (m_1, m_2, \cdots, m_\ell)$ and $\mathbf{M}' = (m_1', m_2', \cdots, m_\ell')$ are such that they coincide except possibly for one (or more) component which satisfies $m_{n+1} > m_{n+1}'$ for some value of n. In this case, we say that the weight $\mathbf{M}$ is higher than the weight $\mathbf{M}'$ and denote this by $\mathbf{M} > \mathbf{M}'$ and also $\phi > \phi'$. Among all possible weights $\mathbf{M}$, there exists the highest weight denoted by (see discussion on weight vectors in section 11.1)

$$\boldsymbol{\Lambda} = (\lambda_1, \lambda_2, \cdots, \lambda_\ell), \tag{11.94}$$

satisfying

$$(A^\mu{}_\mu - A^{\mu+1}{}_{\mu+1})\phi_\Lambda = \lambda_\mu \phi_\Lambda, \tag{11.95}$$

such that $\boldsymbol{\Lambda} > \mathbf{M}$ for any other weight vector $\mathbf{M}$ (we are discussing only finite dimensional representations here). As we will show shortly, the highest weight $\boldsymbol{\Lambda}$ is unique for any irreducible representation. We note that any λ_μ takes nonnegative integer values, namely, $\lambda_\mu \geq 0$.

To keep the discussion parallel to section 11.1, we note that it is more convenient to identify

$$H_\mu = A^\mu{}_\mu - A^{\mu+1}{}_{\mu+1}, \quad \mu = 1, 2, \cdots, \ell, \tag{11.96}$$

so that we can write (11.92) as

$$H_\mu \phi = m_\mu \phi, \tag{11.97}$$

for any eigenvector ϕ with weight $\mathbf{M} = (m_1, m_2, \cdots, m_\ell)$.

Using (11.88), we next note that

$$[H_\mu, A^\alpha{}_\beta] = [A^\mu{}_\mu - A^{\mu+1}{}_{\mu+1}, A^\alpha{}_\beta]$$

$$= (\delta^\alpha{}_\mu A^\mu{}_\beta - \delta^\mu{}_\beta A^\alpha{}_\mu) - (\delta^\alpha{}_{\mu+1} A^{\mu+1}{}_\beta - \delta^{\mu+1}{}_\beta A^\alpha{}_{\mu+1})$$

$$= \left(\delta^\alpha{}_\mu - \delta^\mu{}_\beta - \delta^\alpha{}_{\mu+1} + \delta^{\mu+1}{}_\beta\right) A^\alpha{}_\beta. \tag{11.98}$$

Therefore, for the highest weight state, we have

$$H_\mu \left(A^\alpha{}_\beta \phi_\Lambda\right) = \left([H_\mu, A^\alpha{}_\beta] + A^\alpha{}_\beta H_\mu\right) \phi_\Lambda$$

$$= \left(\delta^\alpha{}_\mu - \delta^\mu{}_\beta - \delta^\alpha{}_{\mu+1} + \delta^{\mu+1}{}_\beta + \lambda_\mu\right) \left(A^\alpha{}_\beta \phi_\Lambda\right). \tag{11.99}$$

We note from (11.99) that if $\mu = \alpha < \beta$, then we have

$$H_\mu(A^\alpha{}_\beta \phi_\Lambda) = (1 + \delta^{\mu+1}{}_\beta + \lambda_\mu) \left(A^\mu{}_\beta \phi_\Lambda\right), \tag{11.100}$$

so that, for $\beta > \mu$, the eigenvector $(A^\alpha{}_\beta \phi_\Lambda)$ has a weight which is higher than Λ itself. This wil be incompatible with our starting assumption that Λ defines the highest weight. Consequently, we must have $A^\mu{}_\beta \phi_\Lambda = 0$ for $\beta > \mu$. Since μ is arbitrary, this implies that

$$A^\alpha{}_\beta \phi_\Lambda = 0, \quad \alpha < \beta. \tag{11.101}$$

Since $A^\alpha{}_\beta, \alpha < \beta$, raises the weight of the state, we call this the raising operator. Similarly, it can be seen, from (11.99), that $A^\alpha{}_\beta, \alpha > \beta$ lowers the weight of the state and, therefore is called the lowering operator. For example, for $\mu = \beta < \alpha$, (11.99) leads to

$$H_\beta(A^\alpha{}_\beta \phi_\Lambda) = (-1 - \delta^\alpha{}_{\beta+1} + \lambda_\beta) \left(A^\alpha{}_\beta \phi_\Lambda\right), \tag{11.102}$$

so that $A^\alpha{}_\beta \phi_\Lambda, \alpha > \beta$, corresponds to an eigenstate with a lower weight (than ϕ_Λ). We have already seen in chapter 3 that the operators $J_+ = A^1{}_2$ and $J_- = A^2{}_1$ correspond respectively to raising and lowering operators of $su(2)$.

We will next show that the highest weight vector ϕ_Λ is unique in any irreducible space and that the irreducible representation is completely specified by its highest weight vector as follows. Using lowering operators $A^\alpha_\beta, \alpha > \beta$, we can construct successively from the highest weight vector states such as

$$\phi_\Lambda, \ A^\alpha_\beta \phi_\Lambda \ (\alpha > \beta), \ A^{\alpha_1}_{\beta_1} A^{\alpha_2}_{\beta_2} \phi_\Lambda \ (\alpha_1 > \beta_1, \alpha_2 > \beta_2), \cdots$$

$$(11.103)$$

Let V_0 denote the vector space spanned by all these vectors obtained from the highest weight vector through the application of the lowering operators. Let us assume that V_0 corresponds to a sub-vector space of the irreducible space V. Apart from the vector ϕ_Λ, the rest of the vectors in the space V_0 have weights lower (smaller) than the highest weight Λ. Therefore, we have

$$A^\alpha_\beta V_0 \subseteq V_0, \quad \text{for } \alpha > \beta, \tag{11.104}$$

since in this case A^α_β corresponds to a lowering operator. Similarly, $H_\mu V_0 \subseteq V_0$ since each vector of V_0 is a weight vector. Therefore, we need to prove

$$A^\alpha_\beta V_0 \subseteq V_0, \tag{11.105}$$

even for the case of raising operators for which $\alpha < \beta$.

We have already shown in (11.101) that $A^\alpha_\beta \phi_\Lambda = 0 \subseteq V_0$ for $\alpha < \beta$. Clearly, any number of raising operators applied to the highest weight vector will also lead to the null vector which is contained in V_0. We next note that, for $\alpha < \beta, \mu > \nu$,

$$A^\alpha_\beta (A^\mu_\nu \phi_\Lambda) = [A^\alpha_\beta, A^\mu_\nu] \phi_\Lambda + A^\mu_\nu (A^\alpha_\beta \phi_\Lambda) = [A^\alpha_\beta, A^\mu_\nu] \phi_\Lambda$$

$$\left(\delta^\mu_\beta A^\alpha_\nu - \delta^\alpha_\nu A^\mu_\beta\right) \phi_\Lambda \subseteq V_0. \tag{11.106}$$

(If $\alpha < \beta$ and $\mu < \nu$, they are both raising operators and the left hand side trivially belongs to V_0 as argued above.) The final identity follows because if the right hand side in (11.106) contains a raising operator, it will be trivially contained in V_0. However, if the right hand side contains a lowering operator, it will be contained in V_0 because of (11.104). Similarly, we can write, for $\alpha < \beta, \mu_i > \nu_i, i = 1, 2$,

$$A^\alpha_\beta \left(A^{\mu_1}_{\nu_1} A^{\mu_2}_{\nu_2} \phi_\Lambda\right) = [A^\alpha_\beta, A^{\mu_1}_{\nu_1} A^{\mu_2}_{\nu_2}] \phi_\Lambda + A^{\mu_1}_{\nu_1} A^{\mu_2}_{\nu_2} \left(A^\alpha_\beta \phi_\Lambda\right)$$

$$= [A^\alpha_\beta, A^{\mu_1}_{\nu_1}] A^{\mu_2}_{\nu_2} \phi_\Lambda + A^{\mu_1}_{\nu_1} [A^\alpha_\beta, A^{\mu_2}_{\nu_2}] \phi_\Lambda \subseteq V_0. \tag{11.107}$$

This follows because if the commutator $[A^\alpha{}_\beta, A^{\mu_1}_{\nu_1}]$ in the first term contains a raising operator, then we can reduce it further using (11.106) to show that it is contained in V_0. If it contains a lowering operator, it is also contained in V_0 by (11.104). Similarly, if the commutator $[A^\alpha{}_\beta, A^{\mu_2}_{\nu_2}]$ in the second term contains a raising operator, it is trivially contained in V_0 while if it contains a lowering operator, it is contained in V_0 by (11.104). This procedure can be generalized to a product of the form $A^\alpha{}_\beta (A^{\mu_1}_{\nu_1} \cdots A^{\mu_n}_{\nu_n} \phi_\Lambda)$ to show that it is also contained in V_0. Therefore, (11.104) and (11.105) together show that

$$A^\alpha{}_\beta V_0 \subseteq V_0, \tag{11.108}$$

for any $A^\alpha{}_\beta \in su(\ell + 1)$. (We note here that the elements of the form $A^\alpha{}_\alpha$ are diagonal, as we have mentioned earlier, and leave the space invariant.) Therefore, we have shown that V_0 is an invariant subspace of the irreducible space V. Then, by the definition of the irreducibility of a space, we must have $V_0 = V$ since $V_0 \neq 0$ (namely, since $\phi_\Lambda \in V_0$). Furthermore, this implies that the irreducible space V is completely determined by its highest weight vector ϕ_Λ by applying the lowering operators. Therefore, it must be unique since any other highest weight vector, obtained by a different lexicographic ordering of $1, 2, \cdots, \ell + 1$, must lead to the same space. In order to make contact with the method of Young tableaux, we recall from (3.54)-(3.55) that we can write the eigenvalues of $A^\mu{}_\mu$ in the form

$$a_\mu - \frac{1}{\ell + 1} \sum_{\lambda=1}^{\ell+1} a_\lambda, \tag{11.109}$$

for some integers a_λ. This follows from the structure (3.54) as well as because $(A^\mu{}_\mu - A^{\mu+1}_{\mu+1})$ has integer eigenvalues and that $\sum_\mu A^\mu{}_\mu = 0$ (see (3.55)). Therefore, for the highest weight vector, we can write

$$A^\mu{}_\mu \phi_\Lambda = \tilde{f}_\mu \phi_\Lambda, \tag{11.110}$$

where we have identified

$$\tilde{f}_\mu = f_\mu - \frac{1}{\ell + 1} \sum_{\lambda=1}^{\ell+1} f_\lambda, \tag{11.111}$$

with the parameterization of f_μ in the form

$$
\begin{aligned}
f_1 &= \lambda_1 + \lambda_2 + \lambda_3 + \cdots + \lambda_\ell + \lambda_{\ell+1}, \\
f_2 &= \lambda_2 + \lambda_3 + \cdots + \lambda_\ell + \lambda_{\ell+1}, \\
&\vdots \\
f_\ell &= \lambda_\ell + \lambda_{\ell+1}, \\
f_{\ell+1} &= \lambda_{\ell+1}.
\end{aligned}
\tag{11.112}
$$

We note that the reason for parameterizing f_μ in this manner is to have a natural ordering in the eigenvalues, namely, $f_1 \geq f_2 \geq \cdots \geq f_{\ell+1}$ as well as to have $f_\mu - f_{\mu+1} = \lambda_\mu$, namely the components of the highest weight. We note from (11.92) or (11.94) that the weights have only ℓ components while in (11.112) the eigenvalues have $(\ell+1)$ components. This has its origin in the fact that although there are $(\ell+1)$ diagonal generators, only ℓ of them are independent because of the traceless condition (3.55). Therefore, we can think of $\lambda_{\ell+1} = -(\lambda_1 + \lambda_2 + \cdots + \lambda_\ell)$. Furthermore, because the eigenvalues in (11.111) are invariant under an arbitrary constant shift $f_\mu \to f_\mu + \Delta$, we can always choose $\lambda_{\ell+1} = 0$ if we wish.

For the tensor representation of $su(N)$ $(N = \ell + 1)$ with the Young tableaux index $(f_1, f_2, \cdots, f_N)$ with $f_N \geq 0$, the highest weight vector ϕ_Λ can be identified with the tensor

$$
\phi_\Lambda = T_{\underbrace{11\cdots1}_{f_1}\ \underbrace{22\cdots2}_{f_2}\cdots\ \underbrace{NN\cdots N}_{f_N}}.
\tag{11.113}
$$

For example, let us compute the eigenvalue of the second order Casimir invariant of $su(N)$

$$
\begin{aligned}
I_2 &= \sum_{\mu,\nu=1}^{N} A^\mu{}_\nu A^\nu{}_\mu \\
&= \sum_{\mu=1}^{N} \left(A^\mu{}_\mu\right)^2 + \sum_{\mu<\nu}^{N} \left(A^\mu{}_\nu A^\nu{}_\mu + A^\nu{}_\mu A^\mu{}_\nu\right).
\end{aligned}
\tag{11.114}
$$

Using the commutation relation (11.88), we note that we can write

$$
\begin{aligned}
\sum_{\mu<\nu}^{N} A^\mu{}_\nu A^\nu{}_\mu &= \sum_{\mu<\nu}^{N} \left(A^\nu{}_\mu A^\mu{}_\nu + [A^\mu{}_\nu, A^\nu{}_\mu]\right) \\
&= \sum_{\mu<\nu}^{N} A^\nu{}_\mu A^\mu{}_\nu + \sum_{\mu<\nu}^{N} \left(\delta^\nu{}_\nu A^\mu{}_\mu - \delta^\mu{}_\mu A^\nu{}_\nu\right)
\end{aligned}
$$

$$= \sum_{\mu<\nu}^{N} A^\nu{}_\mu A^\mu{}_\nu + \sum_{\mu<\nu}^{N} \left(A^\mu{}_\mu - A^\nu{}_\nu \right)$$

$$= \sum_{\mu<\nu}^{N} A^\nu{}_\mu A^\mu{}_\nu + \sum_{\mu=1}^{N} (N-\mu) A^\mu{}_\mu - \sum_{\nu=1}^{N} (\nu-1) A^\nu{}_\nu$$

$$= \sum_{\mu<\nu}^{N} A^\nu{}_\mu A^\mu{}_\nu + \sum_{\mu=1}^{N} (N-2\mu+1) A^\mu{}_\mu. \qquad (11.115)$$

Substituting this into (11.114) we obtain

$$I_2 = \sum_{\mu=1}^{N} \left((A^\mu{}_\mu)^2 + (N-2\mu+1) A^\mu{}_\mu \right) + 2 \sum_{\mu<\nu}^{N} A^\nu{}_\mu A^\mu{}_\nu. \qquad (11.116)$$

As we know, the highest weight vector satisfies (see (11.101))

$$A^\mu{}_\nu \, \phi_\Lambda = 0, \qquad \mu < \nu,$$

so that we see from (11.116) and (11.110) that

$$I_2 \phi_\Lambda = \sum_{\mu=1}^{N} \left((\tilde{f}_\mu)^2 + (N-2\mu+1) \tilde{f}_\mu \right) \phi_\Lambda. \qquad (11.117)$$

As a result, we obtain the eigenvalue $I_2(\rho)$ for the particular representation (given by the highest weight vector ϕ_Λ) to be

$$I_2(\rho) = \sum_{\mu=1}^{N} \left(\tilde{f}_\mu + \frac{N+1}{2} - \mu \right)^2 - \sum_{\mu=1}^{N} \left(\frac{N+1}{2} - \mu \right)^2$$

$$= \sum_{\mu=1}^{N} (\tilde{\sigma}_\mu)^2 - \frac{N}{12} (N^2 - 1), \qquad (11.118)$$

where we have identified

$$\tilde{\sigma}_\mu = \tilde{f}_\mu + \frac{N+1}{2} - \mu = f_\mu + \frac{N+1}{2} - \mu - \frac{1}{N} \sum_{\lambda=1}^{N} f_\lambda, \qquad (11.119)$$

which satisfies

$$\sum_{\mu=1}^{N} \tilde{\sigma}_\mu = 0. \qquad (11.120)$$

This reproduces the earlier result (see (10.148) and the first relation in (10.150)) except for an overall normalization constant N.

Let us recall that the generators $A^\mu{}_\nu$ of $su(\ell + 1)$, for $\mu < \nu$ are known as raising operators. However, among all such raising operators, there exist some very special raising operators given by $A^1{}_2, A^2{}_3, A^3{}_4, \cdots, A^\ell{}_{\ell+1}$. The reason is that all other raising operators can be obtained from these through the commutation relations of $su(\ell + 1)$. For example, we note from (11.88) that we can write

$$A^1{}_3 = [A^1{}_2, A^2{}_3], \quad A^1{}_4 = [A^1{}_3, A^3{}_4] = [[A^1{}_2, A^2{}_3], A^3{}_4], \quad (11.121)$$

and so on. We call these ℓ raising operators to be simple. Let us next introduce $(\ell+1)$ orthonormal unit vectors $\mathbf{e}_1, \mathbf{e}_2, \cdots, \mathbf{e}_{\ell+1}$ in an $(\ell+1)$ dimensional Euclidean space satisfying the inner product

$$(\mathbf{e}_j, \mathbf{e}_k) = \delta_{jk}, \quad j, k = 1, 2, \cdots, \ell + 1. \quad (11.122)$$

Let us also define ℓ independent vectors from these as

$$\boldsymbol{\alpha}_1 = \mathbf{e}_1 - \mathbf{e}_2, \quad \boldsymbol{\alpha}_2 = \mathbf{e}_2 - \mathbf{e}_3, \quad \cdots, \quad \boldsymbol{\alpha}_\ell = \mathbf{e}_\ell - \mathbf{e}_{\ell+1}. \quad (11.123)$$

We call $\boldsymbol{\alpha}_1, \boldsymbol{\alpha}_2, \cdots, \boldsymbol{\alpha}_\ell$ to be simple roots of the Lie algebra (see discussion on simple roots in section 11.1). Let us next identify

$$H_{e_j} = H_j = A^j{}_j - A^{j+1}{}_{j+1},$$

$$E_{e_j - e_k} = A^j{}_k, \quad (11.124)$$

where $j, k = 1, 2, \cdots, \ell + 1$. (We may set $H_{\ell+1} = 0$.) The second definition in (11.124) leads to

$$A^1{}_2 = E_{\alpha_1}, \quad A^2{}_3 = E_{\alpha_2}, \quad \cdots, \quad A^\ell{}_{\ell+1} = E_{\alpha_\ell},$$

$$A^2{}_1 = E_{-\alpha_1}, \quad A^3{}_2 = E_{-\alpha_2}, \quad \cdots, \quad A^{\ell+1}{}_\ell = E_{-\alpha_\ell}, \quad (11.125)$$

while we have

$$A^1{}_3 = E_{\alpha_1 + \alpha_2}, \quad A^1{}_4 = E_{\alpha_1 + \alpha_2 + \alpha_3}, \quad (11.126)$$

and so on, which can be compared with (11.121). Let us make the identification

$$\boldsymbol{\alpha}_{jk} = \mathbf{e}_j - \mathbf{e}_k, \quad (11.127)$$

and call it a positive or a negative root depending on whether $j < k$ or $j > k$. Recall that any positive root can be expressed as a linear combination of simple roots $\alpha_1, \alpha_2, \cdots, \alpha_\ell$ with some nonnegative integer coefficients (see (11.44)).

From the commutation relations in (11.88), we can now find relations such as (see (11.29) and (11.32))

$$[E_\alpha, E_\beta] = N_{\alpha,\beta} E_{\alpha+\beta}, \quad \text{for } \alpha + \beta \neq 0, \text{ and } \alpha + \beta \text{ a root},$$

$$[E_\alpha, E_{-\alpha}] = H_\alpha,$$

$$[H_\alpha, E_\beta] = (\alpha, \beta) E_{\alpha+\beta}, \quad \text{for } \alpha + \beta \text{ a root}. \tag{11.128}$$

$$\alpha_1 \qquad \alpha_2 \qquad \alpha_3 \qquad \cdots\cdots \qquad \alpha_\ell$$

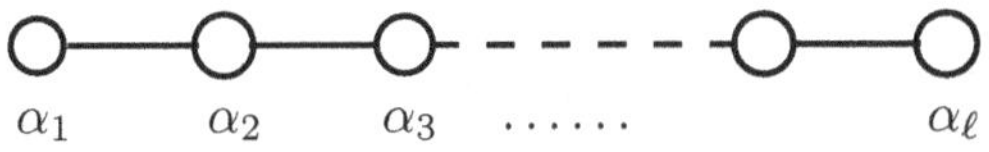

Figure 11.2: The Dynkin diagram for the simple roots of $A_\ell = su(\ell + 1)$.

From the definitions in (11.122) and (11.123), we see that

$$(\alpha_j, \alpha_j) = 2, \quad j = 1, 2, \cdots, \ell,$$

$$(\alpha_j, \alpha_{j+1}) = -1,$$

$$(\alpha_j, \alpha_k) = 0, \quad \text{if } |j - k| \geq 2, \tag{11.129}$$

for simple roots. These relations are generally depicted by the Dynkin diagram as shown in Fig. 11.2.

Finally, we simply introduce the notion of the fundamental weight system $(\Lambda_1, \Lambda_2, \cdots, \Lambda_\ell)$ given by

$$\begin{aligned}
\Lambda_1 &= \mathbf{e}_1 - \tfrac{1}{\ell+1} \sum_{k=1}^{\ell+1} \mathbf{e}_k, \\
\Lambda_2 &= \mathbf{e}_1 + \mathbf{e}_2 - \tfrac{2}{\ell+1} \sum_{k=1}^{\ell+1} \mathbf{e}_k, \\
\vdots &= \vdots \\
\Lambda_\ell &= \mathbf{e}_1 + \mathbf{e}_2 + \cdots + \mathbf{e}_\ell - \tfrac{\ell}{\ell+1} \sum_{k=1}^{\ell+1} \mathbf{e}_k,
\end{aligned} \tag{11.130}$$

which satisfy

$$\frac{2(\Lambda_j, \alpha_k)}{(\alpha_k, \alpha_k)} = \delta_{jk}, \quad j, k = 1, 2, \cdots, \ell. \tag{11.131}$$

If we replace $\mathbf{e}_j$ by $\boldsymbol{\alpha}_j$ in (11.130) using (11.56), we can also write

$$\boldsymbol{\Lambda}_j = \frac{\ell + 1 - j}{\ell + 1} (\boldsymbol{\alpha}_1 + 2\boldsymbol{\alpha}_2 + \cdots + j\boldsymbol{\alpha}_j)$$

$$+ \frac{j}{\ell + 1} ((\ell - j)\boldsymbol{\alpha}_{j+1} + (\ell - j - 1)\boldsymbol{\alpha}_{j+2} + \cdots + \boldsymbol{\alpha}_\ell).$$

$$(11.132)$$

With this, one can show, using the general Cartan-Dynkin theory (see section 11.1), that the highest weight $\boldsymbol{\Lambda}$ of an irreducible representation characterized by $\boldsymbol{\Lambda} = (\lambda_1, \lambda_2, \cdots, \lambda_\ell)$ is given by

$$\boldsymbol{\Lambda} = \lambda_1 \boldsymbol{\Lambda}_1 + \lambda_2 \boldsymbol{\Lambda}_2 + \cdots + \lambda_\ell \boldsymbol{\Lambda}_\ell, \quad \lambda_j \geq 0. \tag{11.133}$$

11.3 Lie algebra $D_\ell = so(2\ell)$:

As we have discussed in (3.23) and (3.24), the Lie algebra of $so(N)$ is given by

$$J_{jk} = -J_{kj}, \quad j, k = 1, 2, \cdots, N,$$

$$[J_{jk}, J_{\ell m}] = -\delta_{j\ell} J_{km} - \delta_{km} J_{j\ell} + \delta_{jm} J_{k\ell} + \delta_{k\ell} J_{jm}. \tag{11.134}$$

This will describe the Lie algebra $so(2\ell)$ with the identification $N = 2\ell$. However, this form is not suitable for the discussion of its irreducible representations. Therefore, we change the basis of generators from $J_{jk}, j, k = 1, 2, \cdots, 2\ell$ to a set of linearly independent generators consisting of $R^{\mu\nu}, R_{\mu\nu}$ and $X^\mu{}_\nu$ with $\mu, \nu = 1, 2, \cdots, \ell$ defined by

$$R^{\mu\nu} = \frac{1}{2} \left((J_{\mu\nu} - J_{\mu+\ell,\nu+\ell}) - i(J_{\mu,\nu+\ell} + J_{\mu+\ell,\nu}) \right) = -R^{\nu\mu},$$

$$R_{\mu\nu} = -\frac{1}{2} \left((J_{\mu\nu} - J_{\mu+\ell,\nu+\ell}) + i(J_{\mu,\nu+\ell} + J_{\mu+\ell,\nu}) \right) = -R_{\nu\mu},$$

$$X^\mu{}_\nu = \frac{1}{2} \left((J_{\mu\nu} + J_{\mu+\ell,\nu+\ell}) + i(J_{\mu,\nu+\ell} - J_{\mu+\ell,\nu}) \right). \tag{11.135}$$

Relations (11.135) can also be inverted to obtain

$$J_{\mu\nu} = \frac{1}{2} \left(R^{\mu\nu} - R_{\mu\nu} + X^\mu{}_\nu - X^\nu{}_\mu \right),$$

$$J_{\mu+\ell,\nu+\ell} = \frac{1}{2} \left(-R^{\mu\nu} + R_{\mu\nu} + X^\mu{}_\nu - X^\nu{}_\mu \right),$$

$$J_{\mu,\nu+\ell} = \frac{i}{2} \left(R^{\mu\nu} + R_{\mu\nu} - X^\mu{}_\nu - X^\nu{}_\mu \right),$$

$$J_{\mu+\ell,\nu} = \frac{i}{2} \left(R^{\mu\nu} + R_{\mu\nu} + X^\mu{}_\nu + X^\nu{}_\mu \right), \tag{11.136}$$

for $\mu, \nu = 1, 2, \cdots, \ell$.

The commutation relations for the new basis can now be obtained from (11.134) and take the forms

$$[X^\mu_\nu, X^\alpha_\beta] = \delta^\alpha_\nu X^\mu_\beta - \delta^\mu_\beta X^\alpha_\nu,$$

$$[X^\mu_\nu, R^{\alpha\beta}] = \delta^\alpha_\nu R^{\mu\beta} + \delta^\beta_\nu R^{\alpha\mu},$$

$$[X^\mu_\nu, R_{\alpha\beta}] = -\delta^\mu_\alpha R_{\nu\beta} - \delta^\mu_\beta R_{\alpha\nu},$$

$$[R_{\mu\nu}, R^{\alpha\beta}] = \delta^\alpha_\nu X^\beta_\mu + \delta^\beta_\mu X^\alpha_\nu - \delta^\alpha_\mu X^\beta_\nu - \delta^\beta_\nu X^\alpha_\mu,$$

$$[R_{\mu\nu}, R_{\alpha\beta}] = 0 = [R^{\mu\nu}, R^{\alpha\beta}], \tag{11.137}$$

for $\mu, \nu, \alpha, \beta = 1, 2, \cdots, \ell$.

The advantage of the transformation (11.135) or (11.136) can be understood as follows. Since $so(2N)$ generates rotations, the generators J_{jk} have the coordinate representation given by

$$J_{jk} = x_j \frac{\partial}{\partial x_k} - x_k \frac{\partial}{\partial x_j}, \quad j, k = 1, 2, \cdots, 2\ell. \tag{11.138}$$

If we now introduce a complex coordinate z_μ through the relation

$$z_\mu = x_\mu + ix_{\mu+\ell}, \quad \mu = 1, 2, \cdots, \ell, \tag{11.139}$$

then, we can express

$$R_{\mu\nu} = z_\mu \frac{\partial}{\partial z^*_\nu} - z_\nu \frac{\partial}{\partial z^*_\mu},$$

$$R^{\mu\nu} = -z^*_\mu \frac{\partial}{\partial z_\nu} + z^*_\nu \frac{\partial}{\partial z_\mu},$$

$$X^\mu_\nu = -z_\nu \frac{\partial}{\partial z_\mu} + z^*_\mu \frac{\partial}{\partial z^*_\nu}, \tag{11.140}$$

in terms of the new variables. Here "star" denotes complex conjugation.

We note (from the first relation in (11.137)) that the generators X^μ_ν define a Lie algebra $u(\ell)$. We also recall that the unitary representation of $so(2\ell)$ requires

$$J^\dagger_{jk} = -J_{kj} = J_{jk}. \tag{11.141}$$

This determines that

$$(X^\mu_\nu)^\dagger = X^\nu_\mu, \quad (R_{\mu\nu})^\dagger = R^{\mu\nu}, \quad (R^{\mu\nu})^\dagger = R_{\mu\nu}, \tag{11.142}$$

for a unitary representation. Let us note that, for our discussion, we can restrict ourselves to the case $\ell \geq 4$ for the following reason. The representation for $so(2)$ corresponding to $\ell = 1$ is trivial (it is an Abelian algebra) and for $\ell = 2$, we note that $so(4) = su(2) \oplus su(2)$. Indeed, for $\ell = 2$, the two commuting $su(2)$ algebras can be identified with the generators

$$(i): \quad X^1_1 - X^2_2, \ X^1_2, \ X^2_1,$$

$$(ii): \quad X^1_1 + X^2_2, \ R^{12}, \ R_{12}. \tag{11.143}$$

For the case of $\ell = 3$ or the algebra $so(6)$, we can define

$$A^\mu_\nu = X^\mu_\nu - \frac{1}{2} \delta^\mu_\nu \sum_{\lambda=1}^{3} X^\lambda_\lambda, \quad A^4_4 = \frac{1}{2} \sum_{\lambda=1}^{3} X^\lambda_\lambda,$$

$$A^4_\mu = \frac{1}{2} \sum_{\alpha,\beta=1}^{3} \epsilon_{\mu\alpha\beta} R^{\alpha\beta}, \quad A^\mu_4 = \frac{1}{2} \sum_{\alpha,\beta=1}^{3} \epsilon^{\mu\alpha\beta} R_{\alpha\beta}, \tag{11.144}$$

with $\mu, \nu, \alpha, \beta = 1, 2, 3$ and $\epsilon^{\mu\alpha\beta} = \epsilon_{\mu\alpha\beta}$ denoting the Levi-Civita tensor in 3-dimensions. These generators can be seen to satisfy the algebraic relations

$$\sum_{a=1}^{4} A^a_a = 0,$$

$$[A^a_b, A^c_d] = \delta^c_b A^a_d - \delta^a_d A^c_b, \tag{11.145}$$

for $a, b, c, d = 1, 2, 3, 4$. We recognize this as the $su(4)$ Lie algebra and, therefore, we conclude that $so(6) = su(4)$. Therefore, these cases need not be studied again.

Let us assume that all the generators belong to some $d \times d$ matrix representation of some irreducible space $V = \{\rho\}$. First we note from (11.137) that

$$[X^\mu_\mu, X^\nu_\nu] = 0, \tag{11.146}$$

for $\mu, \nu = 1, 2, \cdots, \ell$ so that if we identify

$$H_\mu = X^\mu_\mu, \tag{11.147}$$

we can write

$$[H_\mu, H_\nu] = 0. \tag{11.148}$$

These define ℓ commuting generators and, therefore, can be simultaneously diagonalized. We will now show that the (simultaneous) eigenvalues of these generators are either all integers or half integers depending respectively on whether we have a tensor representation or a spinor representation.

Let us recall that the generators

$$J_3 = \frac{1}{2}\left(X^1{}_1 - X^2{}_2\right), \; J_+ = X^1{}_2, \; J_- = X^2{}_1, \tag{11.149}$$

satisfy the familiar $su(2)$ algebra so that the eigenvalues of $X^1{}_1 - X^2{}_2$ is always an integer. Let us also note from (11.137) that

$$[R_{12}, R^{12}] = -\delta^1{}_1 X^2{}_2 - \delta^2{}_2 X^1{}_1 = -(X^1{}_1 + X^2{}_2),$$
$$[X^1{}_1 + X^2{}_2, R^{12}] = 2R^{12}, \quad [X^1{}_1 + X^2{}_2, R_{12}] = -2R_{12}. \tag{11.150}$$

As a result, we can identify

$$J'_3 = \frac{1}{2}\left(X^1{}_1 + X^2{}_2\right), \; J'_+ = R^{12}, \; J'_- = R_{12}, \tag{11.151}$$

with the generators of a second $su(2)$ algebra. We conclude from this that $X^1{}_1 + X^2{}_2$ also always has integer eigenvalues. Since $X^1{}_1 - X^2{}_2$ as well as $X^1{}_1 + X^2{}_2$ have integer eigenvalues, it follows that both $X^1{}_1$ and $X^2{}_2$ can have either integer or half integer eigenvalues. Carrying this analysis for $\mu = 1, 2, \cdots, \ell$, we conclude that each of the diagonal generators $X^\mu{}_\mu$ must have simultaneously integer or half integer eigenvalues.

We can now introduce the notion of highest weight vector as in the last section (for $su(\ell + 1)$) given by

$$H_\mu \phi_\Lambda = f_\mu \phi_\Lambda, \tag{11.152}$$

where $f_1, f_2, \cdots, f_\ell$ are now either all integers or half integers. Moreover, we can now parameterize the eigenvalues in an ordered manner, as in the case of $su(\ell + 1)$, such that

$$f_1 \geq f_2 \geq f_3 \geq \cdots \geq |f_\ell|, \tag{11.153}$$

as we will shortly in (11.161). The root system for this case is a bit more complicated. We note from (11.137) that the $(\ell - 1)$ operators $X^1{}_2, X^2{}_3, \cdots, X^{\ell-1}{}_\ell$ are raising operators just as in the case of $su(\ell + 1)$. In fact, we see from the commutations relations (see the second relation in (11.137)) that $R^{\mu\nu}$ s are also raising operators. As a simple root system of $so(2\ell)$, we can choose the ℓ raising operators to correspond to

$$X^1_{\ 2}, X^2_{\ 3}, X^3_{\ 4}, \cdots, X^{\ell-1}_{\ \ell}, R^{\ell-1,\ell}. \tag{11.154}$$

In analogy with the $su(\ell+1)$ case, we introduce ℓ unit orthogonal vectors $\mathbf{e}_1, \mathbf{e}_2, \cdots, \mathbf{e}_\ell$ in an ℓ-dimensional Euclidean space (see also (11.122)) and write

$$
\begin{aligned}
X^\mu_{\ \nu} &= E_{(e_\mu - e_\nu)}, & \mu \neq \nu, \\
R^{\mu\nu} &= E_{(e_\mu + e_\nu)}, & \mu < \nu, \\
R_{\mu\nu} &= E_{-(e_\mu + e_\nu)}, & \mu < \nu.
\end{aligned}
\tag{11.155}
$$

Defining, as in (11.123),

$$\boldsymbol{\alpha}_1 = \mathbf{e}_1 - \mathbf{e}_2, \quad \boldsymbol{\alpha}_2 = \mathbf{e}_2 - \mathbf{e}_3, \quad \cdots, \boldsymbol{\alpha}_{\ell-1} = \mathbf{e}_{\ell-1} - \mathbf{e}_\ell, \tag{11.156}$$

but with

$$\boldsymbol{\alpha}_\ell = \mathbf{e}_{\ell-1} + \mathbf{e}_\ell, \tag{11.157}$$

they constitute the desired ℓ simple root system of $D_\ell = so(2\ell)$ with

$$E_{\alpha_\mu} = X^\mu_{\ \mu+1}, \quad \mu = 1, 2, \cdots, \ell - 1, \quad \text{and } E_{\alpha_\ell} = R^{\ell-1,\ell}. \tag{11.158}$$

Correspondingly, we define

$$\lambda_1 = f_1 - f_2, \quad \lambda_2 = f_2 - f_3, \quad \cdots \quad \lambda_{\ell-1} = f_{\ell-1} - f_\ell, \tag{11.159}$$

as in (11.112), but with

$$\lambda_\ell = f_{\ell-1} + f_\ell, \tag{11.160}$$

so that all components of $\boldsymbol{\Lambda} = (\lambda_1, \lambda_2, \cdots, \lambda_{\ell-1}, \lambda_\ell)$ are non-negative integers. Explicitly, we can write (compare with (11.112))

$$
\begin{aligned}
f_1 &= \lambda_1 + \lambda_2 + \lambda_3 + \cdots + \lambda_{\ell-2} + \tfrac{1}{2}(\lambda_{\ell-1} + \lambda_\ell), \\
f_2 &= \lambda_2 + \lambda_3 + \cdots + \lambda_{\ell-2} + \tfrac{1}{2}(\lambda_{\ell-1} + \lambda_\ell), \\
&\ \vdots \\
f_{\ell-2} &= \lambda_{\ell-2} + \tfrac{1}{2}(\lambda_{\ell-1} + \lambda_\ell), \\
f_{\ell-1} &= \tfrac{1}{2}(\lambda_{\ell-1} + \lambda_\ell), \\
f_\ell &= \tfrac{1}{2}(\lambda_\ell - \lambda_{\ell-1}),
\end{aligned}
\tag{11.161}
$$

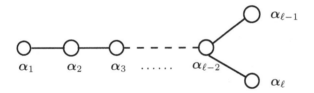

Figure 11.3: The Dynkin diagram for the simple roots of $D_\ell = so(2\ell)$.

which are all integers or half integers satisying (11.153). The corresponding Dynkin diagram is depicted now as shown in Fig. 11.3.

The dimensional formula for the irreducible representation $\{\rho\} = (f_1, f_2, \cdots, f_\ell)$ is given as follows. First we define

$$\sigma_\mu = f_\mu + \frac{N}{2} - \mu = f_\mu + \ell - \mu, \quad \mu = 1, 2, \cdots, \ell, \qquad (11.162)$$

which satisfy $\sigma_1 > \sigma_2 > \cdots > \sigma_{\ell-1} > |\sigma_\ell|$. Then the formula for the dimension can be written as

$$d(\rho) = \frac{2^{\ell-1}}{(2!)(4!)\cdots((2\ell-4)!)} \sum_{j<k}^{\ell} \left((\sigma_j)^2 - (\sigma_k)^2\right), \qquad (11.163)$$

which is due to Weyl.

We can also compute eigenvalues of symmetrical orthonormalized Casimir invariants $I_p(\rho)$ to be

$$I_2(\rho) = \sum_{\mu=1}^{\ell} (\sigma_\mu)^2 - \frac{1}{24} N(N-1)(N-2),$$

$$I_4(\rho) = \frac{1}{2}(N^2 - N + 4) \sum_{\mu=1}^{\ell} (\sigma_\mu)^4 - (2N-1) \left(\sum_{\mu=1}^{\ell} (\sigma_\mu)^2\right)^2$$

$$+ \frac{1}{2880} N(N-3)(N-8)(N^2-1)(N^2-4),$$

$$I_6(\rho) = \frac{1}{4}(N^2 - N + 8)(N^2 - N + 16) \sum_{\mu=1}^{\ell} (\sigma_\mu)^6$$

$$- \frac{3}{2}(2N-1)(N^2 - N + 8) \left(\sum_{\mu=1}^{\ell} (\sigma_\mu)^2\right) \left(\sum_{\nu=1}^{\ell} (\sigma_\nu)^4\right)$$

$$+ 7(N^2 - N - 2) \sum_{\mu=1}^{\ell} ((\sigma_\mu)^2)^3$$

$$- \frac{2}{35} \left(\frac{1}{24}\right)^3 N(N-5)(N-32)(N^2-1)(N^2-4)$$

$$\times (N^2 - 9)(N^2 - 16), \tag{11.164}$$

with a suitable normalization. Of course, we have to identify $N = 2\ell$ in (11.164). However, we note that the same result for the Casimir invariants holds even for $so(2\ell + 1)$ with $N = 2\ell + 1$ if we define σ_μ suitably as is discussed in the later section.

As an example, let us consider the tensor representations with signatures $\{\rho\} = (f_1, f_2, \cdots, f_\ell) = (1, 1, \cdots, 1)$ and $(1, 1, \cdots, 1, -1)$. They are realized as completely anti-symmetric tensors $T_{\mu_1 \mu_2 \cdots \mu_\ell}$ of ℓ indices which take values over $\mu_1, \mu_2, \cdots, \mu_\ell = 1, 2, \cdots, 2\ell$, such that the representation is self-dual for the former and anti-self-dual for the latter. Namely,

$$^*T_{\mu_1 \mu_2 \cdots \mu_\ell} = \epsilon_{\mu_1 \mu_2 \cdots \mu_\ell \nu_1 \nu_2 \cdots \nu_\ell} T_{\nu_1 \nu_2 \cdots \nu_\ell} = \pm T_{\mu_1 \mu_2 \cdots \mu_\ell}, \tag{11.165}$$

where the star denotes the dual operation. For the spinor representations $(\frac{1}{2}, \frac{1}{2}, \cdots, \frac{1}{2})$ and $(\frac{1}{2}, \frac{1}{2}, \cdots, \frac{1}{2}, -\frac{1}{2})$, they are realized as $2^{\ell-1} \times 2^{\ell-1}$ matrices (see section 6.3, in particular, (6.100) as well as (6.77)-(6.78))

$$J_{jk} = \frac{1}{4} \left(\sigma_j \sigma_k^\dagger - \sigma_k \sigma_j^\dagger\right),$$

$$\tilde{J}_{jk} = \frac{1}{4} \left(\sigma_j^\dagger \sigma_k - \sigma_k^\dagger \sigma_j\right), \tag{11.166}$$

where $j, k = 1, 2, \cdots, 2\ell$ and the matrices σ_j and $\sigma_j^\dagger$ satisfy

$$\sigma_j \sigma_k^\dagger + \sigma_k \sigma_j^\dagger = 2\delta_{jk} \mathbb{1} = \sigma_j^\dagger \sigma_k + \sigma_k^\dagger \sigma_j, \tag{11.167}$$

as discussed in chapter 6 on Clifford algebra where the Dirac-Clifford matrix is written as

$$\gamma_j = \begin{pmatrix} 0 & \sigma_j \\ \sigma_j^\dagger & 0 \end{pmatrix}, \quad j = 1, 2, \cdots, \ell. \tag{11.168}$$

Finally, we note that all even dimensional Casimir invariants $I_p(\rho)$ do not distinguish between the two representations $(f_1, f_2, \cdots, f_\ell)$ and $(f_1, f_2, \cdots, f_{\ell-1}, -f_\ell)$ if $f_\ell \neq 0$ since then it follows from (11.162) that $\sigma_\ell = f_\ell$. However, $so(2\ell)$ does possess an ℓ-th order Casimir invariant $\tilde{I}_\ell(\rho)$ given by

$$\tilde{I}_\ell = \frac{1}{(2\ell)!} \sum_{\mu_1,\nu_1\cdots\mu_\ell\nu_\ell=1}^{2\ell} \epsilon_{\mu_1\nu_1\mu_2\nu_2\cdots\mu_\ell\nu_\ell} J_{\mu_1\nu_1} J_{\mu_2\nu_2} \cdots J_{\mu_\ell\nu_\ell}.$$

(11.169)

It is known that its eigenvalue is given by (σ_μ is defined in (11.162) and are not the matrices discussed in (11.166))

$$\tilde{I}_\ell(\rho) = \sigma_1\sigma_2\cdots\sigma_\ell,$$

(11.170)

which changes sign as $f_\ell \to -f_\ell$.

The highest weight vector can be written in the form (see also (11.133))

$$\mathbf{\Lambda} = f_1\mathbf{e}_1 + f_2\mathbf{e}_2 + \cdots + f_\ell\mathbf{e}_\ell$$
$$= \lambda_1\mathbf{\Lambda}_1 + \lambda_2\mathbf{\Lambda}_2 + \cdots + \lambda_\ell\mathbf{\Lambda}_\ell,$$

(11.171)

in terms of the ℓ fundamental weights $\mathbf{\Lambda}_1, \mathbf{\Lambda}_2, \cdots, \mathbf{\Lambda}_\ell$ given by

$$\mathbf{\Lambda}_1 = \mathbf{e}_1, \quad \mathbf{\Lambda}_2 = \mathbf{e}_1 + \mathbf{e}_2, \quad \cdots, \mathbf{\Lambda}_{\ell-2} = \mathbf{e}_1 + \mathbf{e}_2 + \cdots + \mathbf{e}_{\ell-2},$$

(11.172)

and

$$\mathbf{\Lambda}_{\ell-1} = \frac{1}{2}\left(\mathbf{e}_1 + \mathbf{e}_2 + \cdots + \mathbf{e}_{\ell-2} + \mathbf{e}_{\ell-1} - \mathbf{e}_\ell\right),$$

$$\mathbf{\Lambda}_\ell = \frac{1}{2}\left(\mathbf{e}_1 + \mathbf{e}_2 + \cdots + \mathbf{e}_{\ell-2} + \mathbf{e}_{\ell-1} + \mathbf{e}_\ell\right).$$

(11.173)

11.3.1 $D_4 = so(8)$ and the triality relation. In order to explain the triality relation, let us first introduce the notion of inner and outer automorphisms of a Lie algebra. For any Lie algebra L of dimensionality M and with the commutation relation

$$[X_\mu, X_\nu] = \sum_{\lambda=1}^{M} C_{\mu\nu}^\lambda X_\lambda,$$

(11.174)

the transformation

$$X_\mu \to X_\mu' = gX_\mu g^{-1} \in L,$$

(11.175)

where g is an arbitrary element of the associated Lie group $g \in G = \exp(L)$, is clearly a symmetry of (11.174). Namely, it satisfies (recall the case of Yang-Mills gauge transformation)

$$[X'_\mu, X'_\nu] = \sum_{\lambda=1}^{M} C^\lambda_{\mu\nu} X'_\lambda. \qquad (11.176)$$

Such a symmetry is known as an inner automorphism of the Lie algebra. In contrast, when transformations $X_\mu \rightarrow X'_\mu$ preserving the Lie algebra structure cannot be written in the form (11.175), then such transformations are known as outer automorphisms of L. Therefore, the outer automorphisms are unanticipated and special.

For example, consider the Lie algebra of $su(N)$ described by (see (11.88))

$$\sum_{\mu=1}^{N} A^\mu_{\;\mu} = 0, \quad [A^\mu_{\;\nu}, A^\alpha_{\;\beta}] = \delta^\alpha_{\;\nu} A^\mu_{\;\beta} - \delta^\mu_{\;\beta} A^\alpha_{\;\nu}. \qquad (11.177)$$

In this case, we note that the transformation

$$A^\mu_{\;\nu} \rightarrow A'^\mu_{\;\nu} = -A^\nu_{\;\mu}, \qquad (11.178)$$

preserves the Lie algebra structure (11.177). Namely, the new generators satisfy

$$\sum_{\mu=1}^{N} A'^\mu_{\;\mu} = 0, \quad [A'^\mu_{\;\nu}, A'^\alpha_{\;\beta}] = \delta^\alpha_{\;\nu} A'^\mu_{\;\beta} - \delta^\mu_{\;\beta} A'^\alpha_{\;\nu}. \qquad (11.179)$$

However, the transformation (11.178) defines an outer automorphism of the $su(N)$ Lie algebra for $N \geq 3$, while it is an inner automorphism for $su(2)$. This can be easily seen as follows. Consider a tensor representation of $su(N)$ of the form $T^{\mu_1\mu_2\cdots\mu_n}_{\nu_1\nu_2\cdots\nu_m}$. With respect to the transformed basis (11.178), the tensor representation transforms to its complex conjugate

$$T^{\mu_1\mu_2\cdots\mu_n}_{\nu_1\nu_2\cdots\nu_m} \rightarrow T'^{\mu_1\mu_2\cdots\mu_n}_{\nu_1\nu_2\cdots\nu_m} = T^{\nu_1\nu_2\cdots\nu_m}_{\mu_1\mu_2\cdots\mu_n} = \left(T^{\mu_1\mu_2\cdots\mu_n}_{\nu_1\nu_2\cdots\nu_m}\right)^*. \qquad (11.180)$$

Since the complex conjugate representation of $su(N)$ is an inequivalent representation for $N \geq 3$, the transformation (11.178) cannot be an inner automorphism. On the other hand, for $su(2)$, any irreducible representation is real or pseudo-real so that the complex conjugate representation is an equivalent representation and correspondingly, the transformation (11.178) is an inner automorphism in this case.

Let us next consider $so(N)$ Lie algebra given by (see (11.134))

$$J_{jk} = -J_{kj},$$

$$[J_{jk}, J_{\ell m}] = -\delta_{j\ell}J_{km} - \delta_{km}J_{j\ell} + \delta_{jm}J_{k\ell} + \delta_{k\ell}J_{jm}, \qquad (11.181)$$

where $j, k, \ell, m = 1, 2, \cdots, N$. It is known that any permutation σ of the N indices $1, 2, \cdots, N$ gives an inner automorphism of $so(N)$ corresponding to the symmetry group S_N of N objects (see section 1.2.1). In contrast, any symmetry of the Dynkin diagram of a Lie algebra is known to give an outer automorphism of L. Let us consider the Lie algebra $D_4 = so(8)$ for which the Dynkin diagram is shown in Fig. 11.4. It is clear that this diagram is invariant under the symmetric group S_3 of interchanging the three simple roots $\alpha_1, \alpha_3, \alpha_4$ (leaving α_2 fixed). As we will see shortly, this leads to an interesting result that, for $so(8)$, there is no essential difference between the vector and the spinor representations, both of which are eight dimensional.

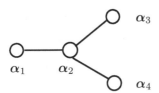

Figure 11.4: The Dynkin diagram for the simple roots of $D_4 = so(8)$.

To see this, let us make a transformation θ given by

$$X^\mu_{\;\nu} \xrightarrow{\theta} X'^\mu_{\;\nu} = -X^\nu_{\;\mu} + \frac{1}{2}\delta^\nu_{\;\mu}\sum_{\lambda=1}^{4} X^\lambda_{\;\lambda},$$

$$R^{\mu\nu} \xrightarrow{\theta} R'^{\mu\nu} = -\frac{1}{2}\sum_{\alpha,\beta=1}^{4} \epsilon_{\mu\nu\alpha\beta}R^{\alpha\beta},$$

$$R_{\mu\nu} \xrightarrow{\theta} R'_{\mu\nu} = -\frac{1}{2}\sum_{\alpha,\beta=1}^{4} \epsilon^{\mu\nu\alpha\beta}R_{\alpha\beta}, \qquad (11.182)$$

for $\mu, \nu = 1, 2, 3, 4$. Here $\epsilon^{\mu\nu\alpha\beta} = \epsilon_{\mu\nu\alpha\beta}$ denotes the completely antisymmetric Levi-Civita tensor in 4-dimensions. It is now straightforward to verify that the commutation relations (11.137) are preserved under the transformations θ given in (11.182). We will see shortly that θ denotes an outer automorphism of $D_4 = so(8)$.

Let P denote an inner automorphism of $so(8)$ which interchanges the indices $1 \leftrightarrow 4$ and $2 \leftrightarrow 3$ and define

$$\tau = P\theta, \tag{11.183}$$

so that we have

$$X^\mu{}_\nu \xrightarrow{\tau} -X^{\bar\nu}{}_{\bar\mu} + \frac{1}{2}\delta^{\bar\nu}{}_{\bar\mu} \sum_{\lambda=1}^{4} X^\lambda{}_\lambda,$$

$$R^{\mu\nu} \xrightarrow{\tau} -\frac{1}{2} \sum_{\bar\alpha,\bar\beta=1}^{4} \epsilon_{\bar\mu\bar\nu\bar\alpha\bar\beta} R^{\bar\alpha\bar\beta},$$

$$R_{\mu\nu} \xrightarrow{\tau} -\frac{1}{2} \sum_{\bar\alpha,\bar\beta=1}^{4} \epsilon^{\bar\mu\bar\nu\bar\alpha\bar\beta} R_{\bar\alpha\bar\beta}, \tag{11.184}$$

where we have identified $\bar\mu = 5 - \mu$ so that $\bar{1} = 4, \bar{2} = 3, \bar{3} = 2, \bar{4} = 1$.

Finally, we introduce another outer automorphism π defined by

1. For $\mu \neq 4, \nu \neq 4$:

$$\pi(X^\mu{}_\nu) = X^\mu{}_\nu, \quad \pi(R^{\mu\nu}) = R^{\mu\nu}, \quad \pi(R_{\mu\nu}) = R_{\mu\nu}. \tag{11.185}$$

2. For $\mu = 4, \nu \neq 4$:

$$\pi(X^4{}_\nu) = R_{4\nu}, \quad \pi(R^{4\nu}) = R^{4\nu}, \quad \pi(R_{4\nu}) = X^4{}_\nu. \tag{11.186}$$

3. For $\mu \neq 4, \nu = 4$:

$$\pi(X^\mu{}_4) = -R^{\mu 4}, \quad \pi(R^{\mu 4}) = -X^\mu{}_4, \quad \pi(R_{\mu 4}) = R_{\mu 4}. \tag{11.187}$$

4. For $\mu = \nu = 4$:

$$\pi(X^4{}_4) = -X^4{}_4, \tag{11.188}$$

where we recall that by anti-symmetry $R^{44} = 0 = R_{44}$.

Let ϕ_0 be the highest weight state of an irreducible representation $\{\rho\}$ specified by (f_1, f_2, f_3, f_4). Then, we can verify that π defined by (11.185)-(11.188) leaves ϕ_0 invariant, but

$$f_j \overset{\pi}{\to} f_j \; (j \neq 4), \qquad\qquad f_4 \overset{\pi}{\to} -f_4, \qquad\qquad (11.189)$$

which implies

$$\sigma_j \overset{\pi}{\to} \sigma_j \; (j \neq 4), \qquad\qquad \sigma_4 \overset{\pi}{\to} -\sigma_4, \qquad\qquad (11.190)$$

where we have identified $\sigma_\mu = f_\mu + 4 - \mu, \mu = 1, 2, 3, 4$. It follows now from (11.159)-(11.160) that under the operation π,

$$\lambda_3 \overset{\pi}{\leftrightarrow} \lambda_4, \qquad\qquad\qquad (11.191)$$

while λ_1 and λ_2 are unaffected. In other words, π interchanges the simple roots $\alpha_3 \leftrightarrow \alpha_4$ in the Dynkin diagram Fig. 11.4.

Similarly, the action of τ defined in (11.183) induces the transformations

$$f_1 \overset{\tau}{\to} \bar{f}_1 = \frac{1}{2}(f_1 + f_2 + f_3 - f_4),$$

$$f_2 \overset{\tau}{\to} \bar{f}_2 = \frac{1}{2}(f_1 + f_2 - f_3 + f_4),$$

$$f_3 \overset{\tau}{\to} \bar{f}_3 = \frac{1}{2}(f_1 - f_2 + f_3 + f_4),$$

$$f_4 \overset{\tau}{\to} = \bar{f}_4 = \frac{1}{2}(-f_1 + f_2 + f_3 + f_4), \qquad (11.192)$$

or equivalently

$$\sigma_1 \overset{\tau}{\to} \bar{\sigma}_1 = \frac{1}{2}(\sigma_1 + \sigma_2 + \sigma_3 - \sigma_4),$$

$$\sigma_2 \overset{\tau}{\to} \bar{\sigma}_2 = \frac{1}{2}(\sigma_1 + \sigma_2 - \sigma_3 + \sigma_4),$$

$$\sigma_3 \overset{\tau}{\to} \bar{\sigma}_3 = \frac{1}{2}(\sigma_1 - \sigma_2 + \sigma_3 + \sigma_4),$$

$$\sigma_4 \overset{\tau}{\to} \bar{\sigma}_4 = \frac{1}{2}(-\sigma_1 + \sigma_2 + \sigma_3 + \sigma_4). \qquad (11.193)$$

Equation (11.193) can also be equivalently written as

$$\sigma_\mu \overset{\tau}{\to} \bar{\sigma}_\mu = \frac{1}{2}(\sigma_1 + \sigma_2 + \sigma_3 + \sigma_4) - \sigma_{5-\mu}, \qquad (11.194)$$

which preserves the ordering

$$\overline{\sigma}_1 > \overline{\sigma}_2 > \overline{\sigma}_3 > |\overline{\sigma}_4|. \tag{11.195}$$

In terms of λ_μs, this can be written as

$$\lambda_1 \overset{\tau}{\leftrightarrow} \lambda_3, \tag{11.196}$$

while leaving λ_2, λ_4 invariant. Therefore, the operation τ implies the permutation of the two simple roots α_1 and α_3 in the Dynkin diagram Fig. 11.4.

It is clear now that we can identify the $3! = 6$ elements of the symmetry group S_3 (see equations (1.13)-(1.16) in section1.2.1) associated with the Dynkin diagram Fig. 11.4 as

1. $P_1 = \mathbb{1} = $ identity,

2. $P_2 = \pi : \alpha_3 \leftrightarrow \alpha_4,$

3. $P_3 = \tau : \alpha_1 \leftrightarrow \alpha_3,$

4. $P_4 = \pi\tau\pi = \tau\pi\tau : \alpha_1 \leftrightarrow \alpha_4,$

5. $P_5 = \tau\pi : \alpha_1 \rightarrow \alpha_3 \rightarrow \alpha_4 \rightarrow \alpha_1,$

6. $P_6 = \pi\tau : \alpha_1 \rightarrow \alpha_4 \rightarrow \alpha_3 \rightarrow \alpha_1.$

Especially, from (11.189) and (11.192) we have

$$\pi : (1,0,0,0) \leftrightarrow (1,0,0,0); \quad (\frac{1}{2},\frac{1}{2},\frac{1}{2},\frac{1}{2}) \leftrightarrow (\frac{1}{2},\frac{1}{2},\frac{1}{2},-\frac{1}{2}),$$

$$\tau : (1,0,0,0) \leftrightarrow (\frac{1}{2},\frac{1}{2},\frac{1}{2},-\frac{1}{2});$$

$$(\frac{1}{2},\frac{1}{2},\frac{1}{2},\frac{1}{2}) \leftrightarrow (\frac{1}{2},\frac{1}{2},\frac{1}{2},\frac{1}{2}). \tag{11.197}$$

In other words, three 8-dimensional irreducible representations, corresponding to a vector, a spinor and a conjugate spinor, transform among themselves under the symmetry group S_3. This is the well known manifestation of the triality principle of $D_4 = so(8)$. We note that the dimension $d(\rho)$ in (11.163) is invariant under S_4 (the permutation group of four elements) and so is the second order Casimir invariant $I_2(\rho)$ in (11.164) since

$$\overline{\sigma}_1^2 + \overline{\sigma}_2^2 + \overline{\sigma}_3^2 + \overline{\sigma}_4^2 = \sigma_1^2 + \sigma_2^2 + \sigma_3^2 + \sigma_4^2, \tag{11.198}$$

as we may easily verify. However, this is not the case for $I_4(\rho)$ in (11.164) as well as for $\tilde{I}_4(\rho)$ in (11.170). For example, using (11.190) as well as (11.193) we obtain from (11.164) and (11.170) that

$$\pi : \; I_4(\rho) \to I_4(\rho), \quad \tilde{I}_4(\rho) \to -\tilde{I}_4(\rho),$$

$$\tau : \; I_4(\rho) \to -\frac{1}{2} I_4(\rho) - \frac{45}{2} \tilde{I}_4(\rho),$$

$$\tilde{I}_4(\rho) \to -\frac{1}{30} I_4(\rho) + \frac{1}{2} \tilde{I}_4(\rho). \tag{11.199}$$

Here we have used the relations

$$\sum_{\mu=1}^{4} (\bar{\sigma}_\mu)^4 = -\frac{1}{2} \sum_{\mu=1}^{4} (\sigma_\mu)^4 + \frac{3}{4} \left(\sum_{\mu=1}^{4} (\sigma_\mu)^2 \right)^2 - 6\sigma_1\sigma_2\sigma_3\sigma_4,$$

$$\bar{\sigma}_1\bar{\sigma}_2\bar{\sigma}_3\bar{\sigma}_4 = -\frac{1}{8} \sum_{\mu=1}^{4} (\sigma_\mu)^4 + \frac{1}{16} \left(\sum_{\mu=1}^{4} (\sigma_\mu)^2 \right)^2 + \frac{1}{2} \sigma_1\sigma_2\sigma_3\sigma_4.$$

$$\tag{11.200}$$

It is clear from (11.199) that $I_4(\rho)$ and $\tilde{I}_4(\rho)$ behave as a two-dimensional irreducible representation of the symmetric group S_3. In contrast, the 6-th order Casimir invariant $I_6(\rho)$ in (11.164) remains invariant under S_3 by the identity

$$4 \sum_{\mu=1}^{4} (\bar{\sigma}_\mu)^6 - 5 \left(\sum_{\mu=1}^{4} (\bar{\sigma}_\mu)^4 \right) \left(\sum_{\nu=1}^{4} (\bar{\sigma}_\nu)^2 \right)$$

$$= 4 \sum_{\mu=1}^{4} (\sigma_\mu)^6 - 5 \left(\sum_{\mu=1}^{4} (\sigma_\mu)^4 \right) \left(\sum_{\nu=1}^{4} (\sigma_\nu)^2 \right). \tag{11.201}$$

This is indeed expected since the sixth order Casimir invariant of $so(8)$, namely, $I_6(\rho)$, is unique. We note that our arguments show that the permutation symmetry group S_3 is an outer automorphism of the Lie algebra $so(8)$ since any inner automorphism will preserve any p-th order Casimir invariant $I_p(\rho)$. (Reference: S. Okubo, J. Math. Physics 23, 8 (1982))

11.4 Lie algebra $B_\ell = so(2\ell + 1)$:

We note here that the Lie algebra $D_\ell = so(2\ell)$ studied in the last section is a sub-Lie algebra of $B_\ell = so(2\ell + 1)$. The number of generators in B_ℓ correspond to $\ell(2\ell+1)$ while that in D_ℓ is $\ell(2\ell-1)$. Therefore, as in the case of $D_\ell = so(2\ell)$ (see (11.134)-(11.137)), we again change basis of the Lie algebra from $J_{jk}, j, k = 1, 2, \cdots, 2\ell+1$, to $X^\mu_\nu, R^{\mu\nu}, R_{\mu\nu}, \mu, \nu = 1, 2, \cdots, \ell$. The remaining 2ℓ generators are identified with

$$B_\mu = \frac{1}{2}\left(J_{\mu+\ell,2\ell+1} - iJ_{\mu,2\ell+1}\right),$$

$$B^\mu = -\frac{1}{2}\left(J_{\mu+\ell,2\ell+1} + iJ_{\mu,2\ell+1}\right), \quad \mu = 1, 2, \cdots, \ell. \tag{11.202}$$

In this case, therefore, in addition to the commutation relations given in (11.137), we also have commutation relations involving B_μ and B^μ. These can be obtained from the definitions in (11.134) and (11.202) and are given by

$$[X^\mu_\nu, B_\lambda] = -\delta^\mu_\lambda B_\nu,$$

$$[X^\mu_\nu, B^\lambda] = \delta^\lambda_\nu B^\mu,$$

$$[R^{\mu\nu}, B_\lambda] = -\delta^\mu_\lambda B^\nu + \delta^\nu_\lambda B^\mu,$$

$$[R_{\mu\nu}, B^\lambda] = \delta^\lambda_\mu B_\nu - \delta^\lambda_\nu B_\mu,$$

$$[R^{\mu\nu}, B^\lambda] = 0 = [R_{\mu\nu}, B_\lambda],$$

$$[B_\mu, B_\nu] = -2R_{\mu\nu},$$

$$[B^\mu, B^\nu] = 2R^{\mu\nu},$$

$$[B_\mu, B^\nu] = -2X^\nu_\mu. \tag{11.203}$$

The unitarity condition for the irreducible representation requires that

$$(B_\mu)^\dagger = B^\mu, \quad (B^\mu)^\dagger = B_\mu. \tag{11.204}$$

Furthermore, we note that the mutually commuting Cartan subalgebra still consists of the generators $H_\mu = X^\mu_\mu, \mu = 1, 2, \cdots, \ell$. However, in this case, we note from (11.203) that the generators B^μ also act as raising operators. On the other hand, we note that the set of raising operators $X^1_2, X^2_3, \cdots, X^{\ell-1}_\ell$ and B^ℓ can now generate all the raising operators. For example, we note from (11.203) that

$$B^{\ell-1} = [X^{\ell-1}_\ell, B^\ell],$$

$$R^{\ell-1,\ell} = \frac{1}{2}[B^{\ell-1}, B^\ell] = \frac{1}{2}[[X^{\ell-1}_\ell, B^\ell], B^\ell], \tag{11.205}$$

and so on. Therefore, the set of raising operators

$$X^1{}_2, X^2{}_3, \cdots, X^{\ell-1}{}_\ell, B^\ell, \tag{11.206}$$

can be thought of as fundamental and the simple root system will change when we take this fact into account.

As before, we introduce ℓ unit orthogonal vectors $\mathbf{e}_1, \mathbf{e}_2, \cdots, \mathbf{e}_\ell$ in an ℓ-dimensional Euclidean space and write (see (11.155))

$$
\begin{aligned}
X^\mu{}_\nu &= E_{(e_\mu - e_\nu)}, & \mu \neq \nu, \\
R^{\mu\nu} &= E_{(e_\mu + e_\nu)}, & \mu < \nu, \\
R_{\mu\nu} &= E_{-(e_\mu + e_\nu)}, & \mu < \nu, \tag{11.207}
\end{aligned}
$$

as well as

$$B^\mu = E_{e_\mu}, \quad B_\mu = E_{-e_\mu}. \tag{11.208}$$

Correspondingly, the simple root system (see (11.156), (11.157) as well as (11.206)), in this case, is given by

$$\boldsymbol{\alpha}_1 = \mathbf{e}_1 - \mathbf{e}_2, \quad \boldsymbol{\alpha}_2 = \mathbf{e}_2 - \mathbf{e}_3, \quad \cdots, \boldsymbol{\alpha}_{\ell-1} = \mathbf{e}_{\ell-1} - \mathbf{e}_\ell,$$

$$\boldsymbol{\alpha}_\ell = \mathbf{e}_\ell. \tag{11.209}$$

As a consequence, the fundamental weight system is given by (compare with (11.172) and (11.173))

$$\boldsymbol{\Lambda}_1 = \mathbf{e}_1, \; \boldsymbol{\Lambda}_2 = \mathbf{e}_1 + \mathbf{e}_2, \; \cdots, \boldsymbol{\Lambda}_{\ell-1} = \mathbf{e}_1 + \mathbf{e}_2 + \cdots + \mathbf{e}_{\ell-1},$$

$$\boldsymbol{\Lambda}_\ell = \frac{1}{2}(\mathbf{e}_1 + \mathbf{e}_2 + \cdots + \mathbf{e}_\ell). \tag{11.210}$$

The highest weight vector can now be written in the form (see (11.171))

$$
\begin{aligned}
\Lambda &= f_1 \mathbf{e}_1 + f_2 \mathbf{e}_2 + \cdots + f_\ell \mathbf{e}_\ell \\
&= \lambda_1 \boldsymbol{\Lambda}_1 + \lambda_2 \boldsymbol{\Lambda}_2 + \cdots + \lambda_\ell \boldsymbol{\Lambda}_\ell, \tag{11.211}
\end{aligned}
$$

with

$$
\begin{aligned}
f_1 &= \lambda_1 + \lambda_2 + \lambda_3 + \cdots + \lambda_{\ell-1} + \tfrac{1}{2}\lambda_\ell, \\
f_2 &= \lambda_2 + \lambda_3 + \cdots + \lambda_{\ell-1} + \tfrac{1}{2}\lambda_\ell, \\
&\;\;\vdots \\
f_{\ell-1} &= \lambda_{\ell-1} + \tfrac{1}{2}\lambda_\ell, \\
f_\ell &= \tfrac{1}{2}\lambda_\ell,
\end{aligned}
\tag{11.212}
$$

The f_ℓs are again all integers or half integers satisfying

$$f_1 \geq f_2 \geq \cdots \geq f_{\ell-1} \geq f_\ell \geq 0. \tag{11.213}$$

We note that f_ℓ does not take negative value in this case consistent with the fact that $so(2\ell+1)$ does not have any self-dual representation in contrast to $so(2\ell)$.

As in (11.162), we define

$$\sigma_\mu = f_\mu + \frac{N}{2} - \mu = f_\mu + \frac{2\ell+1}{2} - \mu, \tag{11.214}$$

with $\mu = 1, 2, \cdots, \ell$. With this, the p-th order Casimir invariant of $so(2\ell + 1)$ can be checked to have the same form as in (11.164) for $so(2\ell)$, but with $N = 2\ell + 1$. However, Weyl's formula for the dimension of a representation now has the form (see (11.163))

$$d(\rho) = \frac{2^\ell}{(1!)(3!)\cdots((2\ell-1)!)} \left(\prod_{\mu=1}^\ell \sigma_\mu\right) \prod_{\mu<\nu}^\ell \left((\sigma_\mu)^2 - (\sigma_\nu)^2\right). \tag{11.215}$$

$$\alpha_1 \qquad \alpha_2 \qquad \alpha_3 \qquad \cdots\cdots \qquad \alpha_{\ell-1} \qquad \alpha_\ell$$

Figure 11.5: The Dynkin diagram for the simple roots of $B_\ell = so(2\ell + 1)$.

The Dynkin diagram for $B_\ell = so(2\ell + 1)$ is shown in Fig. 11.5. Here we have represented the root $\alpha_\ell = e_\ell$ (see (11.209)) by a black vertex (dot) to distinguish it from the other roots since $(\alpha_\ell, \alpha_\ell) = 1$ while $(\alpha_j, \alpha_j) = 2, j \neq \ell$. Furthermore, conventionally the number of lines (bonds) connecting two neighboring roots α_j and α_{j+1} is determined as (see (11.54))

$$n_{j,j+1} = \left|\frac{2(\alpha_j, \alpha_{j+1})}{\sqrt{(\alpha_j, \alpha_j)(\alpha_{j+1}, \alpha_{j+1})}}\right|^2 = n_{j+1,j}. \tag{11.216}$$

Therefore, it follows that the number of lines (bonds) between the roots $\alpha_{\ell-1}$ and α_ℓ in Fig. 11.5 is given by

$$n_{\ell-1,\ell} = \left|\frac{2(\alpha_{\ell-1}, \alpha_\ell)}{\sqrt{(\alpha_{\ell-1}, \alpha_{\ell-1})(\alpha_\ell, \alpha_\ell)}}\right|^2 = \left|\frac{2(-1)}{\sqrt{(2)(1)}}\right|^2 = 2. \tag{11.217}$$

11.5 Lie algebra $C_\ell = sp(2\ell)$:

Let us consider a set of anti-symmetric constants

$$\epsilon_{jk} = -\epsilon_{kj}, \quad j, k = 1, 2, \cdots, 2\ell. \tag{11.218}$$

Then, the Lie algebra $sp(2\ell)$ is defined by the generators J_{jk} satisfying

$$(i) J_{jk} = J_{kj},$$
$$(ii) [J_{jk}, J_{\ell m}] = \epsilon_{j\ell} J_{km} + \epsilon_{km} J_{j\ell} + \epsilon_{jm} J_{k\ell} + \epsilon_{k\ell} J_{jm}, \tag{11.219}$$

where $j, k, \ell, m = 1, 2, \cdots, 2\ell$.

However, the form of the algebra in (11.219) is not very convenient for the discussion of its representations. So, we change basis and define new generators in the Cartan basis

$$S^{\mu\nu} = S^{\nu\mu} = -\frac{1}{2} \left((J_{\mu\nu} - J_{\mu+\ell,\nu+\ell}) - i(J_{\mu,\nu+\ell} + J_{\mu+\ell,\nu}) \right),$$

$$S_{\mu\nu} = -\frac{1}{2} \left((J_{\mu\nu} - J_{\mu+\ell,\nu+\ell}) + i(J_{\mu,\nu+\ell} + J_{\mu+\ell,\nu}) \right),$$

$$X^\mu_{\ \nu} = \frac{i}{2} \left((J_{\mu\nu} + J_{\mu+\ell,\nu+\ell}) + i(J_{\mu,\nu+\ell} - J_{\mu+\ell,\nu}) \right), \tag{11.220}$$

where $\mu, \nu = 1, 2, \cdots, \ell$. Furthermore, let us choose the anti-symmetric constants, ϵ_{jk}, of the form

$$\epsilon_{\mu,\mu+\ell} = -\epsilon_{\mu+\ell,\mu} = 1, \quad \epsilon_{jk} = 0, \quad \text{otherwise.} \tag{11.221}$$

With these, we can now calculate the commutation relations in the new basis using (11.219) which leads to

$$[X^\mu_{\ \nu}, X^\alpha_{\ \beta}] = \delta^\alpha_{\ \nu} X^\mu_{\ \beta} - \delta^\mu_{\ \beta} X^\alpha_{\ \nu},$$

$$[X^\mu_{\ \nu}, S_{\alpha\beta}] = -\delta^\mu_{\ \alpha} S_{\nu\beta} - \delta^\mu_{\ \beta} S_{\alpha\nu},$$

$$[X^\mu_{\ \nu}, S^{\alpha\beta}] = \delta^\alpha_{\ \nu} S^{\mu\beta} + \delta^\beta_{\ \nu} S^{\alpha\mu},$$

$$[S_{\mu\nu}, S^{\alpha\beta}] = -\delta^\alpha_{\ \mu} X^\beta_{\ \nu} - \delta^\alpha_{\ \nu} X^\beta_{\ \mu} - \delta^\beta_{\ \mu} X^\alpha_{\ \nu} - \delta^\beta_{\ \nu} X^\alpha_{\ \mu},$$

$$[S_{\mu\nu}, S_{\alpha\beta}] = 0 = [S^{\mu\nu}, S^{\alpha\beta}], \tag{11.222}$$

where $\mu, \nu, \alpha, \beta = 1, 2, \cdots, \ell$.

Let us next define

$$J_3 = \frac{1}{2} X^1_{\ 1}, \quad J_+ = \frac{1}{2} S^{11}, \quad J_- = \frac{1}{2} S_{11}. \tag{11.223}$$

It follows now from (11.222) that these three generators define a $su(2)$ Lie algebra given by

$$[J_3, J_\pm] = \pm J_\pm, \quad [J_+, J_-] = 2J_3, \tag{11.224}$$

so that the eigenvalues of X^1_1 are all integers. Similarly, we can show that the eigenvalues of each of the generators $H_\mu = X^\mu_\mu, \mu = 1, 2, \cdots, \ell$ are all integers. We can now define the highest weight vector ϕ_Λ to satisfy

$$H_\mu \phi_\Lambda = X^\mu_\mu \phi_\Lambda = f_\mu \phi_\Lambda, \quad \mu = 1, 2, \cdots, \ell, \tag{11.225}$$

with the (integer) eigenvalues satisfying

$$f_1 \geq f_2 \geq \cdots \geq f_\ell \geq 0. \tag{11.226}$$

To obtain the root system, we again introduce an ℓ-dimensional Euclidean space spanned by the orthonormal vectors $\mathbf{e}_1, \mathbf{e}_2, \cdots, \mathbf{e}_\ell$. We note that we can identify the fundamental raising operators of the system to consist of

$$E_{\alpha_1} = X^1_2, \; E_{\alpha_2} = X^2_3, \; \cdots, E_{\alpha_{\ell-1}} = X^{\ell-1}_\ell, \; E_{\alpha_\ell} = \frac{1}{2} S^{\ell\ell}, \tag{11.227}$$

which can be seen (as in (11.205)) to generate all the other raising operators X^μ_ν with $\mu < \nu$ as well as $\frac{1}{2}S^{\mu\nu}$. Here we have identified the simple roots with

$$\alpha_1 = \mathbf{e}_1 - \mathbf{e}_2, \quad \alpha_2 = \mathbf{e}_2 - \mathbf{e}_3, \quad \cdots, \alpha_{\ell-1} = \mathbf{e}_{\ell-1} - \mathbf{e}_\ell,$$

$$\alpha_\ell = 2\mathbf{e}_\ell. \tag{11.228}$$

This allows us to write the fundamental weight system as

$$\Lambda_1 = \mathbf{e}_1, \quad \Lambda_2 = \mathbf{e}_1 + \mathbf{e}_2, \quad \Lambda_{\ell-1} = \mathbf{e}_1 + \mathbf{e}_2 + \cdots + \mathbf{e}_{\ell-1},$$

$$\Lambda_\ell = \mathbf{e}_1 + \mathbf{e}_2 + \cdots + \mathbf{e}_{\ell-1} + \mathbf{e}_\ell, \tag{11.229}$$

so that the highest weight vector can be written in this case in the form (see also (11.211))

$$\Lambda = f_1 \mathbf{e}_1 + f_2 \mathbf{e}_2 + f_3 \mathbf{e}_3 + \cdots + f_\ell \mathbf{e}_\ell$$

$$= \lambda_1 \Lambda_1 + \lambda_2 \Lambda_2 + \lambda_3 \Lambda_3 + \cdots + \lambda_\ell \Lambda_\ell, \tag{11.230}$$

with

$$
\begin{aligned}
f_1 &= \lambda_1 + \lambda_2 + \lambda_3 + \cdots + \lambda_{\ell-1} + \lambda_\ell, \\
f_2 &= \lambda_2 + \lambda_3 + \cdots + \lambda_{\ell-1} + \lambda_\ell, \\
&\vdots \\
f_{\ell-1} &= \lambda_{\ell-1} + \lambda_\ell, \\
f_\ell &= \lambda_\ell,
\end{aligned}
\tag{11.231}
$$

Correspondingly, the Dynkin diagram for $C_\ell = sp(2\ell)$ is shown in Fig. 11.6.

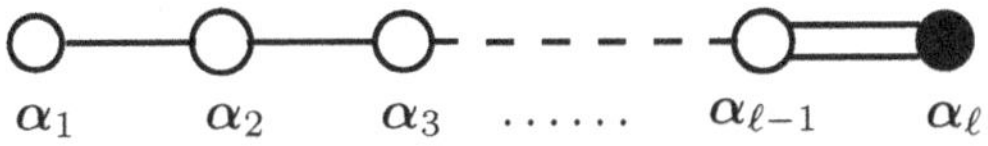

Figure 11.6: The Dynkin diagram for the simple roots of $C_\ell = sp(2\ell)$.

Let us next define

$$
\sigma_\mu = f_\mu + \ell - \mu + 1, \quad \mu = 1, 2, \cdots, \ell.
\tag{11.232}
$$

With this, Weyl's formula for the dimension of a representation (compare with (11.215)) is given by

$$
d(\rho) = \frac{1}{(1!)(3!)\cdots((2\ell-1)!)} \left(\prod_{\mu=1}^{\ell} \sigma_\mu \right) \prod_{\mu<\nu}^{\ell} \left((\sigma_\mu)^2 - (\sigma_\nu)^2 \right).
\tag{11.233}
$$

Moreover the eigenvalues of the Casimir invariants, in a given representation, can also be computed to have the forms

$$
I_2(\rho) = \sum_{\mu=1}^{\ell} \left((\sigma_\mu)^2 - (\sigma_\mu^{(0)})^2 \right),
$$

$$
I_4(\rho) = (2\ell^2 + \ell + 2) \sum_{\mu=1}^{\ell} \left((\sigma_\mu)^4 - (\sigma_\mu^{(0)})^4 \right)
$$

$$
- (4\ell + 1) \left(\left(\sum_{\mu=1}^{\ell} (\sigma_\mu)^2 \right)^2 - \left(\sum_{\mu=1}^{\ell} (\sigma_\mu^{(0)})^2 \right)^2 \right),
$$

$$I_6(\rho) = (2\ell^2 + \ell + 4)(2\ell^2 + \ell + 8) \sum_{\mu=1}^{\ell} \left((\sigma_\mu)^6 - (\sigma_\mu^{(0)})^6 \right)$$

$$- 3(4\ell + 1)(2\ell^2 + \ell + 4) \left[\left(\sum_{\mu=1}^{\ell} (\sigma_\mu)^2 \right) \left(\sum_{\nu=1}^{\ell} (\sigma_\nu)^4 \right) \right.$$

$$\left. - \left(\sum_{\mu=1}^{\ell} (\sigma_\mu^{(0)})^2 \right) \left(\sum_{\nu=1}^{\ell} (\sigma_\nu^{(0)})^4 \right) \right]$$

$$+ 14(2\ell^2 + \ell - 1) \left(\left(\sum_{\mu=1}^{\ell} (\sigma_\mu)^2 \right)^3 - \left(\sum_{\mu=1}^{\ell} (\sigma_\mu^{(0)})^2 \right)^3 \right),$$

$$\tag{11.234}$$

where we have identified (see (11.232))

$$\sigma_\mu^{(0)} = \ell - \mu + 1. \tag{11.235}$$

We note that these formulae are exactly the same as for the Casimir invariants in the tensor representations of $B_\ell = so(2\ell + 1)$ (see discussion around (11.214)) if we let

$$(\sigma_\mu, \sigma_\mu^{(0)}) \to (\sigma_\mu - \frac{1}{2}, \sigma_\mu^{(0)} - \frac{1}{2}), \tag{11.236}$$

in (11.234). The reason for this is not well understood.

Finally, we note that for the case of $C_2 = sp(4)$, the Dynkin diagram in Fig. 11.6 is equivalent to that in Fig. 11.5 for $B_2 = so(5)$ so that we must have $C_2 = B_2$. Therefore, in discussions of C_ℓ, one normally restricts to $\ell \geq 3$. The explicit equivalence of $sp(4)$ and $so(5)$ can be shown as follows. Let $X^\mu_\nu, S_{\mu\nu} = S_{\nu\mu}, S^{\mu\nu} = S^{\nu\mu}, \mu, \nu = 1, 2$, be the generators of $sp(4)$. Let us define a new basis given by

$$\widetilde{X}^1_1 = \frac{1}{2}(X^1_1 + X^2_2), \quad \widetilde{X}^2_2 = \frac{1}{2}(X^2_2 - X^1_1),$$

$$\widetilde{X}^1_2 = -\frac{1}{2}S^{11}, \quad \widetilde{X}^2_1 = -\frac{1}{2}S_{11},$$

$$\widetilde{R}^{12} = -\widetilde{R}^{21} = -\frac{1}{2}S^{22}, \quad \widetilde{R}_{12} = -\widetilde{R}_{21} = -\frac{1}{2}S_{22},$$

$$\widetilde{B}^1 = S^{12}, \quad \widetilde{B}_1 = S_{12}, \quad \widetilde{B}^2 = X^2_1, \quad \widetilde{B}_2 = X^1_2. \tag{11.237}$$

Then, it can be verified that the generators $\widetilde{X}^\mu_\nu, \widetilde{R}^{\mu\nu} = -\widetilde{R}^{\nu\mu}, \widetilde{R}_{\mu\nu} = -\widetilde{R}_{\nu\mu}, \widetilde{B}^\mu$ and $\widetilde{B}_\mu$, for $\mu, \nu = 1, 2$ relaize the commutation relations of the Lie algebra $B_2 = so(5)$ given in (11.137) and (11.203).

11.6 Exceptional Lie algebras

In addition to these classical Lie algebras, there are also Lie algebras known as exceptional Lie algebras conventionally known as the G_2, F_4, E_6, E_7 and E_8. We note here that the subscripts in the case of exceptional algebras correspond to their ranks (or equivalently, the number of simple roots). The standard manner of treating these exceptional algebras is through the Cartan-Dynkin method discussed in section 11.1. However, we will discuss them in the following way.

Let L denote any Lie algebra with a simple or semi-simple Lie sub-algebra L_0, namely,

$$[L_0, L_0] \subseteq L_0, \quad [L_0, L] \subseteq L. \tag{11.238}$$

We can interpret (11.238) as follows. We can think of L as a vector space and denote the action of L_0 on L through the relation

$$a(x) = [a, x], \quad \text{for} \quad a \in L_0, \ x \in L. \tag{11.239}$$

This, then, identifies L as a representation space of L_0. Furthermore, since we have assumed L_0 to be simple or semi-simple and since any arbitrary representation of L_0 is fully reducible by the Peter-Weyl theorem (see section 2.4), we can decompose the representation vector space L as a direct sum of irreducible components of L_0 as

$$L = L_0 \oplus \sum_i V_i, \tag{11.240}$$

where V_i denote the irreducible components of the representations of L_0 so that

$$[L_0, V_i] \subseteq V_i. \tag{11.241}$$

Furthermore, since L is a Lie algebra ($[L, L] \subseteq L$), it follows from the decomposition (11.240) that

$$[V_i, V_j] = x_{ij} + \sum_k \beta_{ijk} V_k, \quad x_{ij} \in L_0, \tag{11.242}$$

for some constant coefficients x_{ij}, β_{ijk}. The commutation relations of any Lie algebra L can be written in this way. Here we apply this method to exceptional Lie algebras since their Lie sub-algebras have been well studied in mathematical literature (for example, see the book by McKay, Patera and Rand).

11.6.1 G_2. As a vector space, the Lie algebra $L = G_2$ is known to be 14 dimensional. Furthermore, it is known that G_2 is known to have a maximal Lie sub-algebra $L_0 = A_2 = su(3)$ such that $L = G_2$ can be reduced to a direct sum of irreducible representations of $L_0 = A_2 = su(3)$ as

$$14 = 8 + 3 + \overline{3}. \tag{11.243}$$

As we have seen, the octet is the adjoint (tensor) representation of $su(3)$ while the other two correspond to the fundamental representation and its conjugate representation. As a result, symbolically we may write

$$G_2 = \{A^\mu{}_\nu\} \oplus \{A_\mu\} \oplus \{A^\mu\}, \quad \mu, \nu = 1, 2, 3. \tag{11.244}$$

Explicitly $A^\mu{}_\nu$, A_μ and A^μ satisfy the $su(3)$ relations (see, for example, (11.88))

$$\sum_{\mu=1}^{3} A^\mu{}_\mu = 0,$$

$$[A^\mu{}_\nu, A^\alpha{}_\beta] = \delta^\alpha{}_\nu A^\mu{}_\beta - \delta^\mu{}_\beta A^\alpha{}_\nu,$$

$$[A^\mu{}_\nu, A_\lambda] = -\delta^\mu{}_\lambda A_\nu + \frac{1}{3} \delta^\mu{}_\nu A_\lambda,$$

$$[A^\mu{}_\nu, A^\lambda] = \delta^\lambda{}_\nu A^\mu - \frac{1}{3} \delta^\mu{}_\nu A^\lambda. \tag{11.245}$$

Let us next consider the commutator $[A_\mu, A_\nu]$. As far as its tensor property is concerned, we can think of the commutator as coming from the direct product $3 \otimes 3$ of $su(3)$. On the other hand, let us recall the decomposition (see (5.93))

$$\square \otimes \square = \boxed{} \oplus \begin{array}{c}\square\\\square\end{array}, \tag{11.246}$$

for any $su(N)$. For the present case, the horizontal blocks (the first term) on the right hand side represent $(2, 0, 0) = 6$ while the vertical blocks (the second term) correspond to $(1, 1, 0) = (0, 0, -1) = \overline{3}$, coming from the decomposition $3 \otimes 3 = 6 + \overline{3}$ (see, for example, (5.106)). However, for the present problem where $L = G_2$, the exceptional algebra G_2 does not contain a 6-dimensional representation of $su(3)$ (see (11.243)). It follows, therefore, that we can write

$$[A_\mu, A_\nu] = -2 \sum_{\lambda=1}^{3} \epsilon_{\mu\nu\lambda} A^\lambda, \quad \mu, \nu = 1, 2, 3, \tag{11.247}$$

where $\epsilon_{\mu\nu\lambda}$ denotes the three dimensional Levi-Civita tensor. We also note here that the multiplicative constant (-2) on the right hand side in (11.247) has been chosen for later convenience (the normalization is not relevant). Similarly, we can write

$$[A^\mu, A^\nu] = 2 \sum_{\lambda=1}^{3} \epsilon^{\mu\nu\lambda} A_\lambda. \tag{11.248}$$

Finally, we must also determine the commutator $[A^\mu, A_\nu]$. We note that with respect to $su(3)$, this commutator has the behavior coming from the decomposition

$$\overline{3} \otimes 3 = 8 \oplus 1. \tag{11.249}$$

However, G_2 does not contain a singlet component of $su(3)$ (see (11.243)) so that we must have

$$[A^\mu, A_\nu] = 3A^\mu{}_\nu, \quad \mu, \nu = 1, 2, 3. \tag{11.250}$$

We note here that the multiplicative constant (3) on the right hand side is not arbitrary, rather it is determined from the Jacobi identity which must be satisfied by the commutators. This completes the construction of the Lie algebra G_2.

Carrying out the analysis as in earlier sections, we can show that the Dynkin diagram for G_2 takes the simple form shown in Fig. 11.7. (Recall that $su(3)$ has rank $\ell = 2$ and, therefore, there are only two simple roots.)

$$\alpha_1 \qquad\qquad \alpha_2$$

Figure 11.7: The Dynkin diagram for the simple roots of G_2.

For G_2, if we identify

$$J_3 = \frac{3}{2} A^2{}_2, \quad J_+ = A^2, \quad J_- = A_2, \tag{11.251}$$

then we can verify (see (11.245) as well as (11.250)) that they satisfy the $su(2)$ algebra

$$[J_3, J_\pm] = \pm J_\pm, \quad [J_+, J_-] = 2J_3. \tag{11.252}$$

As a result, we may choose the two generators of the Cartan algebra as

$$H_{\alpha_1} = \frac{3}{2}(A^1{}_1 - A^2{}_2), \quad H_{\alpha_2} = \frac{3}{2}A^2{}_2, \tag{11.253}$$

as well as the fundamental raising and lowering operators to be

$$E_{\alpha_1} = \sqrt{\frac{3}{2}}A^1{}_2, \qquad\qquad E_{\alpha_2} = \sqrt{\frac{1}{2}}A^2,$$

$$E_{-\alpha_1} = \sqrt{\frac{3}{2}}A^2{}_1, \qquad\qquad E_{-\alpha_2} = \sqrt{\frac{1}{2}}A_2. \tag{11.254}$$

We can now calculate directly and conclude (see (11.32)) that

$$[H_{\alpha_1}, E_{\alpha_1}] = 3E_{\alpha_1}, \qquad\qquad \text{leading to} \quad (\alpha_1, \alpha_1) = 3,$$

$$[H_{\alpha_2}, E_{\alpha_2}] = E_{\alpha_2}, \qquad\qquad \text{leading to} \quad (\alpha_2, \alpha_2 = 2,$$

$$[H_{\alpha_1}, E_{\alpha_2}] = -\frac{3}{2}E_{\alpha_2}, \qquad \text{leading to} \quad (\alpha_1, \alpha_2) = -\frac{3}{2},$$

$$[H_{\alpha_2}, E_{\alpha_1}] = -\frac{3}{2}E_{\alpha_1},$$

$$[E_{\alpha_1}, E_{-\alpha_1}] = H_{\alpha_1}, \qquad\qquad [E_{\alpha_2}, E_{-\alpha_2}] = H_{\alpha_2}. \tag{11.255}$$

This explains the reason for the peculiar Dynkin diagram for G_2 in Fig. 11.7.

The general irreducible representation of G_2 is described by two non-negative integers $\lambda_1, \lambda_2 \geq 0$. We define

$$\sigma_1 = \lambda_1 + \frac{1}{3}\lambda_2 + \frac{4}{3} = (\lambda_1 + 1) + \frac{1}{3}(\lambda_2 + 1),$$

$$\sigma_2 = \frac{1}{3}\lambda_2 + \frac{1}{3} = \frac{1}{3}(\lambda_2 + 1),$$

$$\sigma_3 = -\lambda_1 - \frac{2}{3} - \frac{5}{3} = -(\lambda_1 + 1) - \frac{2}{3}(\lambda_2 + 1), \tag{11.256}$$

so that $\sigma_1 \geq \sigma_2 \geq \sigma_3$ and $\sigma_1 + \sigma_2 + \sigma_3 = 0$. With these, the formula for the dimension of a representation is given by

$$d(\rho) = -\frac{9}{40}\sigma_1\sigma_2\sigma_3(\sigma_1 - \sigma_2)(\sigma_2 - \sigma_3)(\sigma_1 - \sigma_3)$$

$$= -\frac{9}{40}\left(-\frac{1}{27}(\lambda_2 + 1)(3\lambda_1 + \lambda_2 + 4)(3\lambda_1 + 2\lambda_2 + 5)\right)$$

$$\times (\lambda_1 + 1)(\lambda_1 + \lambda_2 + 2)(2\lambda_1 + \lambda_2 + 3)$$

$$= \frac{1}{120}(\lambda_1 + 1)(\lambda_2 + 1)(\lambda_1 + \lambda_2 + 2)(2\lambda_1 + \lambda_2 + 3)$$

$$\times (3\lambda_1 + \lambda_2 + 4)(3\lambda_1 + 2\lambda_2 + 5). \tag{11.257}$$

The second and the sixth order Casimir invariants can also be written as

$$I_2(\rho) = \sum_{\mu=1}^{3}(\sigma_\mu)^2 - \frac{14}{3},$$

$$I_6(\rho) = -3\sum_{\mu=1}^{3}(\sigma_\mu)^6 + 10\left(\sum_{\mu=1}^{3}(\sigma_\mu)^3\right)^2 + \frac{286}{9}. \tag{11.258}$$

We recall here that G_2 does not have any third, fourth and fifth order Casimir invariants (see discussion in the paragraph following (10.101)). The only nontrivial invariants are I_2, I_6 given in (11.258).

11.6.2 F_4. The exceptional Lie algebra $L = F_4$ is known to be 52-dimensional and contains the 36-dimensional Lie sub-algebra $L_0 = B_4 = so(9)$. As the representation space of $so(9)$, it decomposes as

$$F_4 = so(9) \oplus \Lambda_4, \quad \text{equivalently}, \quad 52 = 36 \oplus 16, \tag{11.259}$$

where Λ_4 represents the 16-dimensional spinor representation of $so(9)$.
Let X_{jk} denote the generators of $so(9)$ satisfying (see (11.134))

$$X_{jk} = -X_{kj},$$

$$[X_{jk}, X_{\ell m}] = -\delta_{j\ell}J_{km} - \delta_{km}J_{j\ell} + \delta_{jm}J_{k\ell} + \delta_{k\ell}J_{jm}, \tag{11.260}$$

with $j, k, \ell, m = 1, 2, \cdots, 9$. Let $\psi_\mu, \mu = 1, 2, \cdots, 16$ represent the spinor of $so(9)$. Then, the rest of the commutation relations of F_4 are given by

$$[X_{jk}, \psi_\mu] = \sum_{\nu=1}^{16}\psi_\nu(\sigma_{jk})_{\nu\mu},$$

$$[\psi_\mu, \psi_\nu] = 2\sum_{j,k=1}^{9}(C\sigma_{jk})_{\mu\nu}X_{jk}, \tag{11.261}$$

for $j, k = 1, 2, \cdots, 9$ and $\mu, \nu = 1, 2, \cdots, 16$. Here σ_{jk} denote the generators of $so(9)$ (see (6.103) where it is labelled $J_{\mu\nu}$)

$$\sigma_{jk} = \frac{1}{4}[\gamma_j, \gamma_k], \tag{11.262}$$

in terms of the generators of Clifford algebra γ_j satisfying (see (6.1))

$$\gamma_j \gamma_k + \gamma_k \gamma_j = 2\delta_{jk}\, \mathbb{1}, \quad j,k = 1,2,\cdots,9. \tag{11.263}$$

Furthermore, C represents the charge conjugation matrix associated with the Clifford algebra in 9-dimensions satisfying (see (6.70) as well as (6.72))

$$C^{-1}\gamma_j C = \gamma_j^T, \quad C^T = C, \tag{11.264}$$

with T denoting the transpose of the matrix. The validity of the Jacobi identity

$$[X_{jk}, [\psi_\mu, \psi_\nu]] + [\psi_\mu, [\psi_\nu, X_{jk}]] + [\psi_\nu, [X_{jk}, \psi_\mu]] = 0, \tag{11.265}$$

is guaranteed by the anti-symmetry under charge conjugation

$$(C\sigma_{jk})_{\mu\nu} = -(C\sigma_{jk})_{\nu\mu}, \tag{11.266}$$

while

$$[\psi_\mu, [\psi_\nu, \psi_\lambda]] + [\psi_\nu, [\psi_\lambda, \psi_\mu]] + [\psi_\lambda, [\psi_\mu, \psi_\nu]] = 0, \tag{11.267}$$

follows from the identity

$$\sum_{j,k=1}^{9} [(C\sigma_{jk})_{\mu\nu}(\sigma_{jk})_{\alpha\lambda} + (C\sigma_{jk})_{\nu\lambda}(\sigma_{jk})_{\alpha\mu} + (C\sigma_{jk})_{\lambda\mu}(\sigma_{jk})_{\alpha\nu}] = 0, \tag{11.268}$$

for $\mu, \nu, \alpha, \lambda = 1, 2, \cdots, 16$. We note here that the identity (11.268) can be obtained from the Fierz identity

$$2\sum_{j,k=1}^{9} (C\sigma_{jk})_{\mu\nu}(\sigma_{jk})_{\alpha\lambda}$$

$$= \sum_{j=1}^{9} [(C\gamma_j)_{\lambda\nu}(\gamma_j)_{\alpha\mu} - (C\gamma_j)_{\lambda\mu}(\gamma_j)_{\alpha\nu}] + 3\,(C_{\lambda\nu}\delta_{\alpha\mu} - C_{\lambda\mu}\delta_{\alpha\nu}), \tag{11.269}$$

which, in turn, comes from the ortho-normal relations of the γ_j matrices. To conclude this discussion, we simply note that the Dynkin diagram for F_4 is given by Fig. 11.8. (Recall that $B_4 = so(9)$ is a rank $\ell = 4$ algebra.)

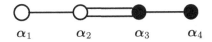

$$\alpha_1 \qquad \alpha_2 \qquad \alpha_3 \qquad \alpha_4$$

Figure 11.8: The Dynkin diagram for the simple roots of F_4.

11.6.3 E_6. The exceptional Lie algebra $L = E_6$ is known to be 78-dimensional and contains $L_0 = A_1 \otimes A_5 = su(2) \otimes su(6)$ as its maximal Lie sub-algebra. Therefore, as a representation space, E_6 decomposes as

$$78 = 1 \otimes 35 + 3 \otimes 1 + 2 \otimes 20 = 35 + 3 + 40, \qquad (11.270)$$

and it can be thought of as being spanned by the generators $X^j{}_k, j, k = 1, 2, \cdots, 6$, $B^\mu{}_\nu, \mu, \nu = 1, 2$ and $A_{jk\ell,\mu}$ where

1. $X^j{}_k \subset A_5 = su(6)$ and belongs to the space $1 \otimes 35$ in (11.270). These generators satisfy the relations (see (11.88))

$$\sum_{j=1}^{6} X^j{}_j = 0,$$

$$[X^j{}_k, X^\ell{}_m] = \delta^\ell{}_k X^j{}_m - \delta^j{}_m X^\ell{}_k, \qquad (11.271)$$

with $j, k, \ell, m = 1, 2, \cdots, 6$.

2. $B^\mu{}_\nu \subset A_1 = su(2)$ and belongs to the space $3 \otimes 1$ in (11.270). These generators satisfy the relations (see (11.88))

$$\sum_{\mu=1}^{2} B^\mu{}_\mu = 0,$$

$$[B^\mu{}_\nu, B^\alpha{}_\beta] = \delta^\alpha{}_\nu B^\mu{}_\beta - \delta^\mu{}_\beta B^\alpha{}_\nu, \qquad (11.272)$$

with $\mu, \nu, \alpha, \beta = 1, 2$.

3. the tensor $A_{jk\ell,\mu}$ is totally anti-symmetric in $j, k, \ell = 1, 2, \cdots, 6$ for any $\mu = 1, 2$ and belongs to the space $2 \otimes 20$ in (11.270).

The rest of the commutation relations between the generators are given by

$$[B^\mu_{\ \nu}, A_{jk\ell,\lambda}] = -\delta^\mu_{\ \lambda} A_{jk\ell,\nu} + \frac{1}{2} \delta^\mu_{\ \nu} A_{jk\ell,\lambda},$$

$$[X^j_{\ k}, A_{\ell mn,\mu}] = -\delta^j_{\ \ell} A_{kmn,\mu} - \delta^j_{\ m} A_{\ell kn,\mu} - \delta^j_{\ n} A_{\ell mk,\mu}$$
$$+ \frac{1}{2} \delta^j_{\ k} A_{\ell mn,\mu},$$

$$[A_{jk\ell,\mu}, A_{abc,\nu}] = -\epsilon_{jk\ell abc} \sum_{\lambda=1}^{2} (\epsilon_{\mu\lambda} B^\lambda_{\ \nu} + \epsilon_{\nu\lambda} B^\lambda_{\ \mu})$$

$$- \epsilon_{\mu\nu} \sum_{p=1}^{6} \left(\epsilon_{jk\ell abp} X^p_{\ c} + \epsilon_{jk\ell apc} X^p_{\ b} + \epsilon_{jk\ell pbc} X^p_{\ a} \right.$$

$$\left. - \epsilon_{jkpabc} X^p_{\ \ell} - \epsilon_{jp\ell abc} X^p_{\ k} - \epsilon_{pk\ell abc} X^p_{\ j} \right), \quad (11.273)$$

for $\mu, \nu = 1, 2$ and $j, k, \ell, a, b, c = 1, 2, \cdots, 6$. Here $\epsilon_{\mu\nu}$ denotes the two dimensional totally anti-symmetric Levi-Civita tensor with $\epsilon_{12} = 1$ while $\epsilon_{jk\ell abc}$ stands for the six dimensional totally anti-symmetric Levi-Civita tensor. We note here that the Jacobi identity

$$[A_{abc,\lambda}, [A_{jk\ell,\mu}, A_{def,\nu}]] + \text{cyclic sum} = 0, \quad (11.274)$$

holds because

$$\epsilon_{\mu\nu} A_{jk\ell,\lambda} + \epsilon_{\nu\lambda} A_{jk\ell,\mu} + \epsilon_{\lambda\mu} A_{jk\ell,\nu} = 0, \quad (11.275)$$

which reflects the fact that there cannot be a third rank anti-symmetric tensor in a two dimensional space. We conclude this discussion by noting that the second order Casimir invariant, in this case, is given by

$$I_2(\rho) = \sum_{j,k=1}^{6} X^j_{\ k} X^k_{\ j} + \sum_{\mu,\nu=1}^{2} B^\mu_{\ \nu} B^\nu_{\ \mu}$$

$$- \frac{1}{72} \sum_{\mu,\nu=1}^{2} \sum_{j,k,\ell,a,b,c=1}^{6} \epsilon^{\mu\nu} \epsilon^{jk\ell abc} A_{jk\ell,\mu} A_{abc,\nu}, \quad (11.276)$$

and that the Dynkin diagram for E_6 has the form shown in Fig. 11.9.

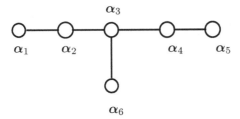

Figure 11.9: The Dynkin diagram for the simple roots of E_6.

11.6.4 E_7. The exceptional Lie algebra $L = E_7$ is known to be 133-dimensional and has a maximal Lie sub-algebra $L_0 = A_7 = su(8)$. Therefore, as a representation space, we can decompose E_7 as

$$133 = 63 + 70, \tag{11.277}$$

where the 63-dimensional representation can be identified with the adjoint representation $X^j{}_k$, while the 70-dimensional representation corresponds to the totally anti-symmetric tensor $A_{jk\ell m}$ with $j, k, \ell, m = 1, 2, \cdots, 8$. The explicit commutation relations for these generators are given by (see (11.88) and note that the other commutation relations are obtained from the transformation of tensors discussed, say, in (3.92))

$$\sum_{j=1}^{8} X^j{}_j = 0,$$

$$[X^j{}_k, X^\ell{}_m] = \delta^\ell{}_k X^j{}_m - \delta^j{}_m X^\ell{}_k,$$

$$[X^j{}_k, A_{\ell mnp}] = -\delta^j_\ell A_{kmnp} - \delta^j_m A_{\ell knp} - \delta^j_n A_{\ell mkp} - \delta^j_p A_{\ell mnk}$$

$$+ \frac{1}{2}\delta^j_k A_{\ell mnp},$$

$$[A_{jk\ell m}, A_{npqr}] = \sum_{s=1}^{8} \left[\epsilon_{sk\ell mnpqr} X^s{}_j + \epsilon_{js\ell mnpqr} X^s{}_k \right.$$

$$+ \epsilon_{jksmnpqr} X^s{}_\ell + \epsilon_{jk\ell snpqr} X^s{}_m + \epsilon_{jk\ell mspqr} X^s{}_n$$

$$\left. + \epsilon_{jk\ell mnsqr} X^s{}_p + \epsilon_{jk\ell mnpsr} X^s{}_q + \epsilon_{jk\ell mnpqs} X^s{}_r \right], \tag{11.278}$$

where $\epsilon_{jk\ell mnpqr}$ denotes the totally anti-symmetric Levi-Civita tensor in 8-dimensions. The Dynkin diagram for E_7 is shown in Fig. 11.10.

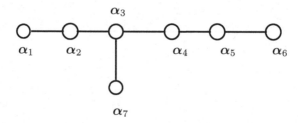

Figure 11.10: The Dynkin diagram for the simple roots of E_7.

11.6.5 E_8. The exceptional Lie algebra $L = E_8$ is known to be 248-dimensional and contains as maximal Lie sub-algebra $L_0 = D_8 = so(16)$. Correspondingly, E_8 decomposes as a representation space as

$$248 = 120 + 128, \qquad (11.279)$$

in terms of the 120-dimensional adjoint representation $X_{jk}, j, k = 1, 2, \cdots, 16$ and the $2^7 = 128$ dimensional (irreducible) spinor representation $\psi_\mu, \mu = 1, 2, \cdots, 128$ (which can be a Weyl/Majorana spinor).

The commutation relations between X_{jk} and ψ_μ are given by (see also (11.260) and (11.261))

$$\sum_{j=1}^{16} X^j{}_j = 0,$$

$$[X_{jk}, X_{\ell m}] = -\delta_{j\ell} X_{km} - \delta_{km} X_{j\ell} + \delta_{jm} X_{k\ell} + \delta_{k\ell} X_{jm},$$

$$[X_{jk}, \psi_\mu] = \sum_{\nu=1}^{128} \psi_\nu \left(\sigma_{jk}\right)_{\nu\mu},$$

$$[\psi_\mu, \psi_\nu] = 2 \sum_{j,k=1}^{16} \left(C\sigma_{jk}\right)_{\mu\nu} X_{jk}. \qquad (11.280)$$

Here $j, k, \ell, m = 1, 2, \cdots, 16$ while $\mu, \nu = 1, 2, \cdots, 128$. Furthermore, σ_{jk} (and $\overline{\sigma}_{jk}$) are defined as (see discussions from (6.100)-(6.106) where σ_{jk} and $\overline{\sigma}_{jk}$ are labelled as $J_{\mu\nu}^{(+)}$ and $J_{\mu\nu}^{(-)}$ respectively)

$$\sigma_{jk} = \frac{1}{4} \left(\sigma_j \sigma_k^\dagger - \sigma_k \sigma_j^\dagger\right),$$

$$\overline{\sigma}_{jk} = \frac{1}{4} \left(\sigma_j^\dagger \sigma_k - \sigma_k^\dagger \sigma_j\right), \qquad (11.281)$$

in terms of the 128×128 dimensional matrices σ_j satisfying

$$\sigma_j \sigma_k^\dagger + \sigma_k \sigma_j^\dagger = 2\delta_{jk} \mathbb{1} = \sigma_j^\dagger \sigma_k + \sigma_k^\dagger \sigma_j, \tag{11.282}$$

which define the generators of the Clifford algebra γ_j as in (6.101)

$$\gamma_j = \begin{pmatrix} 0 & \sigma_j \\ \sigma_j^\dagger & 0 \end{pmatrix}. \tag{11.283}$$

Moreover, we note that C in (11.280) denotes the 128-dimensional charge conjugation matrix defined by

$$(i) \quad C\sigma_j C^{-1} = \left(\sigma_j^\dagger\right)^T,$$

$$(ii) \quad C\sigma_j^\dagger C^{-1} = (\sigma_j)^T,$$

$$(iii) \quad C^T = C. \tag{11.284}$$

And, as in (11.265)-(11.266), here also the Jacobi identity

$$[X_{jk}, [\psi_\mu, \psi_\nu]] + [\psi_\mu, [\psi_\nu, X_{jk}]] + [\psi_\nu, [X_{jk}, \psi_\mu]] = 0, \tag{11.285}$$

holds because of the anti-symmetry under charge conjugation

$$(C\sigma_{jk})_{\mu\nu} = - (C\sigma_{jk})_{\nu\mu}, \tag{11.286}$$

while the Jacobi identity

$$[\psi_\mu, [\psi_\nu, \psi_\lambda]] + [\psi_\nu, [\psi_\lambda, \psi_\mu]] + [\psi_\lambda, [\psi_\mu, \psi_\nu]] = 0, \tag{11.287}$$

for $\mu, \nu, \lambda = 1, 2, \cdots, 128$ follows from the identity

$$\sum_{j,k=1}^{16} [(C\sigma_{jk})_{\mu\nu}(\sigma_{jk})_{\alpha\lambda} + (C\sigma_{jk})_{\nu\lambda}(\sigma_{jk})_{\alpha\mu} + (C\sigma_{jk})_{\lambda\mu}(\sigma_{jk})_{\alpha\nu}] = 0. \tag{11.288}$$

As we have pointed out in (11.268), this identity results from the orthonormality relations for the γ_j matrices in even dimensions. In fact, let us consider a slightly more general case of $so(N)$ Clifford algebra with

$$d = 2^M, \qquad \qquad \text{for} \quad N = 2M + 1,$$

$$d = 2^{M-1}, \qquad \qquad \text{for} \quad N = 2M, \tag{11.289}$$

denoting the dimensions of the σ_{jk} matrices. Let us suppose that

$$\sum_{j,k=1}^{N}[(C\sigma_{jk})_{\mu\nu}(\sigma_{jk})_{\alpha\lambda} + (C\sigma_{jk})_{\nu\lambda}(\sigma_{jk})_{\alpha\mu} + (C\sigma_{jk})_{\lambda\mu}(\sigma_{jk})_{\alpha\nu}] = 0.$$

(11.290)

for $\mu,\nu,\lambda = 1,2,\cdots,d$. This condition can be checked as follows to be satisfied only for three cases. Let us multiply (11.290) by $(\sigma_{\ell m})_{\lambda\alpha}$ and sum over the indices λ,α. This leads to

$$\sum_{j,k=1}^{N}\left[(C\sigma_{jk})_{\mu\nu}\text{Tr}(\sigma_{jk}\sigma_{\ell m}) + (C\sigma_{jk}\sigma_{\ell m}\sigma_{jk})_{\nu\mu} - (C\sigma_{jk}\sigma_{\ell m}\sigma_{jk})_{\mu\nu}\right] = 0,$$

(11.291)

where we have used the anti-symmetry in (11.286). Furthermore, using the identities

$$\sum_{j,k=1}^{N}\sigma_{jk}\sigma_{\ell m}\sigma_{jk} = -\frac{1}{8}\left(N^2 - 9N + 16\right)\sigma_{\ell m},$$

$$\text{Tr}\left(\sigma_{jk}\sigma_{\ell m}\right) = -\frac{1}{8}d\left(\delta_{j\ell}\delta_{km} - \delta_{jm}\delta_{k\ell}\right),$$

(11.292)

in (11.291), we obtain

$$N^2 - 9N + 16 = d.$$

(11.293)

This has only three solutions given by

(i) $N = 16,$ $M = 8,$ as in $E_8 \rightarrow so(16),$

(ii) $N = 9,$ $M = 4,$ as in $F_4 \rightarrow so(9),$

(iii) $N = 8,$ $M = 4,$ for $so(9) \rightarrow so(8).$ (11.294)

We conclude the discussion on E_8 by noting that the Dynkin diagram for the simple roots is shown in Fig. 11.11.

We conclude this section on exceptional Lie algebras by noting that there are several ways to construct these algebras, especially in terms of the octonion algebras. Let us mention this briefly here. Let A be an algebra (which may be non-associative) with a symmetric bilinear non-degenerate form $\langle x|y\rangle = \langle y|x\rangle$. Here, non-degeneracy simply implies that if $\langle a|x\rangle = 0$ for all $x \in A$, then

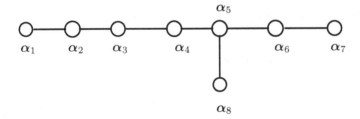

Figure 11.11: The Dynkin diagram for the simple roots of E_8.

$a = 0$. Furthermore, let us suppose that we have the composition law $\langle xy|xy \rangle = \langle x|x \rangle \langle y|y \rangle$. Then A is called a composition algebra. If we assume that A, in addition, is unital, namely, it possesses the unit element e, then A is known as a unital composition algebra or simply a Hurwitz algebra. In this case, the Hurwitz theorem states that there are only four possibilities for such an algebra of dimensionalities $1, 2, 4, 8$, corresponding respectively to the real and complex number fields, quarternion and octonion algebras.

The classical construction of the exceptional Lie algebras F_4, E_6, E_7 and E_8 is due to Tits (see the book by Schafer), where one uses both Hurwitz algebra together with Jordan algebra consisting of 3×3 matrices over the Hurwitz algebra. This leads to Freudenthal's magic square. Another more symmetric approach is due to Elduque which utilizes two independent Hurwitz algebras to obtain essentially the same result. The method also uses the triality relations associated with Hurwitz algebras. However, these are quite technical and we will not go into the details of this.

11.7 References

M. Konuma, K. Shima and M. Wada, *Simple Lie algebras of rank 3 and symmetries of elementary particles in the strong interaction*, Supplement of Progress in Theoretical Physics 28 (1963).

R. D. Schafer, *An introduction to nonassociative algebras*, Academic Press (1966).

J. E. Humphreys, *Introduction to Lie algebras and representation theory*, Springer-Verlag, NY (1972).

S. Okubo, J. Math. Physics 23, 8 (1982).

W. G. McKay, J. Patera and D. W. Rand, *Tables of Branching Rules for Representations of Simple Lie Algebras*, Université de Montréal (1990).

A. Elduque, Rev. Mat. Iberoamericana 20, 847 (2004).

Index

Printed in the United States
By Bookmasters